THE MAKING OF THE
NORTHWEST FOREST PLAN

The Making of the Northwest Forest Plan

The Wild Science of Saving Old Growth Ecosystems

K. NORMAN JOHNSON
JERRY F. FRANKLIN
GORDON H. REEVES

WITH ASSISTANCE IN LEGAL ANALYSIS BY SUSAN JANE M. BROWN
AND SPATIAL ANALYSIS BY DEBORA L. JOHNSON

Oregon State University Press   Corvallis

Generous support from the John and Shirley Byrne Fund for Books on Nature and the Environment helped make publication of this book possible.

Library of Congress Cataloging-in-Publication Data

Names: Johnson, K. Norman, author. | Franklin, Jerry F., author. | Reeves, Gordon H., author. | Johnson, Debora L., contributor. | Brown, Susan Jane M., contributor.

Title: The making of the Northwest Forest Plan : the wild science of saving old growth ecosystems / K. Norman Johnson, Jerry F. Franklin, Gordon H. Reeves, with legal analysis by Susan Jane M. Brown and spatial analysis by Debora L. Johnson.

Description: Corvallis, Oregon : Oregon State University Press, 2023. | Includes bibliographical references and index.

Identifiers: LCCN 2022059659 | ISBN 9780870712241 (trade paperback) | ISBN 9780870712258 (ebook)

Subjects: LCSH: Northwest Forest Plan (U.S.). | Forest reserves—Northwest, Pacific—Management—Planning. | Forest policy—Northwest, Pacific. | Forest conservation—Northwest, Pacific.

Classification: LCC SD413.N67 J64 2023 | DDC 333.75/16091644—dc23/eng/20230215

LC record available at https://lccn.loc.gov/2022059659

♾ This paper meets the requirements of ANSI/NISO Z39.48-1992 (Permanence of Paper).

Oregon State University
OSU Press

Oregon State University Press
121 The Valley Library
Corvallis OR 97331-4501
541-737-3166 • fax 541-737-3170
www.osupress.oregonstate.edu

Oregon State University Press in Corvallis, Oregon, is located within the traditional homelands of the Mary's River or Ampinefu Band of Kalapuya. Following the Willamette Valley Treaty of 1855, Kalapuya people were forcibly removed to reservations in Western Oregon. Today, living descendants of these people are a part of the Confederated Tribes of Grand Ronde Community of Oregon (grandronde.org) and the Confederated Tribes of the Siletz Indians (ctsi.nsn.us).

DEDICATION

We dedicate this book to two beloved colleagues, Jack Ward Thomas (1935–2016) and Jim Sedell (1944–2012). They played major roles in shaping the concepts and ideas we cover here. More than that, they were an almighty good time to be around and made our efforts so much more fun. Guys, we miss you.

Jack Ward Thomas, preeminent wildlife biologist and master communicator, worked as a scientist in the research branch of the US Forest Service and then as its chief. He led many scientific efforts that developed innovative conservation strategies, including the first scientifically credible plan for the northern spotted owl and the science assessment underlying the Northwest Forest Plan. Jack demonstrated that it was possible to provide strong ethical leadership while still being able to modify one's views and adapt to new knowledge. Gruff, outspoken, and loyal to the core, he was mentor and friend to us and to many, many others.

Jim Sedell devoted his professional life to helping save wild fish and their habitat, while working first for Weyerhaeuser, then for the research branch of the US Forest Service, and finally for the National Fish and Wildlife Foundation. Jim pioneered a series of fish conservation strategies and was a major force in development of the Aquatic Conservation Strategy of the Northwest Forest Plan. He brought a contagious

enthusiasm to the resolution of natural resource problems. To those who knew him, Jim seemed a whirlwind of energy, pulling in anything and everything in his path and sending it all skyward in a fountain of creativity. Jim pursued his goal of conserving more fish for all with a broad smile, a hearty laugh, and a pure heart.

Contents

PART 5: SIGNIFICANCE AND FUTURE

Acknowledgments

We wish to thank Mary Elizabeth Braun (Oregon State University Press acquisitions editor, now retired), who kept after us over the last 20 years to write this history. Her view undoubtedly is "Well, it is about time." On the bright side, our procrastination gave us more history to tell.

Susan Jane M. Brown provided invaluable legal advice on key court cases on the need for, and the development and implementation of, the Northwest Forest Plan. Interpretations of what these cases may mean for federal forestry, though, remain ours alone.

Debora Johnson created the maps shown throughout the book and also organized the other graphics. In many ways, her work brings the book to life.

Numerous scientists volunteered to review some or all of this manuscript, perhaps in part because they played a role in the history covered here. Many told us they saw their work on development of the Northwest Forest Plan (NWFP) as the most important of their professional lives. The events of 25–30 years ago still burn brightly in their memories. Key reviewers here include Fred Swanson, Kelly Burnett, Eric Forsman, Marty Raphael, Bruce Marcot, and Dick Holthausen. With their insights and corrections, this is a much better book.

Members of the Clinton administration who were deeply involved in creation of the NWFP gave generously of their time in interviews and reviews, including Katie McGinty, Jim Pipkin, Jim Lyons, Tom Tuchmann, Tom Collier, and David Cottingham. Their reflections helped create a much stronger and well-grounded story.

Others who reviewed pieces or all of the manuscript include Richard Hardt, James Johnston, Ray Davis, Darius Adams, Cristina Eisenberg, and Mindy Crandall. Deanne Carlson redid the FEMAT charts and also summarized the BLM 2016 land allocations. All made important contributions.

Eric Forsman, James Johnston, Cheryl Freisen, Jim Furnish, and Emily Platt all contributed personal stories of the forty-year journey to fundamentally shift federal forest management. Their reminiscences greatly enrich our story. Also, Kathy Thomas allowed us to quote from Jack Ward Thomas's journal. Thank you, all.

Many of the scientists mentioned in this book are part of the Northwest Forest Plan Oral History Collection in the Oregon State University Library's Special Collections and Archives Research Center.[1] Their recollections convey a richness of knowledge, perspective, and sense of importance about the plan they helped craft. These oral histories make a useful and highly entertaining complement to this book.

We also utilized excerpts from the excellent seven-part Oregon Public Broadcasting podcast by Aaron Scott.[2] We recommend it as a companion to the book. For those who wish to dig deeper into the legal struggles underlying the creation and implementation of the NWFP, we strongly recommend Blumm, Brown, and Stewart-Fusek's law review article "The World's Largest Ecosystem Management Plan."

We were fortunate to work with archivists at the William J. Clinton Presidential Library in Little Rock, Arkansas, especially Shannon Lausch. They helped us identify internal memos to the president, which enabled us to understand the inner workings and decision making in the White House during creation of the Northwest Forest Plan.

Two friends from worlds beyond forestry and science, John Rose and Giana Bernardini, reviewed the complete manuscript for readability and coherence. Thank you, it helped.

Michael Furness provided us with the photo of Jim Sedell and Kathy Thomas provided us with the photo of Jack Ward Thomas for the dedication. Thank you, those photos mean a lot to us.

OSU sent an initial draft manuscript out for review. One reviewer, a retired US Forest Service manager, gave back largely positive comments. The other, an anonymous reviewer with experience in environmental history, thought the material was of historical significance but felt that organizationally the book was a mess. He was especially critical that the book lacked a smooth chronological sequence. In addition, he thought it read too much like a Forest Service technical report, among other pithy suggestions. We took those words to heart in our revisions and hope we did them justice.

Kim Hogeland, current OSU Press acquisitions editor, supervised development of our book from beginning to end. She immediately encountered scientists used to producing articles for technical journals and writing textbooks, so she faced real challenges in reorienting them to writing a history book that would interest a wider audience than our colleagues and students. Kim offered constructive criticism throughout her chapter reviews, much of which took hold, and substantially reshaped and improved this book. We thank her for her tireless efforts. Micki Reaman efficiently supervised editing and production for OSU Press, and Laurel Anderton copyedited the entire manuscript and saved us from many embarrassing typos and garbled thoughts.

One final note: most of the statements in this book are deeply referenced and cited, as they should be. However, part of what we report here comes from our personal experiences. Thus, some chapters or statements are shockingly (for scientists) without many citations. Citing ourselves repeatedly in the endnotes, though, seemed foolish. Also, photos without attribution come from our personal files.

Acronyms and Abbreviations

Environmental and Related Laws

APA   Administrative Procedure Act

ESA   Endangered Species Act

NEPA   National Environmental Policy Act

NFMA   National Forest Management Act

O&C Act   Oregon and California Revested Lands Sustained Yield
            Management Act

NEPA Documents

DEIS   Draft Environmental Impact Statement

FEIS   Final Environmental Impact Statement

ROD   Record of Decision

SEIS   Supplemental Environmental Impact Statement

Agencies

BLM	Bureau of Land Management
BLM O&C lands	A commonly used description for BLM's western Oregon lands
National forests or US national forests	United States national forests
NMFS	National Marine Fisheries Service
PNW Research Station	US Forest Service Pacific Northwest Research Station, headquartered in Portland, OR
US Forest Service or Forest Service or USFS	United States Forest Service
USFWS	US Fish and Wildlife Service

Science Assessments That Provided the Scientific Foundation of the Northwest Forest Plan, in Chronological Order

ISC	Interagency Scientific Committee. Purpose: develop a scientifically credible conservation strategy for the northern spotted owl
G-4	Gang of Four. Purpose: develop strategies to conserve old-growth and related aquatic ecosystems and the species within them and evaluate their timber harvest effects
SAT	Scientific Assessment Team. Purpose: develop conservation strategies that would enable species associated with old-growth and related aquatic ecosystems to achieve a high likelihood of persistence
FEMAT	Forest Ecosystem Management Assessment Team. Purpose: prepare a series of options that would keep the timber harvest level from federal lands as high as possible consistent with environmental laws
SAT II	Species Assessment Team. Purpose: develop mitigations for species that did not achieve a high likelihood of persistence under the plan the president selected from the options developed by FEMAT.

Key Descriptors of Different Conservation Strategies and Their Components

Late-successional/ old-growth forest	Also called mature and old-growth forest or older forest. Sometimes shortened to late-successional forest, or LS/OG forest.
ACS	Aquatic Conservation Strategy
AMA	Adaptive Management Area
HCA	Habitat Conservation Area in the ISC's conservation strategy
LS/OG Areas	Forests containing concentrations of LS/OG forest often mixed with younger forest
LSR	Late-Successional Reserve
NWFP	Northwest Forest Plan
PSQ	probable sale quantity, an estimate of average annual timber sale levels likely to be achieved over a decade from Matrix and Adaptive Management Areas under the NWFP
RMPs	BLM 2016 Resource Management Plans
SOHA	Spotted Owl Habitat Areas in the Forest Service's spotted owl conservation strategy

Introduction

Every story needs a place to start.

This story begins when Eric Forsman heard a spotted owl.[1]

Eric Forsman became an avid birder while a high school student in Springfield, Oregon, in the mid-1960s. Owls fascinated him because they hunted at night and were hard to see. He was particularly intrigued by northern spotted owls (*Strix occidentalis caurina*), as little was known about them and they were thought to be very rare.

Eric, though, had to wait until he was a student at Oregon State University (OSU) and working as a temporary employee on the Willamette National Forest to see a spotted owl. One evening while he was sitting on the front porch of the Box Canyon Guard Station on the south fork of the McKenzie River, he heard something barking up on the hill near the station. He thought, "That's a weird-sounding dog," and then it began to dawn on him that the sound seemed similar to the spotted owl calls that he had read about in bird books.

Figure 0.3. A northern spotted owl mother and baby (photo copyright Jared Hobbs.)

Eric hiked down the road to where the sound was coming from. That's where he found a male spotted owl sitting in a tree right next to the road. He started trying to hoot back at him, and a male and a female spotted owl eventually followed him all the way back to the guard station. After that, Eric kept hearing them off and on all summer.

Eric graduated from OSU in spring 1970, was drafted into the army, and served in Germany for over a year. During that time, he corresponded with Howard Wight, a professor in the Fish and Wildlife Department and leader of the US Fish and Wildlife Service (USFWS) Cooperative Fish and Wildlife Unit at OSU, who had encouraged his interest in the northern spotted owl as an undergraduate. Wight agreed to take him on as a master's student to continue his work on the northern spotted owl.

After leaving the army and returning to Corvallis in April 1972, Eric immediately began to document as much as he could about the owl's distribution, habitat, and basic biology for his master's thesis. He found that spotted owls were more common than previously believed and were most abundant in old forests. They nested primarily in cavities in large old trees, and their diet was dominated by small tree-dwelling mammals, especially northern flying squirrels (*Glaucomys sabrinus*), red tree voles (*Arborimus longicaudus*), and wood rats (*Neotoma* spp.). Ominously, Eric also found many of the owl's old-forest sites scheduled for harvest.

Eric's findings generated much interest from the Forest Service and Bureau of Land Management (BLM) in western Oregon once the agencies realized that the habitat needs of the owl could impact their timber program. Wight encouraged Eric to continue as a PhD student, and Eric was soon in the field in the Cascades of western Oregon putting radio collars on spotted owls to track them and determine the size of the home range around their nests, where they would look for small mammals to eat.

Eric finished his fieldwork in 1976 and started to analyze his data. In an interview for the Northwest Forest Plan Oral History Project, he briefly summarized his thinking at that time:

Although I had yet to complete the analysis, I already knew from following the owls around for a year that they were unlike any species of owls that people had studied before. For a relatively small owl, they had huge home ranges that encompassed large amounts of old forest. Annual home ranges of the adults that I studied on the Andrews [Experimental Forest] averaged about 3,000 acres and included more than 1,000 acres of old forest. These data suggested that, in all of our previous recommendations to managers and policy makers, we had horribly underestimated the amount of habitat that spotted owls need.[2]

Eric Forsman's finding that northern spotted owls needed "huge home ranges that encompassed large amounts of old forest" set off a titanic struggle in the Pacific Northwest to find a way to accommodate both conservation of the owl and continuance of the old-growth logging that provided employment for tens of thousands of people. The struggle to find a solution to this problem involved years of controversy and debate, the federal courts, five science assessments, Congress, and eventually the president of the United States, and it led to creation of the Northwest Forest Plan (NWFP)—the first large-scale ecosystem plan for federal forests in the United States.

The creation and implementation of the NWFP is perhaps the most important development in management of federal forests in the range of the northern spotted owl since their establishment. It challenged the classical forest paradigm of good forestry, opened up new ideas about how these forests can best contribute to the public welfare, and sent these precious lands on a new pathway. This book is the story of that effort as the authors understand it.

What type of history is this story? John Burrow, in *A History of Histories*, describes different types of histories that have been written through the ages. He notes that many early writings provide a historical account of events in chronological order, and this book largely follows that model. He describes the evolution over time into histories that attempted to explain these events in terms of broader themes, something that we certainly tried to do here. However, more than anything else, this book covers what Burrow calls *intellectual history*—the history of ideas and how they shape the way in which we view the world, the policies leaders make, and the actions people take. It describes how the interaction of a set of people dedicated to toppling the existing order, laws that required the incorporation of new values, learned jurists who were willing to disrupt established practices, and scientists who challenged old assumptions created new ways to think about forests and their conservation.

Part 1 covers essential background on characteristics of the landscapes, forests, streams, and watersheds within the owl's range, the important role that Indigenous people historically played in their conservation and management, and the current ownership pattern. It also covers early management of the federal forests there by forestry professionals, in which the steady logging of old-growth forests for local wood products industries became the primary way that the federal government sought to contribute to the public welfare through forest use.

Part 2 describes major legislative and scientific developments that would enable a challenge to logging old growth in the name of the public good. Congress passed a series of environmental laws in the 1970s that would revolutionize federal decision-making and give legal standing to those who wished to block what they viewed as environmentally destructive actions. However, wielding these laws in that fight to stop the harvest of old-growth forests depended on the parallel development of new science on the ecological importance of these forests. Such knowledge came on like thunder in the 1970s and 1980s.

Part 3 covers the tumultuous 1980s and 1990s, when science, law, and social activism combined to fundamentally alter the goals of federal forest management. Science that challenged long-standing assumptions of forest management connected to the requirements in the new environmental laws and provided the leverage that environmental groups needed to wreak havoc on agency old-growth harvest plans and, just as importantly, to solidify public support for their efforts. Much of the litigation focused on convincing the courts to prohibit federal timber sales within northern spotted owl habitat (the source of most federal harvest in the area) until federal forest agencies developed conservation plans that would ensure the owl's survival. Once the courts granted injunctions prohibiting sales in spotted owl habitat, the Pacific Northwest had a full-scale economic and political crisis at hand, with tens of thousands of jobs at stake.

Unable to develop a conservation strategy for the owl that would satisfy the courts, the chief of the Forest Service and other agency heads asked an independent group of scientists to step in. Release of the scientists' strategy for spotted owl conservation shook managers and politicians of the Pacific Northwest to their core as they realized that implementing the strategy would require major reductions in fed eral timber harvest. While the courts praised the scientists' efforts as the first scientifically credible spotted owl strategy, they condemned political maneuvering over its adoption, and the injunction on timber sales in owl habitat stayed in place. More science assessments followed, broadening the species of concern to include other species that relied on old-growth forests and the streams that ran through them.

This story took a major turn when presidential candidate William Jefferson Clinton promised, against the advice of his aides, that his administration would solve the owl/timber conflict. And sure enough, the newly elected president held a Forest Conference in Portland, Oregon, and commissioned yet another science assessment. More than anything, the new administration sought a plan that would satisfy the courts. Toward that end, administration leadership made sure there was no political interference in the scientists' effort.

The resulting choices the scientists developed for the president portrayed a stark trade-off between timber production and species conservation. Still, the president pushed on, selecting the plan that best met administration goals and putting it out for public review. That plan reserved approximately 80 percent of the remaining older forest on federal land in the range of the spotted owl and lowered federal harvest there by 75 percent. Intense criticism poured in from all sides. After adjusting the plan to increase its legal sufficiency and lowering federal harvest still further, the administration obtained the long-sought court approval. The Northwest Forest Plan was a reality.

Part 4 describes the first 25 years of the NWFP. The early days of NWFP implementation were a story of immediate difficulties, painful adjustments, and creative adaptation. The harvest strategy in the plan in the near term relied in large part on harvest of the older forest that the NWFP did not place in reserves. Congress stepped in and forced that harvest for a few years by limiting public access to the courts. By 2000, though, that limitation no longer applied, and logging of older forest floundered because of protest, litigation, and other difficulties. Federal timber sales collapsed. In desperation, the Forest Service and BLM turned to thinning plantations established after previous clearcuts as their primary forestry activity, and harvests slowly began to rise.

Many timber workers and their families went through wrenching adjustments as a way of life and a source of employment they thought would last forever disappeared almost overnight. The Clinton administration committed financial resources to help the workers and families to adjust to this new reality, but achieving that goal proved difficult. Also, the federal workforce was slashed. Many communities affected by the loss of this employment were able to adjust without severe dislocation; many others were not so fortunate.

Over time, the amount of older forest on federal land stabilized and watershed and stream conditions improved. Still, many of the problems that the NWFP sought to address remained, with the peril of climate change looming over all other considerations. Another owl, the barred owl (*Strix varia*), invaded much of the northern spotted owl's habitat, threatening its existence. Many salmon populations throughout the owl's range remained at risk. Firestorms in 2020 threatened many small west-side communities as well as suburban and city dwellers in ways they had rarely experienced, burning large areas of older forest. Politicians, land managers, scientists, environmentalists, and others proposed new plans and policies along the way, highlighted by the BLM adopting its own plan for its western Oregon lands.

Part 5 provides our reflections on the significance of the NWFP and the path forward. The NWFP is important in many ways—a scientifically credible approach to protecting biodiversity over a large region, a plan deemed legally sufficient by the courts, and a demonstration of how a multitude of federal agencies could rally around a single plan. Perhaps most importantly, the NWFP represents a fundamental turning point in the role of the national forests in the lives of the people of the United States, from primarily a source of raw material to a place where the conservation of species and ecosystems would be emphasized.

Finally, we address five key issues (older forest, spotted owls, salmon, timber production, and conservation planning) that drove development of the NWFP along with the need for increased recognition of the goals and treaty rights of tribal nations within the area. Looking into the future, we suggest policy changes that we believe will enable the NWFP to address these issues more successfully in the context of a changing and unpredictable world.

PART 1

The Place

Figure 1.1. Cities, towns, mountain ranges, and volcanoes within the range of the northern spotted owl in the United States.

1
Range of the Northern Spotted Owl

The northern spotted owl's range covers much of the Pacific coast of the United States and southwestern British Columbia. In the United States, that range stretches from the Canadian border in the north to San Francisco in the south, and from the Pacific Ocean in the west to the eastern slopes of the Cascade Range (fig. 1.1). That area became a major focus of federal forest policy development in the late 1980s with the realization that the logging of old-growth forest on federal lands there threatened the owl's existence, and new policies to protect the owl were needed. Thus, the range of the northern spotted owl in Washington, Oregon, and California was chosen to define the geographic scope of the Northwest Forest Plan—the world's largest forest ecosystem management plan.

This owl's range is primarily forested and mountainous (figs. 1.1 and 1.2). The Cascade Range runs along its eastern edge, regularly punctuated with active volcanoes, such as Mount Baker, Mount Saint Helens, Mount Rainier, and Mount Shasta. Closer to the Pacific Ocean is a series of mountain ranges displaying diverse geologic histories and topographies—the Olympic Mountains in northwestern Washington, the Coast Range in western Oregon, and the Klamath and Siskiyou Mountains in southwestern Oregon and northwestern California. Most human developments, including cities, industrial areas, and farms, occupy the lowlands that separate the coastal mountains and Cascade Range, such as the Willamette Valley and the Puget Sound Region.

Historically, Indigenous people were part of these landscapes, especially in the lowlands, having lived there for millennia in a multitude of tribes and bands. They actively managed ecosystems, with fire as a primary tool, to stimulate growth and fruit production of plants they favored, clear vegetation, create habitat for animals, and herd game, among other goals. Thus, early explorers of the Pacific Northwest often

Figure 1.2. The Glacier Peak Wilderness of the Cascade Range in northern Washington.

encountered an ecologically heterogeneous landscape of open woods, spacious meadows, and extensive prairies that bore the marks of Indigenous land management practices, especially burning.[1] As an example, David Douglas, for whom the Douglas-fir is named, came down an Indigenous trail into the Willamette Valley in 1826 and encountered a member of the Kalapuya Tribe setting fire to the prairie while he walked through extensive amounts of burned landscape.[2] Yet when he returned in November of that year, he found a transformed landscape with a lush, green prairie broken up by occasional marshes and lakes full of waterfowl.[3]

Not surprisingly, Indigenous peoples' use of fire had a profound effect on shaping ecosystems in many places, especially oak woodlands, prairies, and wetlands at lower elevations in western Oregon and northern California.[4] In addition, the higher, wetter forests, while perhaps less hospitable, would also have drawn attention as a source of food and other supplies. It is well documented that they managed huckleberry fields within the forests using fire and also used wood fires to dry the berries; these activities were the occasional source of larger fires.[5]

By the mid-1800s, though, colonization had brought their longtime conservation of these ecosystems to a sudden and often violent end, as part of the process of settler colonization. First, early European contact decimated Indigenous people through the introduction of deadly diseases. As an example, epidemics greatly reduced the Kalapuya population in western Oregon.[6] Then the US government subjugated them, forcing those who survived onto reservations largely away from the rivers, marshes, prairies, and oak woodlands that had historically provided their sustenance. Those resources were coveted by Euro-American homesteaders for their farms and by a federal government that wanted fertile land to accommodate a growing population and to solidify claims of the United States to those lands.[7] In the end, that was what really mattered, not equitable treatment of peoples who were already there.

As Indigenous people died from disease and conflict or were removed, forests spread and filled in under an emerging policy of fire exclusion, and much of their effect on ecosystems within the range of the northern spotted owl was lost. As an example, a significant part of Oregon State University's McDonald-Dunn Forest, which was once prairie and oak woodland, is now dense forest.[8] It should not be forgotten that some of the forest seen today in the owl's range is a fairly recent development and that other people took care of these lands in sophisticated ways long ago—ways that offer many lessons for current forest management.

A complex ownership pattern now covers the forests in the range of the northern spotted owl (plate 1.1). Almost half the forest was retained in federal ownership, with all the special responsibilities for environmental protection and recovery of endangered species that come with that status. Large and medium-sized corporations own much of the adjacent or intermingled forestland, with a focus on maximizing financial returns from forestry operations for their owners and shareholders. Smaller family forests are numerous near communities, along river bottoms, and around valley fringes, while states and "tribal nations" (on their reservations) control significant blocks of forestlands in several parts of the owl's range.

Also, tribal nations (collectives of tribes and bands recognized as sovereign governments) often retain sovereignty rights that include fishing, hunting, and gathering on their ancestral lands outside their reservations as part of the treaties signed during the 1800s, in which the tribes ceded most of these lands (usually under duress) to the federal government (fig. 1.3).[9] Tribal nations count on their ability to exercise their treaty rights on national forests, BLM lands, and other special places.

Other tribal nations, especially in western Oregon and northern California, thought they also would receive reservations and sovereignty rights on lands they ceded to the federal government, but the agreements were not put into law or were taken back from them, as described by Berg in *First Oregonians*, White in *It's Your Misfortune*, and Robbins in *Landscapes of Promise*. In total, the ancestral lands of Indigenous people cover the owl's range.[10]

With that tragic history as background, this book details the policy and political struggles over the last 40 years of charting a new path for the national forests and BLM lands in the range of the northern spotted owl (fig. 1.4), where logging old-growth forests has long been a dominant forest use. A central issue will be how to best conserve the spotted owl, Pacific salmon, and other creatures that inhabit old-growth forest ecosystems and the streams that run through them. In this chapter, we provide a brief introduction to the physical and biological characteristics of the region and its history of land use, as a context for the political and legal battle to come over this issue.

Climate, Topography, and Geology

The range of the northern spotted owl encompasses a wide array of climatic, topographic, and geologic conditions that interact in ways that help produce the contrasting ecoregions of the Pacific Northwest. Owing to the presence of the Pacific Ocean and prevailing onshore (southwest to northeast) weather patterns, a strong climatic gradient goes from west to east across the area. Winter storms moving onshore interact with the region's topography, producing greater precipitation on the west side of the mountain ranges and rain shadows on the east side.

The Pacific Northwest coastal regions have high levels of precipitation, often in excess of 100 inches of precipitation annually, while the lowlands immediately east of these mountains typically receive less than half this amount. Similarly, precipitation increases as Pacific storms approach the crest of the Cascades from west to east and markedly decreases east of the Cascade Range. The HJ Andrews

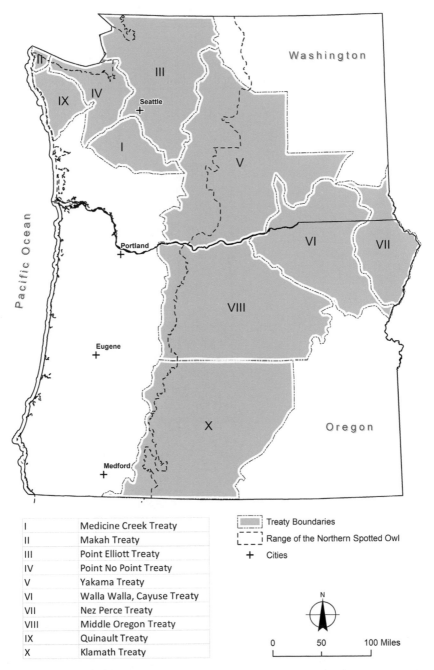

I	Medicine Creek Treaty
II	Makah Treaty
III	Point Elliott Treaty
IV	Point No Point Treaty
V	Yakama Treaty
VI	Walla Walla, Cayuse Treaty
VII	Nez Perce Treaty
VIII	Middle Oregon Treaty
IX	Quinault Treaty
X	Klamath Treaty

Treaty Boundaries

Range of the Northern Spotted Owl

+ Cities

N

0   50   100 Miles

Figure 1.3. Treaty boundaries for tribal nations in Oregon and Washington. (Source: FEMAT, *Forest Ecosystem Management*, 7-84.) Ancestral lands of other tribal nations and unrecognized tribes and bands cover the remainder of the NWFP area. (Long, et al.,"Tribal Ecocultural Resources and Engagement," 856-857). They also were forced to cede much of their lands to the federal government but often got very little in return.

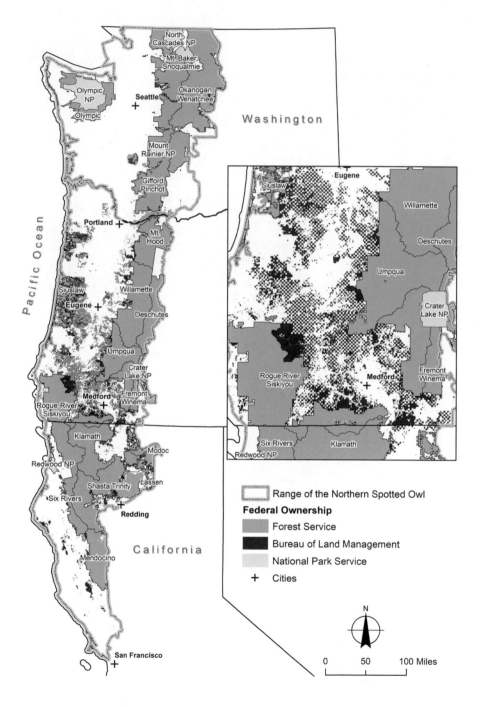

Figure 1.4. National forests, BLM lands, and national parks in the range of the northern spotted owl in the United States.

Experimental Forest in the western Cascades of Oregon, for example, receives approximately 100 inches of precipitation annually, much in the form of snow. In contrast, the town of Redmond, Oregon, about 50 miles to the east on the Cascade Range's leeward side, receives only 8 inches of precipitation annually.

Overall, the climate is cooler and wetter in the coastal and inland mountains toward the north, transitioning to a more Mediterranean climate with warmer, drier summers to the south. In the western Cascades, the decline in both total and summer precipitation south of Eugene helps explain the transition in the character of the forest there.[11]

The amount and form of precipitation strongly affect the timing and quantity of streamflow. Most precipitation falls during the late fall and winter months. In the coastal areas, where elevations rarely exceed 3,000 feet, that precipitation usually falls as rain. Here, the highest streamflows occur during the winter months, coincident with seasonal cyclonic storms sweeping in from the Pacific Ocean. The most intense of these storms bring flooding to the coastal lowlands and initiate landslides in steep headwater streams that create debris flows. At higher elevations and inland from the moderating influence of the Pacific Ocean, such as in the Cascades, an increasing fraction of annual precipitation comes in the form of snow. Melting winter snowpack there supports high streamflow into the spring and early summer months.

In colder years, winter climatic patterns can result in substantial snow accumulation at low to middle elevations of the Cascades, and sometimes the coastal mountains. When warm, intense Pacific storms follow, rapid snowmelt and flooding result.[12] Floods and debris flows caused by these rain-on-snow events are a major source of disturbance in streams of the western Cascades and a primary cause of flooding in lowland agricultural and population centers.

Like the climate and topography, the geology of the area is remarkably variable.[13] The Oregon Coast Range and the Olympic Peninsula are composed largely of uplifted sandstones and mudstones that were once oceanic seafloors, but volcanic intrusions are also present. The most prominent features of the Cascade Range are its active volcanoes, including Mount Saint Helens, which erupted with great force in 1980. South of Snoqualmie Pass in Washington, the Cascade Range can be divided into two geologic formations—the older Western Cascades and the more recent High Cascades. Both are dominantly volcanic in origin but differ in age and character. Rugged topography and streams with steep gradients and confining hillslopes characterize the Western Cascades. In addition to its prominent and still-active stratovolcanoes, the High Cascades has extensive undulating topography. The northern Washington Cascade Range is a also very rugged and geologically complex region that includes extensive intrusive and metamorphosed rock formations as well as several prominent volcanoes. The Klamath Mountains are another very complex geologic region, formed by tectonic processes driving ancient geologic formations of the Pacific Ocean onto the North American landmass. These processes created a unique mosaic of volcanic, sedimentary, and metamorphic rock[14] and resulted in a region of incredible geologic and biological diversity.[15]

Forests

Massive temperate-zone forests of evergreen trees characterize the range of the northern spotted owl (fig. 1.5) and include many of the

Figure 1.5. The massive and structurally complex old-growth forest of the Douglas-fir/western hemlock zone is a wonderous ecosystem that also produces some of the finest structural lumber in the world. Jerry Franklin stands in the center next to an old-growth Douglas-fir on May 14, 2021, at the HJ Andrews Experimental Forest in Oregon.

world's largest and most long-lived conifers.[16] Dominant tree species include Douglas-fir, western hemlock, western redcedar, Sitka spruce, ponderosa pine, Pacific silver fir, noble fir, grand fir, and, in northern coastal California and a small piece of adjacent Oregon, the renowned coast redwood. The dominance of evergreen conifers in a moist temperate region is itself globally notable, as broadleaf or hardwood trees dominate most moist temperate forest regions of the world.

Old-growth forests develop immense biomass accumulations, especially in the wetter portions of the owl's range. A key to the impressive nature of these forests is the prevalence of tree species that survive and grow for many centuries, during which time they develop exceedingly large trunks with decay-resistant heartwood and bark that, in turn, can take several centuries to decay after death of the tree itself. As a result, these forests represent the largest and most stable stores of forest carbon found anywhere in the world.[17]

Quantitative estimates of the historical extent of older forests vary. However, scientists have reconstructed the forests in the owl's range using information gathered in the late 1930s and early 1940s—after the forests around the edges of the major valleys, near ports, and in the flatter topography had already been logged (plate 1.2 left). They found that approximately 60 percent of the forested area of the region was still occupied by late-successional/old growth forests. These estimates include both old-growth forests and mature forests—forests that were developing old-growth characteristics.

Logs from these forests fueled the forest industry of the region, with private forests cut first, followed by federal forests. Most of this forest was gone by the early 1990s on nonfederal lands and was heavily fragmented on much of the federal lands (plate 1.2 right), setting the stage for a ferocious struggle over the future of these forests once it was recognized that the northern spotted owl needed such forests as habitat.

It is not surprising that a region with such an array of climatic, topographic, and geologic conditions also contains a highly diverse flora. Vegetation zones based on dominant plant associations identify areas with similar environmental conditions (plate 1.3), and the distribution of these zones differs greatly among ownerships. The most widespread forest vegetation zone in the area—the Douglas-fir/western hemlock zone—lies at low to moderate elevations and occurs more extensively on nonfederal (mostly private) lands than on federal lands. On the other hand, higher-elevation forests—the Pacific silver fir, mountain hemlock, and subalpine fir zones—are more abundant on federal lands. These patterns resulted from the national forests being created primarily from lands left after lower-elevation forests had been privatized by land grants and homesteading.

In this book, forest vegetation zones in the range of the northern spotted owl are often lumped into two types for discussion—Moist Forests, where infrequent fires that created large areas of high-severity fire were historically a major force shaping their composition and structure (plate 1.4 top); and Dry Forests, where frequent fires with significant areas of low-to-moderate-severity fire were historically a major shaping force (plate 1.4 bottom). Moist Forests occupy most of the coastal ranges and western Cascades, with Douglas-fir/western hemlock and Pacific silver fir the most extensive vegetation zones. Dry Forests occupy the eastern slope of the Cascades, the southern portion of the western Cascades, interior southwestern Oregon, and northern California, with Douglas-fir and white fir/grand fir the most extensive zones (plate 1.3).

Douglas-fir/western hemlock forests initiated by wildfire display many typical attributes of Moist Forest development: (1) high levels of structural legacies (e.g., dead trees and logs) left by the fire; (2) an extended and ecologically important early successional stage in which forbs, shrubs, and grasses dominate the site before conifers once again take over; (3) a young forest stage dominated by a shade-intolerant pioneer tree species (Douglas-fir) that is gradually replaced over time by more shade-tolerant associates; (4) mature and old growth (older forest) stages that develop high levels of structural complexity and accumulations of live and dead biomass; and (5) a requirement for a long period (200–400 years) for full development. Of course, another wildfire or other disturbance could truncate this development at any time and start the process again.[*][18]

---

* Major windstorms, like the Columbus Day storm of 1962, can also reset Douglas-fir/western hemlock forests, in which case the hemlock understory may soon capture the site.

Most large, severe wildfires in the Douglas-fir/western hemlock zone occur during a weather pattern in which east winds bring hot and very dry air from interior continental regions.[19] These wildfires leave complex mosaics of fire effects (severity) at the stand or larger spatial scales, as they burn through diverse topographic and forest conditions and over many days or weeks (plate 1.4 top). They invariably include significant areas (tens to hundreds to thousands of acres) of high tree mortality; that is, they include significant areas of high-severity forest replacement. However, patches of varied size of intermediate and low overstory tree mortality and unburned patches are also part of the burn pattern along with scattered individual and small groups of live trees.* For example, many dominant fire-resistant Douglas-firs may be left alive, providing a source of seed. Partial burns provide suitable conditions for abundant reproduction of shade-tolerant species, such as western hemlock, resulting in two-aged stands.[20] Douglas-fir/ western hemlock stands that live a long life may also experience other, lower-intensity fires, especially in the southern Oregon Cascades.[21] Interactions of development and disturbance create a complex Moist Forest landscape, indeed.

Dry Forests historically experienced frequent and often extensive fires that burned most commonly at low to moderate intensities, killing smaller trees while leaving the larger ones (plate 1.4 bottom). These fires favored species adapted to survive fire, such as ponderosa pine, and disfavored species such as white fir that were easily damaged or killed by fire. Historically, the common outcome of the powerful selective force of frequent fire was a predominantly low-density forest dominated by larger, older trees intermingled with patches of seedlings, poles, and saplings.[22]

Historically, the most extensive frequent-fire ecosystems in the range of the northern spotted owl (Douglas-fir and white fir/grand fir) were dominated by large, old ponderosa pine, Douglas-fir, and white and black oak. With fire suppression and other practices, though,

---

* Spies et al., "Old Growth, Disturbance," 139, similarly describes these fires: "Before the era of fire suppression, a few of these starts likely smoldered for weeks as small fires or as burning snags until a dry east wind event occurred, when those fires could spread rapidly producing large patches of high-severity fire along with patches of moderate- to low-severity fire."

they have filled in with younger Douglas-fir or white fir/grand fir and become more prone to high-severity fire than they were historically.[23]

While this division of forest vegetation zones into two disturbance groups may be oversimplified, it helps illustrate fundamental differences in the owl's range.[24] Certainly, some areas, like the Umpqua

Figure 1.6. (above) Northern spotted owl habitat in a multilayered, old-growth Douglas-fir forest in a Moist Forest in the western Oregon Cascade Range. (Source: Spies et al., *Synthesis of Science*.) (below) Northern spotted owl habitat in an old-growth mixed-conifer forest in a Dry Forest in the eastern Oregon Cascade Range, where a dense, multilayered canopy has developed with fire exclusion.

National Forest in the western Cascades (fig. 1.4), contain mosaics of Moist and Dry Forests where characteristic fire behavior becomes more complex, further complicated by an increase in lightning activity. Still, the general division of forests into Moist and Dry will prove highly useful in this book for policy and management discussions.

While the forest vegetation zones in plate 1.3 all occur within the range of the northern spotted owl, they are not equally desirable as spotted owl habitat. The multilayered old forests of the Douglas-fir/western hemlock zone are recognized as quintessential owl habitat (fig. 1.6 top). On the other hand, spotted owls rarely successfully nest in mountain hemlock and subalpine fir stands, which are too cold and snowy to provide a year-round prey base, although they may occasionally occur there.

Spotted owls can be found in Dry Forests, which historically would have been much more open than Moist Forests and largely unsuitable for them. Although some dense, multistoried stands preferred by spotted owls may have occurred in historical Dry Forests, such as on north-facing slopes or valley bottoms, they would have been a relatively small component of the Dry Forest landscape.[25] However, fire exclusion and logging have shifted large expanses of Dry Forests to dense, multilayered structures that are preferred habitat for both owls and their prey (fig. 1.6 bottom).

Rivers, Streams, and Salmon

A dense network of streams and rivers laces the range of the northern spotted owl, especially in Moist Forests that experience high precipitation levels.[26] Federal lands tend to occupy the headwaters in the steeper, more remote parts of the landscape left over after settlers claimed the lowlands for farms, ranches, towns, and cities.

In the early 1990s, scientists brought evidence of rapidly declining salmon populations throughout the range of the owl that challenged the adequacy of federal conservation of streams, riparian areas, and watersheds there and dramatically broadened the policy debate. While few had heard of the northern spotted owl before Eric Forsman studied it, almost everyone had heard of salmon—an iconic and beloved species of the Pacific Northwest. If salmon were in trouble, things were definitely amiss. This book is partly about the battle to save the federal habitat of salmon populations and other aquatic species within the owl's range.

Major rivers in the range of the northern spotted owl, like the Sacramento, Klamath, Rogue, Umpqua, Siuslaw, lower Columbia, Deschutes, Willamette, and Hoh, have been of ecological, economic,

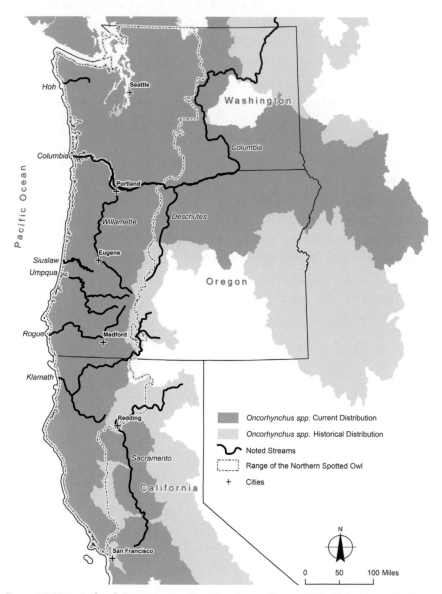

Figure 1.7. Historical and current range of Pacific salmon. (Source: Kelly Christiansen, Pacific Northwest Research Station.)

Figure 1.8. Threatened coho salmon swimming upstream to spawn in the Oregon Coast Range. (Source: NMFS, *Recovery Plan for Oregon Coast Coho*.)

and cultural importance in a multitude of ways for thousands of years, as have many smaller streams (fig. 1.7). Pacific salmon in these streams were a major reason for this importance through the ages as a source of sustenance and life for the people of the Pacific Northwest (fig. 1.8).[27]

Many streams supported abundant runs of coho (*Oncorhynchus kisutch*), Chinook (*Oncorhynchus tshawytscha*), and sockeye (*Oncorhynchus nerka*). Some streams also supported steelhead (*Oncorhynchus mykiss*; a seagoing rainbow trout) and sea-run cutthroat trout (*Oncorhynchus clarkii clarkii*), which are generally considered to be part of the Pacific salmon family. Other seagoing species like candlefish (*Thaleichthys pacificus*; locally known as smelt) and Pacific lamprey

(*Entosphenus tridentatus*) were also present. Thus, anadromous fish—fish that begin life in freshwater, swim to the ocean to grow and mature, and return to freshwater to reproduce—were a dominant component of the fish populations throughout the area.

Within the owl's range, the extent of historical habitat available to Pacific salmon has contracted relatively little over time, although dams in the upper Willamette River and other streams certainly impacted salmon distribution there, and dams on the Klamath River stopped salmon access to the upper watershed. The availability of much of the historical habitat contrasts with other areas of the Pacific Northwest, such as the Columbia River basin east of the owl's range, and the Sierra Nevada in California, where large extents of that habitat have been totally blocked by dams and other migration barriers (fig. 1.7).

Although the historical range of salmon and other aquatic species remains largely intact within the spotted owl's range, many human actions have greatly damaged these species. Conversion of lower watersheds into farms, ranches, towns, and cities; excessive harvest in commercial and recreational fisheries; loss of genetic integrity from the effects of hatchery operations; and water diversion for irrigation have all taken a toll.[28]

Logging and roadbuilding in forests on both federal and nonfederal lands also contributed to the decline of salmon historically through habitat destruction. While this issue was not at the forefront of the spotted owl crisis early on, protection and recovery of aquatic and riparian habitat soon joined protection of spotted owl habitat in conservation battles over federal forest management in the owl's range.

2

Sustained Yield—Pathway to the Greatest Good

Much of the National Forest System was established in the 1890s and early 1900s, and questions soon arose over how the national forests could best contribute to American life. Foresters led by Gifford Pinchot brought forth a vision of "conservative use of these resources [that] in no way conflicts with their permanent value" to assist in the economic and social development of the western United States.[1] In a few decades, that vision translated into providing a sustained yield of timber harvest over time as the best way for the national forests to help stabilize industries, families, and communities. Plans were built and legislation passed to enshrine sustained yield as the central tenet of national forest management for much of the twentieth century. The story of sustained yield forestry briefly covered in this chapter is essential background on how a federal forest management crisis developed in the Pacific Northwest.

Timber, Water, and the Rise of Forest Conservation

In the early days of the United States, most of the nation outside the original 13 colonies along the Eastern Seaboard was part of a public domain that the United States had acquired through purchase from foreign governments or wars with them. Federal policy called for shifting these lands into private hands once the federal government forced the Indigenous people on them onto reservations. Through this process, many fertile farmlands, timberlands, and gold fields across the United States moved into private ownership with little or no environmental restriction on how they were used.[2]

By the late 1800s, scientific and conservation groups became concerned that private ownership was resulting in plundered forests and degraded watersheds. These concerns and interests led Congress to

pass the Forest Reserve Act of 1891, which gave the president authority to reserve forests in the public domain; that is, to retain public
domain forests in federal ownership, preventing their disposition to
private hands. Over the next 16 years, most of the land that is now
national forest was reserved under this authority. In 1897 Congress
passed the Organic Act, which provided the statutory basis for forest
management in the newly created forest reserves. The act established
three purposes for the reserves: "to improve and protect the forest
within the reservation"; "[to secure] favorable conditions of water
flows"; and "to furnish a continuous supply of timber for the use and
necessities of the people of the United States." Thus, the act attempted
to balance protection and use while focusing on wood and water as
the two resources of greatest concern. It also included provisions for
controlling occupancy and use (by private citizens), and for regulating
the sale of timber "for the purposes of preserving the living and growing timber." Further, the act limited harvest to "the dead, matured or
large growth" that had been "marked and designated" and "cut and
removed under the supervision of some person appointed for that
purpose by the Secretary."[3]

Forestry in the Campaign to Save America

In 1905, Congress passed legislation transferring the forest reserves
established by the president from the Department of the Interior to the
Department of Agriculture, where they became the national forests.
The legislation also put these forests in the Division of Forestry, which
was then led by Gifford Pinchot, one of America's first foresters. On
the same day that President Theodore Roosevelt signed the legislation,
Secretary of Agriculture James Wilson sent Pinchot a letter (authored
largely by Pinchot himself) outlining the principles and policies to be
followed in management of the national forests.

In three short paragraphs, Gifford Pinchot expressed principles
that would guide management of the national forests for most of its
first hundred years:

• All resources are for use under such restrictions as will ensure
  their permanence.

- A continued supply of wood, water, and forage from these forests will provide an essential contribution to continued prosperity and economic growth of the West.
- Local questions will be decided on local grounds, with the dominant industry considered first and sudden changes avoided.
- Administration of each reserve will be left in the hands of the local officers, under the guidance of thoroughly trained and competent inspectors.
- Conflicting interests will be reconciled "from the standpoint of the greatest good of the greatest number in the long run."[4]

In implementing these principles, the new US Forest Service endorsed sustained yield forestry, control of grazing, protection of watersheds, and a cooperative relationship between local Forest Service rangers and dominant local industries. Federal forests were for use rather than preservation, as long as that use retained a long-run view in ensuring that these forests could provide wood, water, and forage on a continuing basis.

David Clary, former chief historian of the Forest Service, describes the importance of the Wilson letter in his masterful 1986 history *Timber and the Forest Service*:

> The letter was the Forest Service's Magna Carta; its philosophy and details permeate the agency's policies to this day. The evocation of "the greatest good of the greatest number" reflected the guiding spirit of the Progressive Era, in particular that of Pinchot, who was second only to Theodore Roosevelt as a spokesman—indeed a paragon—of the Progressives. He was on a crusade to transform American Society, and the Forest Service was with him. . . .
>
> The national forests must play their part in a campaign to save America—save it from timber famine, from the depredations of the "interests," and from the sorry social effects of America's way of harvesting timber. Seeking to build stable communities around the national forests, the Forest Service first must determine the social evils against which to guard its neighbors. It found three—communities ruined by "cut out and get out" lumbering, a homeless

underclass created by the nature of the timber industry, and, above all, the threat to the timber economy posed by monopolies.[5]

To better understand the extent of these problems, the Forest Service dispatched Samuel Trask Dana, who would later become dean of the School of Forestry and Conservation at the University of Michigan, on an inspection tour of Pennsylvania and the Great Lakes states in 1915 to examine the social history of traditional lumbering. His report offered many case examples of that history, including that of Cross Fork, Pennsylvania. Cross Fork had a population of a few families when, in 1894, a lumber mill opened there. The community quickly grew to two thousand people, with hotels, churches, a railroad, and an electrical plant. The mill consumed the local timber in 15 years and closed. Cross Fork quickly became a ghost town.

Dana found similar "deserted villages" all around the Lake States, surrounded by barren areas that had once been forest. As Dana recalled years later, "The way to avoid ghost towns was to practice sustained yield forestry."[6]

The Forest Service also commissioned a study of working conditions in the timber industry by Benton MacKaye (father of the Appalachian Trail and cofounder of The Wilderness Society). He reported the findings of a presidential commission that 90 percent of the workforce was unmarried men with incredibly high turnover rates: an industry of homeless men serving a tramp industry. In contrast, he believed that putting forest properties on a sustained yield basis would allow permanent communities of family men to replace the migratory camps.[7]

Progressives of the day, like Pinchot, were quick to identify the cause of devastated communities and hobo workers: the concentration of wealth and power in economic monopolies. Accordingly, Pinchot's first "Use Book" (1905), which covered principles for management of the national forests, promised that monopolies that disadvantaged other deserving applicants would not be tolerated.[8]

In sum, the Forest Service was determined to manage the national forests as instruments of social reform: stave off timber famine, improve the lot of lumberjacks, mill workers, and their families, promote community stability, and fight monopolies. To that end, the agency settled on sustained yield forestry, combined with great care in sale procedures to prevent monopolistic control, as the core policy that would enable the agency to act for the public good. As noted by Clary, "Sustained yield had been at the heart of the entire forestry campaign from the beginning."[9]

Such an approach also fit within the classical notion of a well managed forest, one in which annual harvest equaled annual growth forever. This regulated forest as a forestry ideal would be taught in the forestry schools across the United States and had deep roots in European forestry. However, the national forests were not in a regulated condition; rather, they were wild, old, and growing at a very slow rate as measured by foresters. How should foresters practice sustained yield forestry there?[10]

Prior to World War II, allowable cut calculations on the national forests were largely theoretical, as private timber was able to satisfy most of the demand for logs across the West. In fact, the firms that owned timberland often wanted the Forest Service to restrict its offerings, especially when lumber demand was limited and lumber prices low. That led some foresters and political leaders to urge that sustained yield be thought of as the continuous production of timber harvest volume in a particular geographic area considering both federal *and* nonfederal (mostly private) forestlands.[11]

In the early 1940s, as the demand for national forest timber increased to supply wood for the war effort, another concern surfaced about the ability of national forest timber to contribute to community stability: some way was needed to ensure that national forest timber harvest would be processed locally. To address this issue, and also to make it possible for the Forest Service to combine ownerships in allowable cut calculations, Congress passed the Sustained Yield Forest Management Act in 1944 with the following goal: "To promote sustained-yield forest management in order thereby (a) to stabilize communities, forest industries, employment, and taxable forest wealth; (b) to assure a continuous and ample supply of forest products; and (c) to secure the benefits of forests in regulation of water supply and stream flow, prevention of soil erosion, amelioration of climate, and preservation of wildlife."

With these words, Congress solidified sustained yield forest management as the heart of federal forestry and the key to providing multiple benefits from federal forests.

Specifically, the act authorized the establishment of either federal sustained yield units consisting of federal forestland or cooperative sustained yield units consisting of federal forest and private forest under coordinated management. In both cases, timber processing had to occur in specified local communities. Five federal sustained yield units were established in the United States, along with one cooperative unit—the Shelton Sustained Yield Cooperative Unit—in which the Simpson Timber Company was given a monopoly on timber on the Olympic National Forest near its mill in Shelton, Washington, in exchange for including its lands in the unit in allowable cut calculations.[12] By the early 1950s, sawmill operators and communities who had been excluded from the sustained yield units vigorously protested, which stopped the creation of more units.*[13]

With the coming of World War II and the housing boom that followed the war, demand for national forest timber skyrocketed. By the late 1950s, the Forest Service had settled on a definition of sustained yield that had long been utilized in Europe: each national forest would calculate the amount of timber that it could provide continuously over time and offer that amount for sale each year. The old growth would be metered at such a rate that the young growth, which came in after the initial harvest, could mature sufficiently to sustain the allowable cut into the future.[14] Through this policy, threat of timber famine would be permanently erased and a steady supply of employment provided.[15]

However, few limits would be placed on where the timber could be processed. Mills near the national forests would have to compete for the timber being offered for sale with those far away, weakening the ability to ensure that the local communities dependent on federal timber could be stabilized.

Old-growth forests would fuel the sustained yield engine (fig. 2.1). Those forests, thought to be stagnant, decaying, and wasteful, would then be replaced with rapidly growing young forest that would furnish

---

*After Simpson cut most of the mature timber on its lands in the Shelton Unit, the company turned to the Olympic National Forest for its harvests in the 1970s. The Shelton Unit was disbanded by mutual consent in the late 1980s after much of the national forest in the unit had been logged. Pictures of the large clearcut areas on the Olympic National Forest that resulted from these actions helped fuel public outrage over national forest management that led to the NWFP.

Figure 2.1. Logging old growth on the federal forests. (Source: Bureau of Land Management, Oregon and Washington.)

a continuous wood supply forever. Robbins aptly describes this plan in the second volume of his landmark environmental history of Oregon:

> As the 1950s advanced, industry officials and many public-sector foresters began promoting the idea of conversion, turning "decadent" old growth into fast-growing new stands with projected harvesting cycles of 80–100 years. Although the idea originated earlier in the century, the virtue of converting old-growth forests to fast-growing new stands became an article of faith—especially in Oregon, with its still sizable stands of virgin timber. There was a neat symmetry to the argument posited in an industrial science in which forests were turned into production driven factory regimes or plantation models.[16]

The emphasis on sustained yield, and the associated logging of old growth forests, went beyond the national forests. In fact, some argue that it was first established as a guiding policy for the Bureau of Land Management (BLM), another federal agency.[17] The BLM also controls forestlands in western Oregon, primarily former land grants given to railroad companies to pay for building a rail line from Portland to

Sacramento, which the federal government reclaimed because of fraud-
ulent activities by those companies.* The landscape pattern of much
of these lands reflected the common practice of giving the companies
alternate 640-acre sections for a distance from the rail line as payment
for their work, which would then be sold to settlers. Widespread fraud
in the disposition of those lands by the railroad companies caused the
federal government to reclaim them in the early 1900s. Their ultimate
retention in federal ownership created the checkerboard pattern seen
in western Oregon, especially southwestern Oregon, with much of
the intervening forest owned by timber companies (fig. 1.4). Congress
gave management direction for the reclaimed lands when it passed the
Oregon and California Revested Lands Sustained Yield Management
Act in 1937 (the O&C Act).[18] That act mandated that they be managed
for "permanent forest production . . . in conformity with the principle
of sustained yield . . . providing a permanent source of timber supply,
protecting watersheds, regulating stream flow, and contributing to the
economic stability of local communities and industries, and providing
recreational facilities." Thus, the O&C lands were formally committed
to the principle of sustained yield in statute.[19]

During the 1950s, the Forest Service faced increasing pressures
from the timber industry, livestock operators, recreationists, and wilderness advocates to put a higher priority on their particular interests.
The agency also perceived continuing threats from the Park Service to
take over national forest lands, as had happened in Washington with the
creation of Olympic National Park and demands for creation of North
Cascades National Park.[20] To help stem this tide, the Forest Service
sought congressional clarification of its mission, hoping to strengthen
its hand in balancing different uses and to clearly establish its legislative
authority as the preeminent provider of outdoor recreation.

The Multiple-Use Sustained-Yield Act of 1960, authored by the
Forest Service, authorized the secretary of agriculture to "develop and
administer the renewable surface resources of the national forests for
multiple use and sustained yield of the several products and services
obtained therefrom." The act named the multiple uses as "outdoor

---

* The BLM's western Oregon lands also included public domain lands that had not been
claimed by settlers and Coos Bay Wagon Road lands that were also revested to the
federal government. (These lands in aggregate are generally called the "O&C lands.")

recreation, range, timber, watershed, and wildlife and fish." The Forest Service had explicit legislation for management of the national forests for recreation at last.

The act defined multiple use as "management of all the various renewable surface resources of the national forests so that they are utilized in the combination that will best meet the needs of the American people"; and sustained yield as "the achievement and maintenance in perpetuity of a high-level annual or regular periodic output of the various renewable resources of the national forests." These definitions were interpreted by the courts as allowing the Forest Service largely to balance different uses and set harvest levels as the agency saw fit. In fact, the courts noted that the Multiple-Use Sustained-Yield Act "breathes discretion at every pore."[21]

In practice, the Forest Service's high regard for the potential for wood production on a sustained yield basis to contribute to the public good heavily influenced its interpretation of multiple use. As an example, the timber management plan of the Willamette National Forest in 1965 stated, "Let it be understood, therefore, that all resources on this Forest are of equal importance in the basic concept of multiple use management, and that no one resource shall be allowed to assume an over-riding position." But while no one resource should dominate, good timber management, according to the plan, was the "most useful tool for enhancing other uses."[22]

The Wilderness Movement Challenges Forestry's Merit

Until the 1960s, the Forest Service largely controlled its own destiny, with few laws to impede the agency in implementing its vision. The 1960 Multiple-Use Sustained-Yield Act did little to limit Forest Service discretion; rather, it solidified the agency's ability to use its judgment in pursuing the public good. However, passage of the Wilderness Act in 1964 signaled a changing public attitude toward Forest Service management, a change that would increasingly limit agency discretion.

The Forest Service had afforded special management attention to roadless areas for decades. As early as 1924, the agency managed some forest areas as natural, primitive, or wilderness areas, and regulations it adopted in 1929 gave high priority to the maintenance of primitive conditions in these areas with a view to conserving them for public education and recreation. Under the guidance of Forest Service employ-ees Bob Marshall, Arthur Carhart, and Aldo Leopold—men who went on to become revered advocates for wilderness—the Forest Service had classified almost 14 million acres across the West by 1939 as primitive areas and adopted regulations that prohibited commercial timber cuts, roads, and recreational camps within them.[23]

In the 1950s, the Forest Service conducted a review of classified primitive areas in the context of a new set of administrative wilder-ness and wild area designations it had developed. During this process, it became clear to local preservation groups in the Pacific Northwest, such as the Oregon Cascades Conservation Council, that the Forest Service was attempting to exclude commercial timberland from these classifications—to make more old growth available for the saw. This realization caused wilderness advocates in the Northwest to push hard for legislation to protect the values they held dear, adding their voices to the national clamor for a wilderness law.[24]

The Wilderness Act of 1964 designated 9.1 million acres of national forest lands as wilderness, mostly from portions of existing primitive areas that had little commercial timber on them. Timber harvest and permanent roads were banned. Many high-elevation areas of Oregon, Washington, and California were designated as wilderness, while the status of lower-elevation wild areas that often contained old-growth forest was left unresolved.[25]

The Wilderness Act also established a procedure by which Congress could designate additional wilderness areas in the national forests. It required that the secretary of agriculture review each primitive area on the national forests that remained after passage of the wilderness bill to determine whether those areas too were suitable for preserva-tion as wilderness. When so classified, an act of Congress would be required for those areas to achieve wilderness status.[26] The courts soon expanded that requirement to include all roadless areas contiguous to wilderness or primitive areas.[27]

Implementing the instructions in the Wilderness Act, as inter-preted by the courts, would still leave unresolved the status of millions of roadless acres that were not near other protected areas. Hence, in the late 1960s, the Forest Service began the study of all inventoried roadless areas on the national forests, which totaled 56 million acres or approximately one-third of the national forests, to get the wilderness designation issue behind it once and for all. This inventory included millions of acres of national forests in the owl's range, especially in the Cascade Range and Klamath Mountains.[28]

The commercial forest portions of inventoried roadless areas were, at that time, included in the land base used to calculate allowable cuts. The agency continued that practice during the analysis and debate over which areas should be reserved, under the assumptions that quick decisions would be made and relatively little commercial forest included— assumptions that were wrong on both counts, with dire consequences for the agency in the 1970s and beyond.

Intensive Management to Increase Allowable Cuts

The Forest Service had for decades advocated a shift to more intensive management of the nation's forests, arguing for forest practices that would increase the amount of wood fiber to meet the needs of a growing nation.[29] In the early 1960s, industrial, federal, and university researchers began to investigate technical and scientific solutions to the timber supply problem: refining clearcutting as a harvest practice; increasing the use of chemicals to suppress unwanted vegetation; utilizing precommercial thinning to maximize crop-tree growth; and experimenting with genetically selected supertrees. Weyerhaeuser Company led the way with an extensive research arm that spread the mantra of high-yield forestry. Federal and university researchers added to the chorus through advances that helped accelerate tree and stand growth. The ability to grow trees like corn had arrived.[30]

By the late 1960s, the conversion of natural old-growth forests into young-growth tree farms was well underway (fig. 2.2). Questions increasingly arose about whether intensified management would enable allowable cuts to be increased or at least maintained in the long run, as the agency's timber program exhausted the existing old-growth forests and shifted harvest to the tree farms of the future. The *Douglas-Fir Supply Study* of 1969[31] investigated these questions for the national forests of the Douglas-fir region (western Washington, western Oregon, and northern California) using new computer methods that enabled projections of timber harvest levels far into the future.

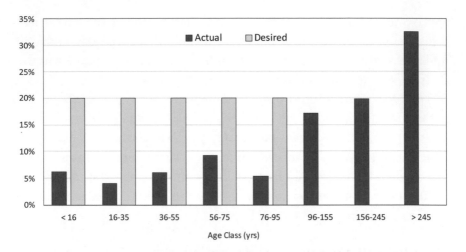

Figure 2.2. Actual and desired age-class distribution of the forest devoted to timber production in the national forests of western Oregon, western Washington, and northern California in the late 1960s, partway through the 100-year conversion of old-growth forests into tree farms. (Source: US Forest Service, *Douglas-Fir Supply Study*.)

The *Douglas-Fir Supply Study* concluded that the national forests would, in fact, need to intensify management to maintain or increase the existing allowable cuts in the long run. The ability to grow wood faster had arrived just in time. The study also showed that intensification would demonstrate extremely high rates of return, and made such actions easier to justify during the federal budgeting process, since it allowed an immediate increase in the rate of old-growth harvest. Using the results of this study, the Forest Service solidified and unified its approach to sustained yield: the national forests would set an allowable cut that it projected would stay level or increase over time (a nondeclining yield), looking out at least a rotation and a half, and would count expected results from intensified management in this calculation.[32]

The BLM similarly pursued management intensification on its forests in western Oregon as a mechanism to maintain and increase its allowable cut. By the early 1970s, federal forest managers in western Washington, western Oregon, and northern California were committed to practicing intensive management on most of their commercial forest. These practices included clearcutting, mechanical and chemical site preparation, genetic improvement, and precommercial thinning to create tree farms that maximized wood production.

Studies like the *Douglas-Fir Supply Study* encouraged the view that increased budgets for the Forest Service (and the BLM) could resolve conflicts over the use of natural resources, a view understandably popular with bureaucratic institutions. More money for timber production, recreation, watershed protection, and wildlife habitat improvement would enable production of a multitude of goods and services from the national forests and, most importantly, continued high timber harvest levels.[33] Perhaps money could, in fact, buy happiness.

Other issues soon arose, though, demonstrating that lack of funds might not be the primary constraint on timber harvest levels. Most importantly, federal managers relied on logging old-growth forests to fuel their mission to contribute to the greater good. This reliance would soon run aground as scientists and citizens began to realize that old-growth forests were highly complex ecosystems that provided many, many benefits beyond board feet.

PART 2

Setting the Stage for Major Change

3

Environmental Laws Reframe Federal Decision-Making

The 1960s were a time of ferment in the United States in many ways, including protest over the Vietnam War and the continuing struggle over civil rights in the South and elsewhere. Also, books like Rachel Carson's *Silent Spring*, which described the deadly effect of pesticides on the natural world, helped create a broad social movement demanding protection of the environment.

Concerns about the effects of unbridled industrial development and air and water pollution on human health and the natural environment came to the forefront of public consciousness in the late 1960s, leading to the first Earth Day in April 1970. Congress responded with the quick passage of a multitude of laws intended to protect the environment. Many of these laws fundamentally changed the goals and processes of federal forest management. Three in particular—the National Environmental Policy Act of 1969, the Endangered Species Act of 1973, and the National Forest Management Act of 1976—provided avenues for environmental groups to successfully challenge federal old-growth timber sales, helped establish a new framework for planning and managing federal forests, and led ultimately to development of the Northwest Forest Plan itself.[1]

An earlier law also played a critical role in changing federal decision-making. The Administrative Procedure Act (APA) of 1946[2] provided direction to federal agencies on rule making and opened federal decision-making processes to public scrutiny and participation for the first time. A key provision in the act prohibited *arbitrary and capricious* decisions by federal agencies. This provision established a standard of review that courts would utilize in evaluating federal decisions—to prohibit actions that were inconsistent with law or fact. Plaintiffs have argued that federal decisions were arbitrary and capricious in many environmental lawsuits that successfully challenged and invalidated proposed federal agency actions. Linked to the specific requirements of environmental laws, this standard became a powerful tool, indeed.

Recognizing Environmental Impacts of Federal Decisions

The National Environmental Policy Act (NEPA),[3] passed in late 1969, kicked off the environmental era. NEPA addresses all aspects of the environment and brings consideration of them into federal decision-making policy: "The purposes of this Act are: To declare a national policy which will encourage productive and enjoyable harmony between man and his environment; to promote efforts which will prevent or eliminate damage to the environment and biosphere and stimulate the health and welfare of man; to enrich the understanding of the ecological systems and natural resources important to the Nation."

Toward that end, NEPA requires that "all agencies of the Federal Government shall . . . utilize a systematic, interdisciplinary approach which will ensure the integrated use of the natural and social sciences and the environmental design arts in planning and in decision making which may have an impact on man's environment."

Before NEPA, primarily foresters and road engineers ruled the federal forestry roost. With NEPA's requirement for interdisciplinary planning, hydrologists, wildlife and aquatic biologists, other environmental specialists, sociologists, archaeologists, and anthropologists joined the planning team. These professionals brought not only technical knowledge but also different perspectives and values to the decision-making process.[4]

NEPA required that their concerns be systematically assessed, directing federal agencies to "identify and develop methods and procedures . . . which will ensure that presently unquantified environmental amenities and values may be given appropriate consideration in decision making along with economic and technical considerations."

To implement this approach to decision-making, NEPA called for a new kind of analysis and reporting before major federal actions were taken: "Include in every recommendation or report on proposals for legislation and other major Federal actions significantly affecting the quality of the human environment, a detailed statement by the

responsible official." This detailed statement became known as an *Environmental
Impact Statement* (EIS).

NEPA embedded the EIS in a structured
decision-making process that provided
multiple points for public involvement (fig.
3.1) and revolutionized the way the federal
government made decisions that affect the
environment. From timber sales to grazing
allotment renewal to culvert replacement
to outdoor concerts on federal lands, a
broad range of federal decisions now had
to go through environmental review.

However, Congress did not provide
guidance on how to decide which decisions are "major Federal actions significantly affecting the quality of the human
environment" and thus require an EIS.
Rather, Congress left that decision to the
new Council on Environmental Quality as
it formulated regulations to guide implementation of NEPA. The council mandated *Environmental Assessments* to help
determine whether an EIS was needed,
and *Categorical Exclusions* for actions
where minimal NEPA documentation was
required because of obvious lack of environmental impact.

NEPA and its implementing regulations
broadened the list of agencies, groups, and
individuals who could comment on, and
challenge, a decision by a federal agency.

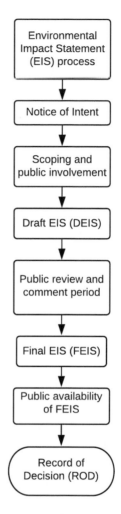

Figure 3.1. The NEPA
decision-making process
for major federal actions
significantly affecting the
quality of the environment.
(Source: J. Franklin et
al., *Ecological Forest
Management*, chap. 7.)

The act called for the proposing agency to seek comments from other
federal agencies with potential jurisdiction over, or special expertise
concerning, the environmental impacts that might occur, such as the
US Fish and Wildlife Service and the National Marine Fisheries Service (both of which administer the Endangered Species Act). These

regulatory agencies thus gained access to the formerly internal affairs of land management agencies.

NEPA regulations also required agencies to make diligent efforts to involve the public in preparing and implementing NEPA procedures and to respond to public comments about proposed actions. These requirements created a source of accountability that changed the relationship between federal agencies and the public. With the public able to question how completely an agency divulged environmental consequences, proposed actions could be scrutinized in ways they had not been before and provided fodder for much NEPA-based litigation.

However, NEPA does not protect the environment through substantive environmental standards or by regulating actions, as do other laws such as the Endangered Species Act. Rather, NEPA is often referred to as a *procedural* or *stop-and-think* law because it requires a certain analysis and documentation process for decisions affecting the environment. Federal courts have repeatedly said that the role of NEPA is to demonstrate that federal agencies have taken a hard look at the potential impacts of their proposed actions to be sure that those impacts are understood well enough to make environmental surprises unlikely.[5] The courts have also noted that it is the responsibility of agency personnel to decide how to use the resulting information: "Other statutes may impose substantive environmental obligations on federal agencies, but NEPA merely prohibits uninformed—rather than unwise—agency action."[6]

Even with these limitations, NEPA had an immediate and major impact on national forest management when the Forest Service published its *Roadless Area Review and Evaluation* (RARE) report in 1972. That review gave the agency's recommendations as to which inventoried roadless areas, covering one-third of the national forests, should become wilderness and which should be available for development. Lawsuits were filed claiming that the agency had not sufficiently complied with NEPA regulations—in fact, the agency had not followed the relatively new NEPA process at all. The courts agreed and halted development (i.e., road construction, timber sales, and other activities) in roadless areas until the Forest Service completed an EIS documenting the environmental impacts of its recommendations.[7]

A second roadless review, *Roadless Area Review and Evaluation II* (RARE II), was published in 1979 utilizing the NEPA process. When a lawsuit challenged the Forest Service's decision to identify a number of RARE II areas in California as unsuitable for wilderness designation, the courts ruled that the RARE II process violated NEPA because it was insufficiently site specific and failed to consider an adequate range of alternatives. The court also affirmed an injunction against release of areas being considered for wilderness designation until the Forest Service met the legal requirements of NEPA.[8]

Congress then began a state-by-state approach to designating roadless areas as wilderness under the provisions in the Wilderness Act of 1964. Northwest legislators, such as Senator Hatfield and Congressman AuCoin and their staffs, worked diligently in the 1970s and 1980s to add particular roadless areas in their states to the wilderness system.[9] However, many roadless areas remained in limbo for decades. In the meantime, concentration of harvest in the portion of the forest that remained accessible created the visual and ecological specter of over-cutting. Jack Usher, the director of timber management in the Pacific Northwest Region (Oregon and Washington), summed up the resulting predicament in which the Forest Service found itself:

> For more than a decade the allowable cut has included timber volumes and anticipated tree growth in some of these roadless areas, even though the actual cutting has been in areas where roads are already established. . . . We have assumed that the political controversy would be resolved quickly, and no area would be heavily impacted by this intense harvesting. . . . But when it goes on for a decade, we think we may be perilously close to violating our soil and watershed protection requirements in some of our roaded areas.[10]

Some of those heavily cut accessible areas became poster children for the "environmental destruction" that Forest Service management had caused in the range of the northern spotted owl during the debates that led to the Northwest Forest Plan.

The RARE/NEPA process gave the Forest Service a clear lesson that times had changed, and that it was no longer solely in charge of its own destiny. NEPA had interjected other federal agencies into evaluation of the environmental impacts of these Forest Service actions. Even more importantly, the public process prescribed in the NEPA regulations enabled environmental groups to gain access to federal agency explanations and justifications for proposed actions—a general requirement for transparency and accountability that would prove extremely useful in litigating proposed federal forest actions in the range of the spotted owl in the late 1980s and early 1990s.

Conserving Threatened Species and Their Habitats

The Endangered Species Act (ESA) of 1973[11] may be the United States' most controversial and powerful environmental law. It is difficult to overestimate its significance or impact. The ultimate goal of the ESA is to stop and reverse the trend toward biological extinction of imperiled native species, and to recover those species to the point where federal protection is no longer required.

The ESA establishes two categories of species: *endangered species*, "any species which is in danger of extinction throughout all or a significant portion of its range"; and *threatened species*, "any species which is likely to become an endangered species within the foreseeable future throughout all or a significant portion of its range." Anyone can petition the federal government to list a species in one of the two categories. The list of species protected under the ESA includes mammals, birds, fish, mollusks, amphibians, reptiles, conifers, flowering plants, grasses, ferns, lichens, and insects. While delisting has happened only occasionally, there have been quite notable success stories: bald eagles, peregrine falcons, and gray whales have all been delisted after increases in their populations.

The ESA calls for the government to use the best available scientific and commercial data in determining whether a species is threatened or endangered. However, if citizens do not think the government used the best information available in assessing a threat, or think authorities misinterpreted this information, they can sue to force the agency to reconsider, claiming that the agency made an arbitrary and capricious decision. Environmental groups from the Pacific Northwest successfully brought such a case in federal court when the US Fish and Wildlife Service (USFWS) declined to list the northern spotted owl in 1988. That directly led to the USFWS decision to list the owl as threatened in 1990.

The ESA contains language with significant consequences for forest policy in the United States: "The purposes of this Act are to provide a means whereby the ecosystems upon which endangered species and threatened species depend may be conserved, [and] to provide a program for the conservation of such endangered species and threatened species." Note the use of the term *ecosystems*—one of the first references to that concept in any congressional legislation. Moreover, Congress put the conservation of threatened and endangered species into the job description of all federal agencies: "All Federal departments and agencies shall seek to conserve endangered species and threatened species and shall utilize their authorities in furtherance of the purposes of this Act." Furthermore, "The terms 'conserve,' 'conserving,' and 'conservation' mean to use and the use of all methods and procedures which are necessary to bring any endangered species or threatened species to the point at which the measures provided pursuant to this Act are no longer necessary."

The ESA calls for critical habitat to be designated at the time of listing—habitat that is essential to survival of the listed species and that requires special management. In actuality, the designation of critical habitat often comes a considerable time after listing, for a variety of reasons. In the case of the northern spotted owl, the USFWS has designated and updated critical habitat several times in response to changes in scientific information, changes in the political environment, and court order.

The ESA makes it unlawful for any person to take an endangered species, defining *take* as "to harass, harm, pursue, hunt, shoot, wound, kill, trap, capture, or collect, or to attempt to engage in any such conduct." The verb *harm* is the key word in this definition for forest management operations. Harm includes habitat modification where actual injury to wildlife can be demonstrated and includes "significant habitat modification or degradation where it actually kills or injures wildlife by significantly impairing essential behavioral patterns, including breeding, feeding or sheltering."[12] The USFWS (and the National Marine Fisheries Service, NMFS) also has broad authority under the ESA to issue take regulations for the conservation of threatened species. Using this authority, the USFWS extended the prohibition of take to the northern spotted owl once it was listed as threatened.

An early legal test established that the ESA was an immensely powerful law in directing the goals and actions of federal agencies. Hiram (Hank) Hill was a second-year law student at the University of Tennessee whose biology professor and friend had discovered the snail darter while scuba diving in the Little Tennessee River. It appeared that the darter existed only in the reach below a dam being constructed on that river. Hill then wrote a paper for his environmental law class in 1975 on the dam's threat to the existence of the darter and how the newly adopted ESA might come to the aid of the tiny fish.[13] This ultimately led to the US Supreme Court's decision in *Tennessee Valley Authority v. Hill* in 1978, in which the snail darter, which had been listed as threatened under the ESA in 1975, prevailed over the economic interests represented by the dam. In that case, the court ruled that saving endangered species was the first priority for federal agencies:

> The plain intent of Congress in enacting this statute [the ESA] was to halt and reverse the trend toward species extinction, whatever the cost. This is reflected not only in the stated policies of the Act, but in literally every section of the statute. . . . In addition, the legislative history . . . reveals an explicit congressional decision to require [federal] agencies to afford first priority to the declared national policy of saving endangered species. The pointed omission of the type of qualifying language previously included in endangered species legislation reveals a conscious decision by Congress to give endangered species priority over the "primary missions" of federal agencies.[14]

In response to this decision, Congress passed legislation in 1978 amending the ESA and establishing the Endangered Species Committee, which is often called the *God Squad* because the committee can determine whether a listed species is conserved or is allowed to go extinct. The God Squad can exempt federal actions from the ESA if it deems the cost of compliance too high. Very few exemptions have been granted, and this amendment to the ESA has generally fallen into disuse. It briefly came back to life in the early 1990s over protection of the northern spotted owl but had little impact on policy outcomes, as described in chapter 6.

Two cabinet-level secretaries are responsible for implementing the ESA. The secretary of the interior, through oversight of the USFWS, oversees threatened and endangered species that are not oceangoing, such as the northern spotted owl and the bull trout. The secretary of commerce, through oversight of the NMFS, oversees threatened and endangered species that spend all or part of their lives in the ocean, such as anadromous salmon. These two secretaries, through their agencies, take care of listing, protection, and delisting processes.

Major categories of responsibilities for the conservation of a listed species differ between federal and nonfederal land managers. Federal land managers have responsibilities related to take, consultation, and recovery of a listed species; nonfederal landowners and managers have responsibilities related only to take.[15]

Each federal agency must ensure that any action it authorizes, funds, or carries out is not likely to jeopardize the continued existence of any endangered or threatened species or result in the destruction or adverse modification of critical habitat of such species. Toward that end, federal agencies must consult with the USFWS or NMFS. If the USFWS or NMFS concludes that jeopardy or adverse modification will occur, that agency will suggest alternative actions consistent with the ESA. If the alternatives are not acceptable to the agency requesting consultation, it will usually call off the proposed action or modify it in some substantial way and send it back for more consultation. In rare cases, the agency may seek relief from the God Squad.

The ESA also calls for the development of recovery plans for listed species. Recovery plans are comprehensive landscape plans that provide a template for both the spatial and temporal recovery of the listed species. For species with a large geographic distribution, such as the northern spotted owl, a recovery plan provides important guidance on federal forest management over a very large landscape.

Reforming National Forest Management

Clearcutting on the national forests became a national issue in the 1960s as the Forest Service implemented intensive timber management across the United States. Congressional action to control clearcutting through discretionary guidelines did little to satisfy critics. In 1973 a local chapter of the Izaak Walton League brought suit to halt that practice on the Monongahela National Forest in West Virginia. The lawsuit was not brought under the recently passed environmental laws; rather, it alleged that Forest Service harvesting practices violated the Organic Act of 1897.[16]

The Organic Act authorized the Forest Service to sell "dead, matured, or large growth of trees" and required that such timber be "marked and designated" and "cut and removed under the supervision of some person appointed for that purpose by the Secretary." The argument before the Fourth Circuit Court of Appeals centered on the meaning that Congress, originally and in subsequent legislation, intended with the words *dead, matured, large growth, marked and designated*, and *cut and removed*.

Interpretation of these words could affect all federal timber harvesting practices, not just clearcutting.[17] The court held that by cutting immature and unmarked trees and also frequently not removing them, which the Forest Service did as part of its clearcutting practices, the agency had exceeded its authority. The decision did not specifically ban clearcutting, but it was difficult to see how it could occur while meeting the provisions of the Organic Act.[18] With much national forest timber harvest coming from clearcutting, the Forest Service predicted that the court's decision could have dire consequences for the economies of rural communities throughout the western United States.

Congressional leaders rejected the notion of a simple elimination of the clauses at issue in the Organic Act and instead entertained comprehensive approaches to addressing persistent problems and ambiguities in management of the national forests relative to timber harvest and timber production. The troublesome clauses from the Organic Act could be eliminated as part of legislation, but at the price of much more congressional direction of national forest management. The floor leader of the legislation in the Senate set the tone for what was to come: "The days have ended when the forest may be viewed only as trees and trees viewed only as timber. The soil and the water, the grasses and the shrubs, the fish and the wildlife, and the beauty of the forest must become integral parts of the resource manager's thinking and actions." Hubert Humphrey, 1976.[19]

The resulting National Forest Management Act (NFMA) of 1976[20] reaffirmed congressional intent that multiple use and sustained yield, as described in the Multiple-Use Sustained-Yield Act of 1960, were to remain foundational purposes of national forest management.[21] Within that context, much of NFMA's language provided further congressional definition of what these terms meant, especially regarding timber production and timber harvest. Many of these provisions dealt with technical issues on which Congress could give only general guidance. Congress therefore deferred to the executive branch, with its substantial expertise embedded in the agencies it controlled, to develop the specifics: "The Secretary [of Agriculture] shall . . . promulgate regulations . . . that set out the process for the development and revision of land management plans, and the guidelines and standards prescribed by this subsection."[22]

NFMA called for an integrated approach to national forest planning, employing many of the same processes prescribed under NEPA, including preparation of land management plans by an interdisciplinary team and public participation in the development, review, and revision of such plans. Each management unit (generally a national forest) would need to develop an integrated multiple-resource plan in a single document, unlike in the past when the agency developed separate plans for timber, range, recreation, and other resources.

In developing that integrated plan, NFMA required the Forest Service to "provide for diversity of plant and animal communities based on the suitability and capability of the specific land area in order to meet overall multiple-use objectives"[23]—the first commitment to protecting biodiversity in federal statute. Regulations implementing NFMA attempted to apply this broad mandate to conserve biodiversity through, in part, the concept of "management indicator species." These species were to be "selected because their population changes are believed to indicate the effects of management activities."[24] Many national forests selected the northern spotted owl as a management indicator species for the health of old-growth forests and of other species within them.

NFMA also required the promulgation of regulations to provide for the public notice of and comment on "the formulation of standards, criteria, and guidelines applicable to Forest Service programs."[25] While this sounds similar to NEPA's public comment process, the Forest Service interpreted this requirement as a need to provide for administrative review of agency decisions, in which higher levels of the agency could be asked to review decisions made at lower levels. That type of review played an important role in challenging agency decisions about conservation of the northern spotted owl.

Committee of Scientists Shapes NFMA Regulations

NFMA mandated that a Committee of Scientists, who were not employees of the Forest Service, be appointed to assist the secretary of agriculture in promulgating regulations.[26] This committee had an enormous impact on the 1979 regulations implementing NFMA, both because the committee broadly interpreted its charge and because the secretary deferred to its judgment.

As an important part of its work, the Committee of Scientists prescribed environmental safeguards to help limit and direct national forest management, with the most important provision stating that "fish and wildlife habitat shall be managed to maintain viable populations of existing native and desired non-native vertebrate species in the planning area."[27] Thus, NFMA regulations issued in 1979, which were called the Planning Rule, mandated a major new goal for the national forests: maintaining habitat for viable populations of native vertebrates in the planning area.

One member of the Committee of Scientists, William Webb, who was an ornithologist and university professor, guided development of the viability requirement. Thomas Wellock used the notes of the Committee of Scientists' meetings in the late 1970s to describe Webb's approach in his story "The Dicky Bird Scientists Take Charge":

> Webb became the chief spokesman for the Service's biologists. He believed NFMA's diversity language could advance the cause of saving wildlife. He complained that the Service had traditionally given "non-timber values of the forest a minor place. . . . Now there is a different direction!" As Webb read the NFMA, Congress intended that the "planning process starts with the assumption that all resources of the forest are equal in value." The Service was "no longer to maximize timber production but to manage public lands for public benefit."*[28]

---

* Legal scholars later confirmed Webb's view: "When the section [on diversity of plant and animal communities] is read in light of the historical context and overall purposes of the NFMA, as well as the legislative history of the section, it is evident that section 6(g)(3)(B) requires Forest Service planners to treat the wildlife resource as a controlling, co-equal factor in forest management and, in particular, as a substantive limitation on timber production." Wilkinson and Anderson, *Land and Resource Planning*, 296.

Wellock then summarized the extraordinary nature of the regulation that Webb helped develop: "With very little debate, Webb and the Forest Service staff created a whole new class of protected wildlife not covered by existing federal law. In effect, the Forest Service required of itself that it safeguard minimum populations of every [vertebrate] species on the national forests."[29]

However, the initial statement on viability in the 1979 Planning Rule gave inadequate direction for forest planning. Most specifically, it did not clarify whether the requirement could be satisfied in only a few isolated areas on a national forest—an issue that national forest planners were starting to ponder when forest planning under NFMA began in earnest in the early 1980s.

The Forest Service tasked a small group of its wildlife biologists to come up with an answer to this puzzler, including Steve Mealey (who had been explaining the Planning Rule to national forest planning teams,) Hal Salwasser (who was the Forest Service's lead wildlife biologist,) and others. They contacted academic conservation biologists such as Michael Soule for help. Given the state of knowledge at the time, though, the Forest Service biologists concluded that only general policy guidance would be possible, such as distributing habitat over the planning area.[30]

At about the same time, newly elected president Ronald Reagan set up a Task Force on Regulatory Relief, chaired by Vice President George H. W. Bush. The task force had the goal of reducing the regulatory burden of federal agencies, and it selected the Planning Rule for review.

The Forest Service asked the Committee of Scientists to assist in the review, which it was glad to do. The committee was proud of its effort on the 1979 Planning Rule and was deeply interested in how this new approach to federal forest planning was working out. Its members also wanted to assist the Forest Service in interpreting the Planning Rule's meaning where the wording and concepts proved confusing, and they readily deferred to each other's expertise on particular issues.

Wildlife requirements in the Planning Rule, especially the viability clause, were the focus of a phone call between Webb and Forest Service planning and wildlife staff (including Mealey) during that review. They discussed the shortcomings of viability metrics and procedures as well as the distribution problem that was puzzling planners. After some discussion. Mealey suggested that the phrase *well distributed in the planning area* be added to the 1979 viability requirement. Webb acknowledged that the suggested addition appeared to meet his intent to provide for viable populations—in the blink of an eye and a few words, the viability requirement was greatly strengthened, all in the name of reducing regulatory burdens.[31]

The updated viability standard appeared as part of a revised Planning Rule in 1982, with a viable population defined as "one which has the estimated numbers and distribution of reproductive individuals to ensure its continued existence is well distributed in the planning area."[32] Meeting that viability standard would become central to litigation over protection of the northern spotted owl and the foundation of the conservation strategy developed for the Northwest Forest Plan.* As Jack Ward Thomas later opined, that obscure clause "shook the agency more than any other legislative or regulatory requirement in its history. It boiled down to two words, 'viable populations.'"[33]

The Committee of Scientists also developed an additional level of planning that had not been mentioned in NFMA: "A regional guide shall be developed for each administratively designated Forest Service region. . . . Regional guides shall provide standards and guidelines for addressing major management concerns that need to be considered at the regional level to facilitate forest planning."[34] Essentially, regional guides provided big picture direction to national forests within a region on management issues best addressed at very large—that is, regional— scales. In turn, forest plans for individual national forests would provide additional direction based on local conditions.

The *Regional Guide for the Pacific Northwest Region* covered appropriate harvest methods by geographic area and forest type as well as processes for deciding the maximum size of clearcuts, among other standards and guidelines. It also prescribed a conservation strategy for the northern spotted owl, but the strategy did not survive legal review. That failure, after repeated attempts, led the agency to hand the

---

* The courts affirmed during the spotted owl litigation that the viability clause was a permissible interpretation of the diversity provision in NFMA. "Consistent with the statute, the implementing regulations provide that 'fish and wildlife habitat shall be managed to maintain viable populations of existing native and desired non-native vertebrate species in the planning area.'" Seattle Audubon Society v. Moseley (No. C92-479WD, July 2, 1992).

problem to independent scientists to resolve, a novel approach to forest planning that sent the agency in new directions.

Planning for Sustained Yield Proves Inadequate to Meet the Law

Perfecting and applying the integrated planning approach mandated by NFMA consumed the Forest Service for much of the 1980s. Toward that end, the forest planning effort under NFMA in the early 1980s focused on estimating the maximum feasible sustained timber yield possible under the many environmental goals and considerations prescribed in NFMA and other recently passed environmental laws. Norm Johnson helped develop models to do such calculations, and one (FORPLAN) became the primary forest planning tool for implementing the National Forest Management Act.[35]

However, the central planning problem had changed to a search for scientifically credible conservation strategies for at-risk species and protection of biodiversity in general. That planning problem, in turn, was wrapped inside a social context of increased demands for public participation in decision-making, a loss of trust in technical solutions to planning problems, and a new appreciation for the wonders of forests, especially old-growth forests, and the creatures within them.

As the Forest Service constructed its forest plans under NFMA in the early 1980s, it could not achieve a high enough harvest level to satisfy the Reagan administration. Thus, draft plans were repeatedly returned to the Forest Service with instructions to evaluate whether a higher harvest level might be possible, delaying plan completion year after year until the late 1980s, when they were finally released.

In the meantime, the allowable cut train rolled on. The Forest Service had large staffs of foresters and engineers skilled at, and committed to, providing the nation with wood products. The economic health of many rural communities throughout the West depended on federal logs being delivered to their mills. The Oregon and Washington congressional delegation, a powerful force in Congress at the time, was dedicated to maintaining the high harvest level and providing the needed budgets for doing so. Old-growth logging in the range of the spotted owl continued unabated.

The Forest Service's release of final versions of NFMA-prescribed regional and national forest plans and associated NEPA documents in the late 1980s, after years of delay, opened the floodgates of appeal and litigation that would ultimately lead to the Northwest Forest Plan. From the prohibition on arbitrary and capricious behavior in the APA, to the need to divulge the environmental effects of actions in NEPA, to the priority on recovery of threatened species and the ecosystems on which they depend in the ESA, to the diversity and viability protections for a multitude of species in NFMA, a revolution was coming in the management of old-growth forests on federal lands. The allowable cut train would be stopped at last.

4

Wild Science Creates New Understandings

Significant research on forest management began with the emergence of the forestry profession and the Forest Service in the United States in the early twentieth century. The agency's research branch had an independent line of authority directly to the chief in order to protect the science program and its publications from direct control by the management personnel administering the national forests. Most of its research was intended to help foresters become more efficient in achieving their timber management goals, such as establishing tree reproduction, growing and harvesting timber, and protecting forests from damaging agents such as fire, insects, and disease. Research in forestry schools similarly focused on timber management, with the support of federal programs and agencies.

Such *domesticated science* was designed largely to assist forest managers to do better what they had already decided to do. Scientists directed little effort toward learning about the natural history of forests and about functions other than timber production, except for some watershed research to improve knowledge of how forest conditions influenced water yields.

Most Forest Service research prior to World War II was field oriented, emphasizing growth and yield studies in forested environments, including putting in large numbers of permanent sample plots in the Pacific Northwest and elsewhere. In the postwar era, the Forest Service deemphasized such research, and programs at experimental forests languished. Instead, the Forest Service built laboratories in many locations (e.g., Olympia, Corvallis, Wenatchee, and Bend) with sophisticated scientific instrumentation, such as electron microscopes and growth chambers, to support physiological and microbial research.

Congress's creation of the National Science Foundation (NSF) in 1950 led to a new source of funding for basic biological (including ecological) research. After a few years of very modest budgets, the challenge of the Russians' Sputnik success led quickly to larger budgets and greater interest in science unconnected to its immediate practical application.

NSF preferred basic, curiosity-driven research to develop and test scientific theories over efforts to solve practical problems, and it provided most of its financial support through competitive grants to individual scientists in traditional university departments, such as biology, botany, zoology, and ecology. NSF initially discouraged research proposals from scientists in applied sciences, such as agriculture, forestry, and wildlife management; peer-review panels considered these practical sciences to have other sources of funding, and proposals that smelled of management applications had little prospect of NSF funding.

Attitudes at NSF began to change in the early 1960s with modest funding for the emerging and controversial topic of ecosystem science. Congress provided NSF with funding for the US International Biological Program, some of which it directed at ecosystem-level research in the Western Coniferous Forest Biome Project.

Funding for the Western Coniferous Forest Biome Project had a profound impact on forestry research in the Pacific Northwest. Scientists receiving the funds were fiscally independent and thereby largely free of control by the Forest Service or its cooperators in the various forestry schools across the West. They could pursue scientific topics that they deemed most important to understanding the region's coniferous forest ecosystems without regard for any limitations on what they studied.

This *wild science*, although methodologically rigorous, can produce consequences for policies and practices that are largely unpredictable.[1] Knowledge generated from such efforts can challenge assumptions underlying management policy and approaches, rather than simply facilitating them. Hence, agencies and management professionals may view such science as subversive and threatening to existing policies and practices. It often takes a bold resource manager to welcome and apply such science.

This chapter focuses on wild science that challenged assumptions underlying classical federal forest management policies and created both the need for the Northwest Forest Plan and the concepts that guided it. Three scientific efforts that took this approach are examined in terms of how they occurred, their major findings, and why they were important.

First, the chapter describes the study of an old-growth forest ecosystem on the HJ Andrews Experimental Forest by a scientific team funded largely by NSF. Second, it considers Eric Forsman's personal and professional effort to understand the habits of a small forest owl that few had ever seen. Finally, the chapter traces how a scientific team, which included Gordie Reeves, helped transform our knowledge of ecological processes underlying aquatic ecosystems in coastal forests.

Discovering an Old-Growth Forest Ecosystem

Funding for the Western Coniferous Forest Biome Project was split between the University of Washington's College of Forest Resources and a team of Oregon State University and Forest Service scientists in Corvallis. After considerable internal debate, the Corvallis team focused their research on sites at the HJ Andrews Experimental Forest, located within the Willamette National Forest. Richard Waring (an OSU scientist) and Jerry Franklin led the research team on the university and agency sides, respectively.

The Andrews was an exception to the deemphasis on experimental forests. Established as the Blue River Experimental Forest in 1948 and renamed in 1953, the forest was the focus of research on how to most effectively convert old-growth Douglas-fir forests to young, productive managed forests. Research conducted there emphasized designing efficient road system and harvest unit layouts to facilitate the dominant Forest Service management strategy of clearcutting old growth followed by prompt tree regeneration to jump-start the next crop of trees. Also, scientists calibrated watersheds for later studies to examine road and harvest impacts on water yield and quality.

The Andrews team felt free to select its own research topics under the Forest Biome Project. Jerry Franklin had a longtime interest in understanding old-growth forests, and at last the funding for such an effort was at hand (see Franklin's story on pages 82–83). After some literature review and discussion about other biome efforts, the team decided to focus on developing carbon, hydrologic, and nutrient budgets for an old-growth forest ecosystem occupying a 25-acre watershed (Watershed 10) on the Andrews. This would enable scientists to trace the flow of carbon, water, and nutrients into, out of, and through that ecosystem. The Andrews ecosystem team was born.

The team chose Watershed 10 on the Andrews for the study area because Forest Service scientist Richard Fredriksen had a long-term research project there that could provide data on streamflows and water chemistry needed for the hydrologic and nutrient budgets. Waring and Franklin, with the urging of Jim Hall who was a fish biologist in the Department of Fish and Wildlife at OSU, committed to including aquatic and fish biology in the project too. However, Watershed 10 lacked fish, so a newly minted Andrews stream team studied both Watershed 10 and a larger fish-bearing stream on the Andrews.

NSF funded the Forest Biome Project for five years of study, a relatively short time for complex and highly focused research. Although the original study group included graduate students, the short-term nature of the project made their involvement impractical, so Waring and Franklin recruited a group of postdoctoral associates to replace the students. Fortunately, they found a group of young postdocs who made major contributions to this effort, including James Sedell (an aquatic biologist) and Stan Gregory (a fish biologist), who illuminated forest/stream interactions; and Fred Swanson (a geomorphologist), who advanced knowledge on geologic disturbance processes in forests. Also, the group included a postdoc (Kemick Cromack) and a graduate student (Mark Harmon) who led investigations into the decomposition and ecological importance of dead wood. In addition, Susan Stafford, a new assistant professor in the Department of Forest Science at OSU, contributed mightily to research information management at the Andrews—to help bring order out of the chaos of scientific discovery.

Research on old-growth forests developed rapidly, albeit with many adjustments to initial plans. For example, a plan to study the canopies of old trees by cutting them down did not work: after a tree was felled, its leaves and branches scattered over hundreds of square feet and defied reconstruction. Climbing trees and making measurements in place would be necessary; two graduate students from the University of

Figure 4.1. Slowly decaying downed logs in an old-growth Douglas-fir/western hemlock forest can persist for several centuries after the death of the trees themselves. Understanding their ecological importance was a major advance in describing old-growth ecosystems.

Oregon pioneered the approach using mountain climbing techniques to assist the ascents.

A critical early epiphany was the simultaneous recognition by both terrestrial and aquatic scientists of the importance of dead wood in old-growth forests and streams. It seems incredible that the importance of dead trees and their several forms, such as snags, logs, and other woody detritus, had gone essentially unrecognized by scientists and resource managers up to that time (fig. 4.1). Even more remarkably, existing syntheses of global forest carbon budgets failed even to have a category for large dead wood, nor did analyses of annual dead organic matter production (e.g., litterfall) recognize the contributions of tree mortality.

Recognition of the importance of dead wood emerged quickly in efforts to quantify the carbon and energy budgets in old-growth forests and streams: the scientists learned that there would be no balancing of stream or terrestrial cycling of carbon without a full accounting of big dead wood. In the northwestern conifer forests, dead trees could not be ignored since they typically totaled 20 percent or more of the existing forest organic matter. Fireside chats and data tabulations highlighted that, in addition to the big old trees themselves, large snags and logs on land and in streams were critical to the massive carbon accumulations characterizing the old forests and to providing habitat for old forest–dependent animals and creating structurally complex streams and forest stands.

The Andrews ecosystem team's analyses of animal, plant, and fungal life-forms dependent on old forests made apparent the central role of dead wood as essential habitat. Scientists had understood the importance of snags to primary and secondary cavity dwellers for some time, but not so the downed logs. The Andrews work helped establish the importance of dead wood structures as homes, sources of food, and protection for a wide variety of animals, and it found that snags and logs often contain more biomass of living organisms than they did as live trees, when decomposers (such as beetles) are included.

The Andrews ecosystem and stream teams also found that dead wood created structural complexity in streams. Logjams, sandbars, and pools, including those associated with wood-generated falls, were important contributors to aquatic habitat and essential to creating channel complexity and to retaining litterfall that provided the base of food webs in small streams.

For the first time, it was possible to describe the central role that large, old trees in all their forms played in an old-growth ecosystem (fig. 4.2).[2] As the Andrews ecosystem and stream teams expanded their studies to younger forests, to rivers, and, most profoundly, to effects of forest disturbances (see the discussion of Mount Saint Helens on page 74), it became clear that dead wood was a critical structural element or building block of essentially all natural forest ecosystems, terrestrial and aquatic. This represented a fundamental reversal of forestry's historical view of dead trees and woody debris as nothing but waste, fuel for future fires, and a hazard to humans.

The Andrews ecosystem team had developed a first approximation of old-growth forest ecosystem structure and function as the end of Forest Biome Project funding approached in 1975. Clearly, such forests had very complex physical structures and high levels of biodiversity; they also had high levels of productivity as measured by the absorption of carbon dioxide through photosynthesis. Foresters, however,

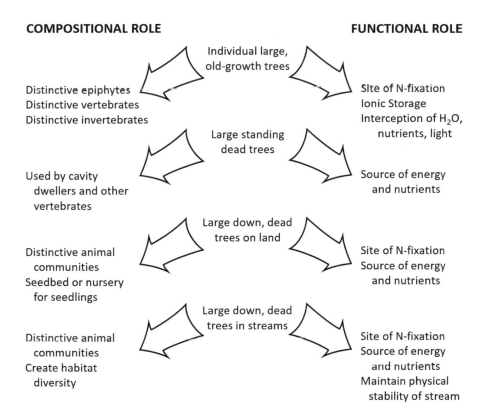

COMPOSITIONAL ROLE

FUNCTIONAL ROLE

Individual large,
old-growth trees

Distinctive epiphytes
Distinctive vertebrates
Distinctive invertebrates

SIte of N-fixation
Ionic Storage
Interception of H_2O,
nutrients, light

Large standing
dead trees

Used by cavity
dwellers and other
vertebrates

Source of energy
and nutrients

Large down, dead
trees on land

Distinctive animal
communities
Seedbed or nursery
for seedlings

Site of N-fixation
Source of energy
and nutrients

Large down, dead
trees in streams

Distinctive animal
communities
Create habitat
diversity

Site of N-fixation
Source of energy
and nutrients
Maintain physical
stability of stream

Figure 4.2. The central roles played by large, old trees in forest ecosystems, in both their live and dead forms, were a key finding in the HJ Andrews scientists' initial synthesis of scientific knowledge about old-growth forests. (Source: J. Franklin et al., *Ecological Characteristics*.)

viewed the productivity of old forests as low because they traditionally measured productivity in terms of annual net additions of live wood. Once the ecosystem team included all the biomass in the forest (live and dead), though, it reached a different conclusion.

The Andrews ecosystem and stream teams also demonstrated important reciprocal relationships between forests and streams as the terrestrial and aquatic scientists documented the influences and flows of structure, energy, and animal and plant life across the forest and stream interface and the importance of wood in streams to all these processes. No previous research group in the PNW had been willing to leave their favored habitats and work across land and water. This work was captured, in part, in the influential 1988 book *From the Forest to the Sea: A Story of Fallen Trees.*[3]

Most importantly, this research showed that an old-growth forest was a complex, interconnected ecosystem and not simply a collection of trees, as many foresters had traditionally viewed it. From now on, forest planners and managers would have to recognize and accept that fact as part of their obligations under such newly created laws as the National Forest Management Act and Endangered Species Act.

Although Forest Biome Project funding ended in 1975, the Andrews teams continued and expanded their work on conifer-dominated, forest and stream ecosystems of the region as their reputation and membership grew. The science was initially largely descriptive (development of carbon, nutrient, and water budgets for an old-growth forest ecosystem), but it quickly generated an array of interesting topics for research proposals. The ecosystem team's increasing fame led to the funding of several proposals, including one supporting establishment of a major field station at the Andrews, and the stream team received significant separate NSF funding for stream and river research. The long-term future (and independence) of the forest ecosystem program at Corvallis was ensured in 1980 when NSF awarded the Andrews research effort one of the first Long-Term Ecological Research grants.

The Andrews Work Gets Noticed

By the late 1970s the Andrews ecosystem team had scientifically documented enough of the critical features of the moist old-growth coniferous forests of the range of the northern spotted owl that they could begin to answer the question, What is an old-growth forest? Forest planners on the Siuslaw National Forest came to the team with exactly that question. The planners were grappling with the emerging issues of old-growth forests and their role in biodiversity, including habitat for the northern spotted owl; they understood that old growth was about more than just big old trees, but what else did they need to recognize to describe an old-growth forest? The Siuslaw query stimulated the Forest Service to ask a group of scientists, resource specialists, and foresters under Franklin's leadership to quantify the characteristics of old-growth forests.

This group developed an interim description of old-growth Douglas-fir forests, and what makes them distinctive and unique, for use in planning and management. Not surprising, given the conceptual framework that the Andrews ecosystem team had developed, large old

Table 4.1. Interim minimum standards for Douglas-fir/western hemlock forests.

Stand characteristic	Species/amount
Live trees	2 or more species with a range of ages and sizes
	Douglas-fir > 8 trees/acre > 32" in diameter or > 200 years old
	Tolerant associates (western hemlock, western redcedar, Pacific silver fir, or big leaf maple > 12 trees/acre > 16" in diameter
Canopy	Deep and multilayered
Snags	Conifer snags > 4/acre > 20" in diameter and > 15" tall
Logs	15 tons/acre including 4 pieces/acre > 24" in diameter and 50" long

Source: J. Franklin et al., *Interim Definitions for Old-Growth Douglas-Fir.*

trees and their derivatives—large snags, and large persistent logs on the forest floor and in streams—were central elements of this first approximation (table 4.1).

The group crafted similar descriptions for Douglas-fir forests on mixed conifer sites and mixed evergreen sites. The resulting publication, popularly known as PNW-RN-447, became a standard reference for description of old-growth Douglas-fir forests and would later be enshrined in congressional legislation.[4]

Many forestry professionals generally, including those charged with managing the national forests, were not enthusiastic about the science emerging from the Andrews. Federal foresters in the Pacific Northwest were in the business of converting old-growth forests into plantations. More broadly, most professional foresters, including academics, advocated or at least accepted the view that productive national forest lands should be managed primarily for wood production and not preserved as de facto national parks. Old-growth forests were often described as cellulose cemeteries by foresters and as biological deserts by wildlife professionals (see box on page 68).

Relationships between experiment station scientists and the Willamette National Forest began to thaw when Zane Smith became forest supervisor, and they continued to improve when Mike Kerrick was named supervisor. Kerrick's appointment of a science-oriented district ranger, Steve Eubanks, on the Blue River Ranger District (which included the Andrews) in 1984 created the first truly collaborative

A MANAGER REACTS:
"YOU ARE EITHER WITH ME OR AGAINST ME!"
(Jerry Franklin)

Federal scientists participating in the Andrews ecosystem studies were targets for criticism from some agency foresters, in part because the results of that research were increasingly cited in appeals and lawsuits against agency old-growth harvest projects. Sometimes the conflicts became personal.

The Willamette National Forest, where the Andrews is located, was working hard to become the first national forest to reach an annual allowable cut of one billion board feet. David Gibney, the Willamette National Forest supervisor during the 1960s and early 1970s, aggressively pushed this agenda and grew greatly agitated by Forest Service scientists conducting research at the Andrews on the ecological value of old-growth forests.

One morning four of us stopped by the Willamette National Forest intending a friendly visit with forest supervisor Gibney about our research. Jack Rothacher, Dick Fredriksen, Ted Dyrness, and I were all Forest Service scientists. We were directed into the supervisor's office and found him sitting red faced at his desk.

Gibney was not pleased to see us, and among his first words were "I know that you are all members of the Sierra Club! I have the list of members right here!" He immediately pounded a stack of papers on his desk. His concluding words were "You are either with me in what we are trying to do here or you are against me, and I know that you are against me!"

Subsequently the word went out that "that boy scout Jerry Franklin is no longer allowed on the Willamette National Forest!" Later, geomorphologist Fred Swanson was told he was no longer welcome on the Sweet Home District of the forest, presumably because of his research on landslides in the Middle Santiam River.

relationship between the federal forest managers and scientists there, and Lynn Burditt, who came after Eubanks, furthered that relationship.

Research at the Andrews on the ecological importance of dead wood—coarse woody debris—in both aquatic and terrestrial ecosystems was the first research from the Andrews to influence national forest management. The Forest Service had developed aggressive programs to remove dead and rotten logs from harvest sites and streams in the 1960s for a wide variety of reasons, including removing

impediments to planting seedlings, reducing fuel for wildfire, easing
fish passage, and reducing the impact of floods on bridges and other
structures. By the early 1980s, though, the agency wound down these
programs, largely because of research from the Andrews documenting
the ecological value of dead wood in the forests and streams, including a widely cited publication by Harmon and others.[5] Also, Swanson
and Sedell published an influential paper based on their work at the
Andrews in 1976 concluding that

> organic debris has historically been an important element in
> small streams of the Pacific Northwest [fig. 4.3]. The debris
> serves to slow the movement of water and inorganic and fine
> organic matter through the channel. Debris may remain in the
> channel for decades or longer. . . . The combination of clearcutting
> and the complete removal of large debris in a channel may

Figure 4.3. Organic
debris in Lookout
Creek, HJ Andrews
Experimental Forest.
Note small figure
standing on downed
tree at center left.
(Source: F. Swanson
et al., *History,
Physical Effects*.)

deprive a stream of this natural feature for a century or longer. The consequences are likely to be downcutting of the stream, accelerated transport of fine organic and inorganic sediment, and a possible decrease in biological productivity of the stream ecosystem. Therefore, stream debris management during logging operations should include leaving undisturbed the natural, stable organic debris in the channel.[6]

Thus, the Forest Service proved willing to listen to suggestions for change that had ecological benefits and reduced cost. It probably helped if such suggestions did not affect the allowable cut.

Concerns over the effects of clearcutting emerged in the 1970s, as clearcuts began to dominate federal forest landscapes in the Douglas-fir region and appear on national forests across the country. Some of the concerns were related to the impact on wildlife, such as different species of birds that required larger trees with decadent features, large dead trees (snags), and large logs on the forest floor. Other concerns were related to the potential mining of nutrients and the resulting effects on site productivity. Also, it became clear that much of the public detested clearcutting on public lands, and reducing clearcutting became a major impetus for the National Forest Management Act of 1976. Something different was badly needed.

And something different was coming from the Andrews. Foresters had long been reluctant to leave scattered trees at final harvest because they thought the trees would probably blow down and their economic value would be permanently lost. That result had been common in earlier decades when cull (rotten) trees were left behind. However, research done on the Andrews by Roy Silen showed that this was not necessarily the case; Silen had marked and harvested an old-growth Douglas-fir forest, leaving behind a portion of the sound trees in the stand, and little blowdown had occurred.[7]

Franklin and his colleagues embarked on demonstrations on the Andrews in the mid-1970s of retaining some green trees at final harvest (fig. 4.4). These demonstrations retained dominant and codominant Douglas-fir trees representing about 15 percent of the preharvest stand. Leave trees were distributed uniformly over the harvest area, and retained trees were to remain through the entire next rotation. By the

Figure 4.4. The first variable-retention harvest at the HJ Andrews Experimental Forest.

end of the 1970s, many foresters had designed timber harvests in the Douglas-fir region with significant retention and survival of dominant and codominant green trees. And field trips to see retention harvests on the Andrews became lively indeed as many foresters took grave exception to the idea that an alternative to clearcutting might be better. Once again work at HJ Andrews challenged conventional wisdom.[8]

By the early 1980s, national forests in western Oregon and elsewhere were developing forest plans under the National Forest Management Act. The Andrews work was well known by that time, and the scientists there had established positive relationships with forest planners on the Willamette National Forest. Thus, the scientists worked collaboratively with the planners to incorporate their findings into the standards and guidelines for the forest plan crafted there, including retention of green trees at final harvest and of downed wood on harvest sites and in streams. Also, work by the Andrews stream team strongly influenced riparian management standards in that plan through the development of the *Riparian Management Guide* by Stan Gregory and Linda Ashkenas.[9] Other national forests in the owl's range, though, made much less use of Andrews-derived forest ecosystem research in forest plan development.

During the 1980s, the Andrews became the place to learn about the wonders of old-growth forests. Creative writers and photographers increasingly extolled the beauty and mystery of ancient forests and included scientists and their science-inspired stories in magazine articles and books such as *Secrets of the Old Growth Forest*.[10] As controversy over continued cutting of public old-growth forest and associated threats to the northern spotted owl heated up in the late 1980s and into the 1990s, the frequency of field tours on the Andrews increased to several dozen per year. Participants included journalists, elected officials, forest managers, students, and the public. The pace of media coverage also picked up substantially in those years in both regional (e.g., Kathie Durbin's many articles in the Portland *Oregonian*) and national outlets (e.g., Jon Luoma's four articles in the *New York Times* plus several in national magazines). Presentation of the new science on old-growth ecosystems, on-the-ground examples of forward-looking practices, and a large cadre of experienced presenters on major issues involved in forestry conflicts of the time proved a considerable draw. The HJ Andrews Forest became a central place for public discussion on the future of forestry, and everyone was learning.[11]

The Owl and Other Wildlife Get Their Due

Wildlife biologists at the same time were developing a separate body of science on a variety of wildlife species—particularly on the northern spotted owl—that would converge with the Andrews ecosystem studies and other efforts in clarifying the importance of old-growth forests to regional biodiversity.

Eric Forsman's chance encounter with a northern spotted owl on the Willamette National Forest in 1969, when he was an undergraduate at Oregon State University, started him on a path of discovery that profoundly affected federal forestry. We briefly summarize his story below; Forsman tells it in his own words at the end of the chapter.

Forsman's master's degree research in the Department of Fish and Wildlife at Oregon State University focused on the distribution, habitat associations, and diet of northern spotted owls in Oregon and included an extensive survey of spotted owls in the state, in which he found the owls in old-growth forests.[12] His major professor (Howard Wight) encouraged Forsman to pursue a PhD as soon as he finished his master's degree and obtained funding from the Forest Service and BLM. This time, the owls of the Andrews and surrounding area became his primary focus for a radiotelemetry study to evaluate habitat use. The mixture of old growth and clearcuts in that forest made it especially valuable for this work:

> When I got up there to do the telemetry study in 1975–76, most of the cutting that would occur had already happened. They had done big clearcuts in many places on the Andrews. One of the things we wanted to do with the radio-collared owls was to find out what kinds of habitat they preferred or selected. And if all we'd had was old growth, we wouldn't have been able to do that. But because we had these contrasting clearcuts and early seral and old forest, and even a little bit of mid-seral stuff, we were able to do those comparisons.[13]

When Forsman returned to OSU in the fall and analyzed his data, two facts stood out: this little owl had a large home range for an owl, and that home range included substantial amounts of old-growth forest. From these humble beginnings, the northern spotted owl became the animal most identified with the Pacific Northwest's old-growth forests, and its conservation became a pivotal factor in the major shift in federal forest management that led to the Northwest Forest Plan.

Scientific interest in the northern spotted owl was indicative of a broadening interest in biodiversity emerging among agencies and academic departments active in wildlife research. Wildlife research had traditionally focused on animal species of interest to hunters, such as deer, elk, and game birds. Forest scientists occasionally found themselves studying nongame animals, but usually because of one of two circumstances—a predator eating a game animal or a rodent or some other animal eating tree seedlings.

Agency-sponsored basic research beyond game or problem animals expanded gradually during the 1970s and rapidly in the 1980s. This included potential candidates for listing under the Endangered Species Act but also extended to a broad range of animals as well as plants, fungi, and other organisms. An outstanding example of this was a major decade-long project on old-growth forest wildlife habitat.

The Old-Growth Forest Wildlife Habitat Research and Development Program was chartered by the Forest Service PNW Research Station in 1981 and carried out from 1983 to 1990. The goal of this program was to increase knowledge about vertebrate animals found in unmanaged west-side Douglas-fir forests, including their habitat relationships. The study included all vertebrate animal groups—birds, small mammals, amphibians, and reptiles. Old-growth forests were a primary focus of the research, but the scientists leading the program recognized the need to also study vertebrate communities in young and mature forest. The resulting book, *Wildlife and Vegetation of Unmanaged Douglas-Fir Forests*, became an important general reference on vertebrates in Douglas-fir forests.[14]

Concurrent with this work, Marty Raphael and Bruce Marcot of the PNW Research Station conducted studies of both vertebrate and nonvertebrate diversity in the forests of northern California.[15] Their studies highlighted the habitat values of litter, duff, coarse downed wood, standing dead and partially dead trees, and the physical structure of older forests. They found that the complex physical structure of these forests—deep, multilayered canopies, deeply furrowed and sloughing bark on tree boles, and so forth—was of vital importance for a wide variety of associated species including bryophytes, lichens, invertebrates, amphibians, reptiles, birds, and mammals.

A Volcanic Explosion Reveals the Resilience of Nature

The catastrophic eruption of Mount Saint Helens on May 18, 1980, provided ecological scientists with a major new opportunity to study the effects of natural disturbances on forest ecosystems and their subsequent development (usually referred to as *recovery*). The stark conditions created by the eruption resulted in surprising discoveries about the biological legacies of disturbances and the importance of habitats that immediately follow a major natural disturbance.

The explosive blasts of the 1980 eruption smashed over 160,000 acres of forested landscape and rained tephra (ash) on hundreds of thousands of acres of additional forest. Much of the affected forest was part of either the Gifford Pinchot National Forest or large timber company holdings. Many of the scientists, technicians, and students associated with the Andrews responded to this research opportunity and began a long-term collaboration with an ecological team from Utah State University led by James A. MacMahon; one of this latter team, Charlie Crisafulli, was critical to sustaining ecological research at Mount Saint Helens for the next four decades.

Initial impressions of the severely impacted blast zone were of an area that had been sterilized by the eruption and would therefore require recolonization by plants and animals from adjacent areas. Reality proved otherwise. This new understanding began when the Franklin-MacMahon teams helicoptered into the blast area a few weeks after the eruption. Eric Wagner tells this dramatic story in his history of the eruption and its aftermath:

> Franklin opened the helicopter's side door and hopped out. His boots sent up little puffs of ash when they hit the ground. He glanced down, but instead of the gray he expected, he saw a bit of green poking up next to him. He knelt. It was a plant shoot, maybe 2 or 3 inches tall. *I'll be damned*, Franklin thought. It was *Chamaenerion angustifolium*, a plant much more widely known by its common name, fireweed. . . .
>
> Franklin stood and took in the landscape again. He realized that this fireweed was one of tens, hundreds, maybe even thousands of little green shoots emerging from the ash. He saw the beginnings of pearly everlasting and thistles. . . . More than plants, he noticed evidence of other forms of life. Beetles were scuttling over the downed trees. Ants trooped along the ground, leaving trails of dimpled footprints. Dark mounds of dirt showed where a pocket gopher must have pushed up from its subterranean tunnels. Roving herds of ungulates—elk, probably, judging by the size of the hooves—had already planted prints in the ash as they picked their way through the tangle of [downed] trees.[16]

An immense array of organisms—plants, animals, fungi, and microbes—survived the eruption to play major roles in the redevelopment of aquatic and terrestrial ecosystems in the varied habitats in the blast zone. Massive amounts of dead organic matter were also present, much of it as shattered fragments of trees, including large logs and snags.

The wide distribution of living organisms and abundant dead organic matter and their important roles in the postdisturbance Mount Saint Helens landscape were an important phenomenon from which the concept of biological legacies of disturbances emerged.[17] For ecologists, the wake-up lesson came with recognition that most natural disturbances of forests leave behind vast legacies of living organisms and dead organic materials—the latter including large snags and logs—that result in continuity of plant and animal life, structure, and function between forest generations. Wildfires, windstorms, and insect epidemics exemplify this, leaving conditions in stark contrast with those following clearcut timber harvests. Importantly, such natural disturbances and their legacies provided models that could be emulated in forest harvesting, and they ultimately provided the scientific underpinning for the practice of *variable retention harvesting*—an alternative to the widespread practice of clearcutting that retains biological structures such as large and old live trees, snags, and logs.[18]

The importance of early successional ecosystems was the second major epiphany for ecologists from research at Mount Saint Helens, although this insight took longer to emerge than the biological legacy concept. There is a period following a severe forest disturbance when the disturbed area is free of dominance by tree canopies, allowing for vigorous development of herb- and shrub-dominated communities. The redevelopment of a closed forest canopy takes decades since trees must regenerate and grow substantially in height.

Foresters traditionally minimized this early successional or preforest period, which lacks full occupancy by growing trees, by planting trees and spraying herbicides. Much of the blast zone escaped this treatment because it was federal forest that eventually became part of Mount Saint Helens National Monument. Natural development of early successional ecosystems was allowed to proceed there, revealing a rich plant and animal diversity. It did take many years, however, for this knowledge to become developed to the point where its implications and potential in management and policy were apparent.[19]

For scientists who would ultimately develop the Northwest Forest Plan, the eruption of Mount Saint Helens also provided an important and humbling reminder about the potential for large natural disturbances in the Pacific Northwest. Any system of ecological reserves would have to account for disturbances of this scale and severity.

Listening to the River and the Fish

While the pioneering research at the Andrews and Mount Saint Helens continued, other scientists were doing seminal work on the relationship of salmon productivity to the condition of aquatic ecosystems, primarily in the Oregon Coast Range. This work proved to be an important foundation for what became the Aquatic Conservation Strategy of the Northwest Forest Plan.

In 1985, the Siskiyou National Forest asked Gordie Reeves and Jim Sedell to examine the distribution of anadromous salmon in the federal portion of Elk River, which is on the southern Oregon coast near Port Orford. They jumped at this chance because the Elk was an important salmon stream, and they wanted to test methods for estimating fish numbers and amount of habitat.

Reeves was responsible for the fieldwork and began to determine what portions of the river were to be surveyed by reviewing previous reports and visiting potential sample areas. On a Friday afternoon, Reeves and Charlie Dewberry, who worked for the Forest Service, went to the North Fork of the Elk River in the uppermost part of the basin. Reeves in his wet suit found the trail hard going as he tumbled down a steep slope through a dense natural stand. Soon after Reeves submerged himself in the stream, though, he made a discovery that reframed how he thought about aquatic ecosystems and helped set up his research agenda for much of his career (see box on page 78 and plate 4.1).

Other colleagues were making similar observations during the early and mid-1980s in Oregon's Coast Range. Lee Benda identified the attributes of stream channels that produce debris torrents[20] that reach fish-bearing streams as part of his PhD work at the University of Washington. Parallel work at the PNW Research Station led by Fred Everest examined the effects of landslides on fish habitat (plate 4.1). Swanson

THE DAY MY THINKING ABOUT RIVERS AND SALMON CHANGED
(Gordon Reeves)

The lower part of the North Fork was not particularly spectacular—it was moderately steep, with small boulders and a few pools. About a half mile up, things changed dramatically. The valley widened and the stream was essentially flat. Pools and pieces of large wood were abundant. We stuck our heads in the water, and hundreds of juvenile salmon and trout happily met us. Certainly not what we expected based on earlier surveys by the Oregon Department of Fish and Wildlife that described this stretch of stream as unproductive and with little suitable habitat for salmon.

The banks along the stream were even more spectacular. They were 50–70 feet tall, with remnants of large trees protruding from stream level to the top. Clearly some large disturbance event had occurred here in the relatively recent past that had set the stage for the conditions we saw today. Charlie and I were flabbergasted and struggled to make sense of what we were seeing. The farther we walked the more awesome it became.

We hustled back to our truck—the trail seemed much easier than when we came up because of our newfound excitement—and got to the nearest phone as soon as possible to call Fred Everest (senior fish biologist at the PNW Research Station) and Jim Sedell. I must have sounded like a little kid who had just been given the most incredible gift he could have imagined.

A few weeks later, we went back to the North Fork with Fred, Jim, and Lee Benda, a geomorphologist. Everyone struggled to understand what they were seeing and why it had occurred. It was clear that something big had happened—an entire hillside had failed and slid to the valley bottom. Given that much of the wood contained in the banks and within the stream showed signs of having been burned, the hillslope failure must have followed a wildfire. Based on the age of the trees on the hillside, we estimated that fire occurred 100–120 years ago. And how did fish respond to this? They loved it! It was clear that moving forward, we needed to consider a new way of thinking about aquatic ecosystems—they were not stable but rather were dynamic—and to develop a research program to further this understanding.

continued his work on the positive effect on stream ecosystems of earth flows and landslides and delivery of large wood.

The idea that landslides initiated by earth movement and debris torrents were necessary to help rejuvenate streams puzzled many people.

The suggestion that wood and boulders delivered by them could create high-quality fish habitat met with surprise and resistance from many managers and regulators, as it questioned prevailing conventional wisdom that these events needed to be prevented.

Reeves and Sedell continued to study fish distribution and habitat along more than 40 miles of the Elk River for many years. They found that fish assemblages varied within and among streams based on the local features of the stream and the valley in which it was set. Chinook and coho salmon were most abundant in low-gradient streams that flowed through wide valleys, which provided for the development of complex habitat in the main channel as well as on the floodplain. In contrast, steelhead were found in steeper-gradient stream reaches that had little to no floodplain because the stream was confined by valley walls or high banks.* Additionally, the research suggested that past large-scale disturbances strongly influenced the stream conditions that existed at any point in time.

Other work by Reeves and his colleagues showed that Pacific salmon are well adapted to natural disturbances. Adults returning to freshwater do not necessarily return to the stream where they were reared prior to their ocean migration. As many as 15–20 percent of returning salmon may stray from their natal streams, allowing populations to colonize newly available habitats. Additionally, female salmon produce many eggs, allowing for rapid population establishment in favorable habitats. Also, juvenile salmon may move to streams other than those where they hatched and then venture to new streams as adults. All these characteristics provide useful evolutionary traits in a dynamic landscape.

The scientists concluded that native fish like salmon could thrive in a dynamic environment in which occasional major disturbances reset the landscape; in fact, they needed the sediments and wood that the disturbances provided to prosper. Such findings called for a rethinking of conservation measures for this iconic species.

Key research was also being done at the University of Washington, particularly by Robert Naiman. Naiman and a number of graduate

---

* Burnett et al., in "Distribution of Salmon-Habitat Potential," popularized the constrained/unconstrained classification that has become important in identifying the "intrinsic potential" of a stream segment for different salmon species.

students including Benda made an especially valuable contribution in a paper on the attributes of an ecologically healthy watershed. They noted that conditions in streams within a watershed were not stable, as suggested by other scientists. Rather, stream conditions could vary widely in the variety of habitat types and features such as amount of large wood, composition of sediments, and depth over time in response to periodic disturbances such as fires and floods. The headwater and middle portions of a stream network, which were where federal lands tended to be in the range of the northern spotted owl, were identified as particularly dynamic and variable over time.[21]

Naiman's paper and others that suggested a similar perspective on aquatic ecosystems resulted in a shift of the focus of fish conservation efforts in two important ways. First, they urged consideration of the entire stream network rather than just the stream reaches that were generally in the middle and lower portions of a watershed where most salmon habitat was located. This shift in perspective was particularly important in considering the dynamic and variable headwater streams, which can make up 70 percent or more of the total stream length in a watershed, and which had previously been largely ignored.

Second, they urged an emphasis on conserving ecological processes that influence aquatic ecosystems rather than simply conserving attributes of habitat units (number of pieces of wood or pools). These processes included the input of large wood, leaves, and other organic materials to the stream over time from riparian vegetation, which provides stream structure and nutrients and energy for aquatic insects. This shift in perspective caused scientists to emphasize processes that were largely a function of forest composition and structure and to utilize tree height as an indicator of the spatial extent of forest impacts on streams. Thus, they began to recommend riparian buffer widths in terms of the maximum height that trees could attain on a site rather than in terms of fixed distances.

There was skepticism among managers and regulators, and within parts of the scientific community, about the new approach to conservation of aquatic ecosystems. The focus on ecological processes and large spatial scales challenged the basic assumption that aquatic ecosystems were static and maintained a narrow range of conditions through time, and that management practices should be tailored to maintain stability.

The new dynamic approach challenged this view by suggesting that habitat features and productivity in aquatic ecosystems were naturally variable in space and time as a result of wildfire, windstorms, and other periodic disturbances, much as in natural forests. This presented a particular dilemma for managers and regulators whose approach was premised on achieving stable conditions within a stream reach and across a watershed.

A Foundation for the Northwest Forest Plan

Ecosystem-oriented scientific study of the forests of the Pacific Northwest proliferated in the 1970s and 1980s. The research began with a goal of understanding and documenting the nature of old-growth coniferous forest and stream ecosystems. These studies quickly deepened from a largely descriptive initial focus to increasingly sophisticated research on forest and stream structure, function, and plant and animal life.

Access to funding from NSF was the critical element in many studies, as it provided interdisciplinary teams of academic and agency scientists the freedom to pursue these pioneering efforts independent of agency or forestry school agendas. The PNW Research Station also made important contributions to ecosystem research by allowing its scientists to participate in such programs and to use its experimental forests and data from its gauged watersheds.

The results of the old-growth ecosystem research converged with the organism-focused science on native species that used these forests (especially the northern spotted owl) to inform the developing controversy over management of federal forestlands, particularly the logging of old-growth forests. This science helped increase public and professional understanding about the unique roles played by natural forest ecosystems, contributed essential information for litigation and policy debates, and ultimately became a major scientific foundation of the Northwest Forest Plan.

In addition, the focus on the dynamic nature of coastal stream systems was pivotal in rethinking how to conserve and restore aquatic ecosystems in the science assessments to come and in the Northwest Forest Plan. The scientific emphasis there would be on ecological processes that underlay functioning aquatic ecosystems.

WHY I STUDIED OLD GROWTH
(Jerry Franklin)

Why study old-growth forests? There are many reasons, and for me they are long-standing and personal.

I was first introduced to old-growth forests as a nearly nine-year-old boy on our family's first major camping trip to the Columbia National Forest's beautiful Government Mineral Springs Campground, and I fell immediately in love with the immense old-growth Douglas-fir trees and monkey-bar vine maples that populated the camp. The following summer we camped in the equally impressive old-growth forest at La Wis Wis Campground along the Cowlitz River. There I learned at the forest guard station that there was a profession called forestry, and people were actually paid to work in such forests. I then started reading books about forestry adventures including Martha Hardy's wonderful diary as a summer lookout, *Tatoosh*, and Montgomery Atwater's exciting 1947 novel *Hank Winton: Smokechaser* (first serialized in *Boy's Life*, I think). These locked in my determination to become a forester.

On entering the School of Forestry at Oregon State College, though, I found that no one could tell me much about old-growth forests. Everyone had something to say, but their comments were about values and not facts. Conservationists commented about their beauty and inspirational value but could offer little on what such forest ecosystems did. Foresters saw them as cellulose cemeteries that needed to be quickly replaced with productive young forests before they rotted and ultimately collapsed. Wildlife specialists called them biological deserts because they lacked significant populations of game species. I was disappointed and dismayed by the lack of knowledge and, even more profoundly, by the lack of interest in the true nature of the old-growth forest.

I was able to spend much time in these forests as a student and then a professional involved in research at the HJ Andrews Experimental Forest. The more I experienced them the more convinced I became that old-growth forests represented an extraordinary ecosystem that carried out important and unique functions for the earth and humankind. We just didn't know what they were.

Administrators of Forest Service research were not interested in studying old-growth forests, although individual forest scientists were. Administrators had other priorities. By 1959 they felt that we knew enough about how such forests could be cut down and regenerated, and that no further investigations of old growth were needed. The only exception was some research on effects of harvesting on streamflows and water quality.

They even proposed disbanding the Andrews Experimental Forest and returning it to the Willamette National Forest for harvesting.

Similarly, there was little interest among academic foresters and ecologists in studying old-growth forests and little financial support for such work. A few Forest Service scientists did work on aspects of old-growth forests. Project leader Dick Miller continued to measure the singular set of permanent old-growth forest plots at the Wind River Experimental Forest. Soil scientist C. T. Dyrness and I bootlegged studies of old-growth forest composition and structure on the Andrews Forest and in research natural areas we visited. But it certainly seemed that old-growth forests would go to the mills without being fully understood and appreciated.

The emergence of the US International Biological Program and funding for a Western Coniferous Forest Biome study offered hope that we could change that. A group of us at Corvallis first battled the College of Forestry at the University of Washington for a piece of that action and then successfully argued at Oregon State that we study an old-growth forest rather than a plantation. We would focus on a 25-acre old-growth forest occupying Experimental Watershed 10 adjacent to the HJ Andrews Experimental Forest.

OK, but what exactly should/would we do? At that point we had no great scientific hypotheses that we wanted to test—we didn't know enough about any forest, let alone an old-growth forest, to pose such questions. The youthful science of ecosystems traditionally focused on ecosystem functions—productivity and carbon cycling, influences on the hydrologic cycle, including water yields, and nutrient cycling. So, we would begin our work by describing our old-growth forest ecosystem in these terms. We would document its carbon/energy, hydrologic, and nutrient budgets. The work already in place by Forest Service researcher Dick Fredriksen on water and nutrient cycles in Watershed 10 would jump-start us.

Within 18 months we had learned enough to have a first approximation of the structure and function of an old-growth coniferous forest ecosystem and scores of working hypotheses that would require decades of further research. The old-growth coniferous forest ecosystems of the Pacific Northwest would finally be described, understood, and ultimately appreciated for their contribution to the planet and humankind.

TRACKING THE NORTHERN SPOTTED OWL
(Eric Forsman)

When I was in high school, I became an avid birder and developed an interest in owls because they hunted at night and were mysterious critters that most people didn't see or particularly relate to. Spotted owls in particular intrigued me because they were thought to be very rare and almost nothing was known about them. When I first got interested in owls, there were only about 28 previous published sightings of spotted owls in Oregon, and no one had ever described a spotted owl nest in Oregon or Washington. Most of the observations in Oregon were cases where somebody stumbled onto a spotted owl in the woods. As an avid birder I really wanted to see a spotted owl, but I was just a high school kid with no reliable transportation, so I had no inkling about the topic that would soon come to dominate the next 50 years of my life.

In 1969, I was a college student in my third year working a summer job for the Forest Service at the Blue River District on the Willamette National Forest. I was stationed at the Box Canyon Guard Station in late summer at the peak of the fire season. My job was to patrol the roads and trails, talking to campers and loggers and keeping an eye out for fires. One evening I was sitting at the guard station and heard something barking up on the hill and thought, "That's a weird-sounding dog." And then it began to dawn on me that it sounded like the spotted owl calls I had read about in bird books.

I had been looking for spotted owls for years not really knowing what I was doing. But there I was, sitting on the front porch of the guard station in the evening, and all of a sudden, they started calling on the hill right above me. And I thought, "Oh, shit, that sounds like a spotted owl!" So I hiked down the road to where the sound was coming from and found the male perched in a tree next to the road just south of the guard station. I started trying to hoot back at him. I was pretty much an amateur, but spotted owls respond to almost anything that's close to their calls, and both the male and female eventually followed me all the way back to the guard station. After that, I kept hearing them all summer off and on. They'd call up on the hills around the guard station, and I practiced calling back at them and gradually got better at it.

So, when I went back to school my senior year in September 1969, I started running around looking for spotted owls with my friend and fellow OSU student Richard Reynolds. Reynolds was an undergraduate student in the Zoology Department and was interested in accipiters (goshawks, Cooper's hawks, sharp-shinned hawks), so we had similar interests regarding forest birds, and we began to search for spotted owls and accipiters in forest areas in western Oregon. In that first year, we found spotted owls at a number of locations, including Tombstone Summit and Indian Creek on Santiam Pass and a couple of sites on McDonald Forest. I also found my first spotted owl nest in a stand of old-growth forest on BLM land at Greasy Creek just a few miles southwest of Philomath. This all happened while I was still an undergraduate and before I was drafted into the army.

I got out of the army in March 1972, and by April 1972, I was in the field doing spotted owl surveys for my master's research project. My assignment was finding spotted owls and documenting as much as I could about their distribution, habitat, and basic biology. And I had to do it fast because the breeding season for spotted owls is from late March through the end of August.

What we found that first summer was that spotted owls were more common than we had previously believed. We still thought they were pretty rare, but they were certainly more common than prior data would have suggested, and they were definitely most abundant in old forests. We also found that they nested primarily in cavities in large old trees and their diet was dominated by arboreal mammals, especially flying squirrels, tree voles, and wood rats. And many of the sites where we found spotted owls were getting cut very rapidly. There were harvest units laid out all over the old forests where these owls occurred.

My goals in that initial study were to find as many spotted owls as I could, survey as broadly as I could in western Oregon to document distribution and abundance, find as many nests as I could, and collect as many pellets as possible so we could describe the diet. I also collected basic behavioral information on when the young came out of the nest, how long they were cared for by the adults, and when they dispersed. But most importantly I tried to determine what kinds of habitat the owls occurred in. It took me three years (1972–1974) to collect all the data and get a thesis written.

While I was wrapping up my master's thesis, Howard Wight (my major professor) asked whether I wanted to continue on and do a PhD project on the spotted owl. I quickly agreed because there was a lot of interest in spotted owls after our initial study. Howard talked to the Forest Service and the BLM and was able to get funding. So I went right from finishing a master's thesis in 1974 to starting a PhD in 1975. In fact, the ink was barely dry on the master's thesis before I was in the field on the Andrews in April 1975, putting radio collars on spotted owls. Unfortunately, Howard passed away from cancer in 1975. After that, Chuck Meslow took over as my major professor.

The PhD project on the Andrews was a radiotelemetry study in which

I put radio transmitters on adult spotted owls so we could learn about the size of their home ranges and what kind of forest they preferred for foraging habitat. Those were the two things we really needed to understand better because up until that time nobody knew for sure how much habitat spotted owls needed in order to survive and reproduce. We went into the telemetry study thinking that spotted owls would be similar to barred owls, which typically had home ranges of about 900–1,200 acres. We even thought that spotted owl home ranges might be smaller than barred owl home ranges, because spotted owls are slightly smaller than barred owls. We also knew that the tawny owl, a similar species in Europe, had home ranges of only about 100 acres or so. So we thought we were going to put transmitters on spotted owls and find that they had home ranges that were similar to those of these other owl species.

I put transmitters on eight adults in April and May 1975 and tracked those owls for the next year. I didn't get much sleep that year because I lived with the owls almost 24/7, tracking them at night and hiking to their roost sites during the day so I could document the kinds of habitat they used for roosting. I survived on about five hours of sleep a day and lived in a little 21-foot travel trailer that Meslow had found on the US Fish and Wildlife Service list of surplus property. At the time, there were no living facilities on the Andrews, so I parked the trailer in the gravel storage area that has since become the main administration site on the Andrews.

When you do this kind of study, you see the owls every day and regularly hear them talking to each other at night. You spend your nights following them from place to place and you put them to bed in the morning before you stagger back to the trailer to get a few hours of sleep. You watch them breed and bring up young, and you watch the young come out of the nest. And you constantly worry about whether the radio-marked owls will survive or whether your radios will reduce their survival rate. You get very attached to your animals; you can't help it. But, as a scientist, you also worry about getting enough data to estimate home ranges. So you bust your ass, burning the candle at both ends to get that data.

I came back to OSU at the end of that year in the woods and started to analyze the data. Although I had yet to complete the analysis, I already knew from following the owls around for a year that they were unlike any owl species that people had studied before. For relatively small owls, they had huge home ranges that encompassed large amounts of old forest. Annual home ranges of the adults I studied on the Andrews averaged about 3,000 acres and included more than 1,000 acres of old forest. These data suggested that, in all of our previous recommendations to managers and policy makers, we had horribly underestimated the amount of habitat spotted owls need.

PART 3

*Social Activism, Law, and Science
Upend Federal Forest Management*

5

The Northern Spotted Owl Takes Center Stage

When Eric Forsman began to study the habitats of the northern spotted owl in 1972 as a master's student in the Department of Fish and Wildlife at Oregon State University, he immediately made an important discovery: spotted owls showed a strong preference for old-growth forests. Furthermore, many of the owl's nesting sites were in areas designated for upcoming timber sales: "Everywhere we went we'd find spotted owls and boundary markers for timber sales. They were right smack in the same area."[1]

In July 1972, Howard Wight, Forsman's major professor and head of the local US Fish and Wildlife Service (USFWS) unit, sent letters to his boss at the USFWS in Washington, DC, and to officials in the regional Forest Service and BLM offices informing them of the early results of Eric Forsman's spotted owl research: "Forsman has located 37 sites where pairs of spotted owls can regularly be seen. All of these pairs are inhabiting old-growth timber. No spotted owls have been located in any other type of habitat. These birds seem to require climax forests and are very sensitive to any alteration of their habitats."[2]

Wight also noted that it was too early to provide a definitive recommendation on the size of the habitat areas needed by the owls; in the meantime, he recommended that agencies proceed cautiously in logging old-growth forest.

As Forsman discovered ever more numerous owl–timber sale conflicts, he began to advocate for owl protection. He notified the district ranger of the Klamath District of the Fremont-Winema National Forest in 1973 that five of the six nesting pairs he had located there were in cutting units, adding that "the Forest Service and the BLM are systematically destroying the Spotted Owl population in your area, and yet I would bet that not a single timber-sale impact statement produced

on the entire Winema National Forest makes any mention of the
destruction of Spotted Owl habitat." He also notified the administrator
of the Corvallis city watershed, where the city cut timber to fund different projects, that he had found a pair of owls 100 feet from the edge of a
planned timber sale and requested that the city reconsider the location
of the harvest unit.[3]

Not surprisingly, Forsman's pleas hit the proverbial stone wall.
The Klamath district ranger responded: "Your letter suggests that
nesting habitat for each pair ranges from about 200–600+ acres and
that pairs are found a minimum of about 1 mile apart. I think we can
both understand the magnitude of the set-aside areas if there are many
nests involved. Owls have nesting demands, but other resources have
demands too."

After the administrators of the Corvallis city watershed pointed at
the risks from old-growth forests in terms of their decay, windthrow,
and potential to spread disease, they stated, "You have suggested that
approximately 300 acres per nesting site be set aside to remain as old-
growth timber. In viewing the economics alone on a 300 acre parcel,
you have placed a value of 2 million dollars per nest on these areas. I
am sure that you can appreciate [our] responsibility . . . to the citizens
of Corvallis to manage the watershed in a manner that provides maximum service to the citizens."[4]

Reserving the forest for owls at the expense of timber harvest and
associated revenue was not something managers were ready to do vol-
untarily. A lot more was needed for them to take such action than some
graduate student yelling about timber harvest impacts on an owl few
had ever even seen.

In 1973, the increased visibility of the spotted owl as a public controversy, the heightened interest in nongame wildlife, and the soon-
to-be-enacted Endangered Species Act (ESA) led to the first public
response to the conflict. The Oregon Endangered Species Task Force
was created to study the issue, with representatives from the Oregon
State Game Commission, the BLM, Forest Service, USFWS, and
Oregon State University.[5] At its first meeting the task force set a high
priority on establishing guidelines to preserve suitable habitat for the
northern spotted owl. Wight indicated that its habitat was restricted to
old-growth forests and estimated that each pair required a minimum

of 300 acres of old growth per pair, with the pairs spaced approximately one mile apart. He also noted that over half of the 70 known pairs of owls were within proposed timber sales.[6]

Thus began a multiyear back-and-forth discussion between the Endangered Species Task Force and the land management agencies: the task force would make recommendations on a framework to conserve the spotted owl and the Forest Service and BLM would either reject the recommendations outright or delay their acceptance because of concerns over the impact on timber harvest. Additionally, as Forsman's work progressed, his estimates of home range size kept increasing, making a conservation strategy a moving and ever more costly target.[7]

In late 1977, the Endangered Species Task Force issued management recommendations that called for the protection of 400 northern spotted owl pairs, with guidelines taken largely from Forsman's master's thesis. The task force recommended clusters of at least three owl pairs, with the pairs separated by approximately one mile; clusters not more than 12 miles apart to allow for dispersal of owls between clusters; a minimum of 1,200 contiguous acres designated for each owl pair, with a core area of at least 300 acres of old growth; and the remaining 900 acres in the designated area managed so that at least half of the acreage contained stands that were at least 30 years old.[8]

Early in 1978, the Forest Service and BLM agreed to follow the task force recommendations, subject to ongoing agency planning processes. Forest Service officials generally planned on accommodating the old-growth reserve for each owl pair (300 acres) in areas already withdrawn from timber production. For example, planners on the Willamette National Forest, which had the largest owl allocation and was the biggest timber producer in the owl's range, expected to locate owl habitat areas within existing wilderness, roadless recreation areas, and research natural areas. On the adjacent Mount Hood National Forest, forest supervisor (and later chief of the Forest Service) Dale Robertson commented, "A reduction in timber yield is not anticipated. Management for the above pairs can be accomplished without entering programmed commercial forest lands proposed in the Timber Management Plan."[9] However, the Forest Service noted that any owl decisions would be temporary; permanent decisions would be made through the forest planning processes established under the National Forest Management Act of 1976.

Protest Begins over Threats to Owls and Old Growth

In the summer of 1978, Cameron LaFollette, a staff member of the Oregon Student Public Interest Research Group (OSPIRG), wrote a report titled *Saving All the Pieces: Old Growth Forest in Oregon*, which was the first public analysis of the Forest Service's implementation of the Endangered Species Task Force's spotted owl recommendations.[10] She summarized her findings in the OSPIRG newsletter: "The decisions of the plan do not have a solid grounding in biology; they are entirely political in nature. Old-growth forests are highly valued by the public land management agencies for their timber production. The spotted owl plan looks very much like an effort of these agencies to have their cake and eat it too."[11]

Prophetic of what was to come, LaFollette made two more claims: that no one really knew what constituted a minimum viable population, so it would be wise to provide for as many owls as possible (also claiming there were many more owl pairs than the 400 that would be protected); and that the owl was not the "crux of the problem, it is the old-growth forest community which is highly complex both in flora and fauna. There is no biological sense in using one species as an indicator for the welfare of all the other old-growth dependent species, each of whose needs are vastly different from those of the spotted owl."[12] LaFollette had been to one of the Andrews workshops on old-growth ecosystems and was putting the ideas she had learned there to good use.

LaFollette next wrote a paper that became the basis for the first administrative appeals directed at federal spotted owl management. That appeal, filed by a cluster of Oregon environmental groups in early 1979, claimed that a major federal action (adoption of the task force recommendations by the Forest Service) was being undertaken without the required preparation of an Environmental Impact Statement (EIS), and that the Forest Service and BLM should suspend harvest in spotted owl habitat until the EIS was completed.

The regional forester, Richard Worthington, and subsequently the chief of the Forest Service, John McGuire, denied the appeal in late summer 1980, concluding that adoption of the owl recommendations was not an action but a restraint on action. However, Chief McGuire also made a commitment that would haunt the Forest Service for the next decade: the *Regional Guide*, which would provide guidance for the development of forest plans under the National Forest Management Act, would include "a proper biological analysis to determine the number and distribution of spotted owls which constitute a viable population."[13]

Since the regional guides were "major Federal actions significantly affecting the quality of the human environment," the guides would be subject to the environmental analysis and public review requirements in the National Environmental Policy Act (NEPA). Therefore, the public would have a chance to comment on the Draft Environmental Impact Statement for the *Regional Guide* and then to administratively challenge the agency's decision to approve it once the regional forester signed the Record of Decision.

While waiting for the Forest Service to complete its NEPA process, local Oregon environmental groups began a public information campaign. They sought and received coverage in the regional press about how continued logging of spotted owl habitat would doom this charismatic animal to extinction, and how that logging would destroy forests of trees that were alive long before the pilgrims landed at Plymouth Rock, Massachusetts, in 1620. The public awakening had begun.

Also, the 1980 election of Ronald Reagan as president changed the nature of the political struggle over protection of owls and old growth. The early 1980s were a period of rapid growth for environmental organizations nationally and regionally. They gained substantial increases in membership and contributions, spurred by the appointment of anti-environmentalist James Watt as secretary of the interior and Reagan administration policies that included calling for doubling the harvest on the national forests.

Even the more cautious national environmental groups such as the National Wildlife Federation swung into action. As stated by Andy Stahl, a National Wildlife Federation staff member in Oregon who focused on developing strategies for protecting the northern spotted owl: "James Watt turned the National Wildlife Federation into an activist organization. It took someone of Watt's outrageous behavior to awaken the slumping giant of these sportsmen."[14]

It also became increasingly clear to environmental groups that the only way to advance environmental goals, given the views and policies of the Reagan administration, was to move aggressively through the courts. Thus, they began to develop analytical and legal expertise to do just that.

Successful litigation of a case over timber harvest on landslide-prone forests on the Mapleton District of the Siuslaw National Forest in 1984 made environmental groups in the Pacific Northwest believe that they could use the courts to overcome timber's historical political dominance in the region.[15] As Stahl explained:

> The first forestry issue that the [National Wildlife Federation] got into in the Northwest was the issue of landslides in the Oregon Coast Range. . . . We were able to get the court to enjoin all timber sales on the Mapleton Ranger District [of the Siuslaw National Forest] . . . because they had never done an EIS. . . . After we won the Mapleton case, we realized that potentially we had the ability to change the world.[16]

The Society of American Foresters (SAF) also got involved through Jim Lyons, policy director of the society and manager of a task force that the SAF had set up on scheduling the harvest of old-growth timber. The task force included federal foresters, members of interest groups, and three scientists: Franklin, Jack Ward Thomas (then a senior Forest Service research scientist), and John Gordon (dean of the Yale School of Forestry). The three scientists successfully turned the discussion from scheduling old-growth harvest to managing an old-growth ecosystem. Thus, the task force's report included recommendations to define old growth; inventory existing old-growth forests on federal lands; determine how much was needed to sustain spotted owls and associated old-growth species; and schedule the sustainable harvest of the remaining old growth on public lands.[17] To the surprise of many, the report recognized old-growth forests as an end in themselves, legitimizing a preservation policy. "At least until substantial research is completed, the best way to manage for old growth is to conserve an adequate supply of present stands and leave them alone."[18] While the task force's report kept the working title "Scheduling the Harvest of

Old Growth Timber," it was much more than that. The times they were a-changin'.

The SAF Task Force effort reflected a science-based approach to the old-growth issue that Lyons would advance in the future. Jack Ward Thomas described Lyons's emergent role in old-growth conservation in a 1998 oral history:

> About this time, a figure enters the picture with influence up to the present. His name is Jim Lyons. His influence goes back . . . to the genesis of the technical consideration of the "old growth" issue. This was the SAF [Society of American Foresters] assessment and subsequent report on scheduling old-growth timber harvest. Lyons was the SAF staffer who organized and guided that effort. When you trace the players in the entire "old growth" issue, there are those who continue a role through the entire drama.[19]

Jim Lyons was one of those people.

Sinking Ever Deeper into Quicksand

During the 1980s, the Forest Service proposed a sequence of management strategies for the northern spotted owl that it hoped would withstand legal scrutiny under the inevitable environmental litigation that would follow (table 5.1).* However, the legal and scientific foundations for that strategy kept shifting in ways that made that outcome ever more difficult to achieve.

By the late 1970s, Eric Forsman's owl work had shown much larger home ranges than previously thought: 2,500–4,000 acres (generally along a south-to-north gradient within the owl's range), with a minimum of 1,000 acres of old growth.[20] As Forsman explained, "They're very much a specialized forest predator. They feed on things that occur at fairly low densities. Flying squirrels, for example, and red tree voles. When you're that much of a specialist, you have to forage over a big

---

* This chapter focuses on the struggles of the Forest Service in crafting a spotted owl management plan, as it took the lead in that effort. The BLM suffered through similar difficulties. See Yaffee, *Wisdom of the Spotted Owl*, for that agency's trials and tribulations. We include the BLM in the discussion of spotted owl litigation in chapter 6.

Table 5.1. Key Events in the Forest Service's Effort to Adopt a Conservation Strategy for the Northern Spotted Owl, 1980–1991.

1980	Forest Service begins work on *Regional Guide* for national forest planning in Pacific Northwest Region.
1982	Viability provision in NFMA Planning Rule revised to require that habitat be well distributed.
1984	Final *Regional Guide* for Pacific Northwest Region appealed: claim of inadequate northern spotted owl plan. Forest Service agrees to try again.
Nov. 1988	Court (Judge Zilly: *Northern Spotted Owl (*Strix occidentalis caurina*) v. Hodel*) rules USFWS must list northern spotted owl as threatened or endangered under the ESA.
Dec. 1988	Supplemental *Regional Guide* for the northern spotted owl litigated: claim of inadequate owl strategy.
May 1989	Court (Judge Dwyer: *Seattle Audubon Society v. Robertson*) enjoins sale of spotted owl habitat until owl case decided.
June 1989	USFWS announces intent to list northern spotted owl as threatened under the ESA.
Oct. 1989	Congress mandates federal old-growth harvest without the possibility of environmental litigation (Sec. 318).
Nov. 1989	Court (Judge Dwyer) lifts injunction on sale of owl habitat because of Sec. 318.
May 1990	Interagency Scientific Committee (ISC) spotted owl conservation strategy released.
Sep. 30, 1990	Sec. 318 ends; owl litigation resumes. US Forest Service embraces ISC's owl strategy, which is litigated: claim that adoption process is illegal.
Mar. 1991	Court (Judge Dwyer: *Seattle Audubon Society v. Robertson*) rejects Forest Service claim that ESA listing of the owl exempts the owl from the NFMA Planning Rule.
May 1991	Court (Judge Dwyer: *Seattle Audubon Society v. Evans*) rejects Forest Service process for adopting the ISC's strategy, enjoins sales of owl habitat until satisfactory owl plan legally adopted; notes that the ISC's strategy is the first scientifically credible approach.

area to get enough food."[21] In addition, the spotted owl was turning out to be relatively common rather than rare, once systematic surveys were undertaken. The owl was found not just in a few isolated places, but rather throughout old-growth forests of western Oregon, western

Washington, and northern California. The stage was set for a titanic political battle.

In the early 1980s, both scientists and agency specialists grew concerned about what criteria could be used to assess adequacy of a spotted owl conservation management plan under the requirement that national forests provide habitat to ensure viable populations. Forest Service wildlife biologist Hal Salwasser was among the first to interpret the emerging science of viable populations: "A viable population is capable of persisting on the forest in the face of anticipated environmental changes, both natural and man caused. It is not a maximum, nor even a desired population level. Rather it is a minimum level, below which the population is considered to be subject to local elimination due to habitat loss, disease, harassment, or some other factor."[22]

The determination of what viability meant for the northern spotted owl came to a head with the preparation of the *Regional Guide for the Pacific Northwest Region* of the Forest Service. The *Regional Guide* would provide a framework for forest planning on individual national forests and would need to spell out what was meant by a viable population of northern spotted owls—a commitment previously made by the chief of the Forest Service. The Forest Service settled on 500 owl pairs distributed across the national forests as a viable population in its draft *Regional Guide* distributed for public comment in May 1981. That number was based on the findings of Michael Soule, a California geneticist, who published one of the first books on conservation biology. He had done pathbreaking work in the late 1970s on the effects of genetic inbreeding on survival, and his experimental work on fruit flies and theoretical modeling led him to conclude that a breeding population of 500 individuals would minimize the effect of inbreeding. He passed this advice on to the NEPA team preparing the draft *Regional Guide*. However, he also noted that "this number does not represent the pairs required to distribute the owl throughout its natural range."[23]

The draft *Regional Guide* also documented the need for at least 1,000 acres of old growth (300 acres in the owl core and 700 acres within 1.5 miles) around each owl pair as the minimum standard for owl habitat protection.[24] After receiving public comment, the Forest Service put out the 1984 final *Regional Guide* with only modest changes from the draft, despite significant intervening developments in owl science and policy.[25] To bolster its case, the Forest Service claimed that conservation biologists had assisted in developing the owl management plan, that the approach had been peer-reviewed by population viability experts, and that it was supported by a scientific monograph on the owl authored by Eric Forsman and Chuck Meslow (Wight's successor as leader of the Cooperative Wildlife Unit at OSU). The Forest Service also claimed that "scientific uncertainty surrounding adequacy is very limited."[26]

The National Wildlife Federation, Oregon Natural Resources Council, and others filed an administrative appeal of the 1984 final *Regional Guide*, arguing that the Forest Service had misrepresented the research of Forsman, Wight, and Meslow. They noted that the Forest Service had used 1,000 acres as the required amount of old growth within an owl home range in its final *Regional Guide*, while the average amount of old growth in home ranges reported in the 1984 monograph published by Forsman, Meslow, and Wight was 2,200 acres; 1,000 acres was the smallest home range size found in the research. They also cited Forsman's presentation on the owl at a recent Forest Service symposium in which he said that the agency's adopted approach "involves a high degree of risk" along with other substantive criticisms. The appellants asked that the amount of old growth protected per pair be raised to 2,200 acres, that the number of owl pairs protected be raised to 1,000, and that old-growth harvest cease until those actions were taken.[27]

The final decision rested with Deputy Assistant Secretary of Agriculture Doug MacCleery, a former staff member of a forest product trade association. Although MacCleery largely supported the *Regional Guide*, he concluded in March 1985 that "new information has become available since the Regional Guide was prepared which may be relevant to regional direction on spotted owl management" and ordered preparation of a Supplemental EIS (SEIS) on owl management. However, he did not stop logging of spotted owl habitat in the meantime.[28]

Andy Stahl, still working for the National Wildlife Federation, began devoting much of his time to understanding the scientific basis of the Forest Service's claim that 500 owl pairs distributed in 1,000-acre patches across the landscape would be sufficient to maintain a viable population.[29] Toward that end, he recruited Russel Lande, a theoretical biologist from the University of Chicago, to study the issue. Lande applied a patch extinction model to the question of owl population dynamics in a patchy forest habitat. The key to Lande's model was the assumption that a juvenile's probability of successfully dispersing and finding its own territory depended on both the fraction of suitable habitat patches that were unoccupied and the density of those patches on the landscape. Because the likelihood of successful dispersal depended on the availability of suitable habitat, Lande's model predicted that owl populations would decline as logging destroyed old-growth forests. Based on his assumptions and modeling, Lande estimated that at least 21 percent of the total landscape needed to be old growth for a stable owl population to persist.[30] In a letter to Andy Stahl summarizing his work in June 1985, Lande concluded that "Based on the available evidence, harvest of any of the remaining old forest in the Pacific Northwest poses a potentially serious threat to the long-term viability of the northern spotted owl."[31]

Stahl promptly introduced Lande's conclusions into the fray by holding a press conference about them and requesting (again) that Douglas MacCleery stop all old-growth harvests while the SEIS was being written. MacCleery did not do that, but he did stop the harvest of six old-growth timber sales that the National Wildlife Federation and other groups found especially egregious.[32] Further challenges would need to await completion of the SEIS.

The Forest Service knew that the issue was bigger than spotted owls, but after some debate, agency planners decided to focus narrowly on the owl issues in the SEIS. Kathy Johnson, a highly respected wildlife biologist from the Willamette National Forest, remembered a key decision made in a strategy meeting in July 1985 on how to proceed with the SEIS:

> The discussion was really focused on owls, although MacCleery's letter did have a lot of reference to old growth forests. There was a discussion about whether we were going to deal with the old growth forest issue, or is this just going to be an SEIS on spotted owl viability. . . . [New regional forester] Tom Coston basically threw something out on the table, "Well, why shouldn't we deal with old growth forests if that's really the issue?" This was indeed the issue at that time. . . . Al Lampi [director of land management planning] spoke first . . . it was important that we move ahead with forest planning, that opening this issue up to old growth

forests when the appeal was specific to spotted owls only would put the rest of the Regional Guide at risk.[33]

While everyone involved in the SEIS knew that old-growth forests were the real issue, the SEIS would not address it.[34] Regional forester Coston "unexpectedly retired after a few months."[35]

The SEIS represented the first systematic effort by the Forest Service to lay out the choices in owl management and their implications. Toward that end, planners within the Forest Service used analytical procedures to estimate the probability of providing a well-distributed population over different time frames, rather than just looking at the data and intuitively coming to conclusions. The agency-proposed alternative (Alternative F) provided for 550 spotted owl habitat areas (SOHAs), with 2,200 acres of old growth set aside (reserved from timber production) per SOHA, which resulted in an estimated high probability of providing habitat for a well-distributed population in the short run, but a low probability in the long run. Also, the agency estimated that its proposed plan would require the annual timber harvest on national forests to be reduced by approximately 5 percent within the range of the owl, which was consistent with Chief Robertson's desire to strictly limit the impact of spotted owl conservation on timber harvest.[36]

Predictable outrage by stakeholder groups followed release of the draft SEIS in spring 1986, along with schemes to advance their respective causes. Environmental groups led by the National Wildlife Federation utilized the work of independent scientists to poke holes in the viability conclusions for Alternative F, especially employing the analysis that Lande had done for Stahl. Environmental groups also organized teams of scientists to propose more protective conservation strategies for the owl, such as Audubon's Blue-Ribbon Panel, which recommended a much more robust plan—1,500 owl pairs across all ownerships, with SOHA sizes more reflective of spotted owl home ranges. Environmental groups also charged that the Forest Service's estimated low level of long-term persistence violated the National Forest Management Act (NFMA).[37]

Industry, labor, and timber community groups put most of their energy into arguing that the preferred alternative was not a balanced approach and that the estimated 5 percent decrease in the allowable cut would lead to economic and social chaos in the Pacific Northwest as sawmills closed and thousands of people lost their jobs. In addition, they claimed that the information on the habitat needs of the spotted owl was too uncertain to make a decision that would be so harmful to communities.

An agency team began to develop a final SEIS for the *Regional Guide* for the northern spotted owl against a backdrop of the interested parties' claims and counterclaims. In the final SEIS, owl protections in Alternative F were slightly improved by varying SOHA size across the range based on empirically derived home range sizes, from 3,200 acres per pair in Washington to 1,250 acres per pair in northern California, but Alternative F Adjusted still achieved only a moderate probability of achieving sufficient habitat for a well-distributed population in the long run. As with the draft plan, the final SEIS projected the harvest level to drop about 5 percent.[38] Chief Robertson was philosophical about adopting Alternative F Adjusted in the Record of Decision. He said, "Tell me why I shouldn't just go ahead and make this decision and let the forces that shape this country react. It's going to go into the court system, it's going to go into the legislative system, that's the way our country is supposed to work."[39]

In late 1988, the Forest Service issued the NEPA documents for this updated *Regional Guide* containing the revised spotted owl conservation strategy. Environmental groups led by Seattle Audubon were not satisfied and challenged the guide in court, alleging—among other claims—that the new requirements were insufficient to provide for diversity of plant and animal communities and maintain habitat for viable populations of the owl as required by NFMA.[40] An important part of the lawsuit was the conclusion from Lande's work, which by now had been published in the peer-reviewed literature, that the Forest Service's conservation strategy would doom the owl.[41]

The presiding judge in the case (*Seattle Audubon Society v. Robertson*) was William Dwyer, a widely respected jurist. He agreed with the plaintiffs that their claims were likely to prevail and granted a temporary restraining order in March 1989 on 139 planned Forest Service timber sales in owl habitat covering thousands of acres. In May of that year, he extended the injunction indefinitely, pending a decision on the merits of these claims.[42]

Judge Dwyer brought an interesting background to this case. He had taken forestry classes as an undergraduate at the University of Washington. His law practice included participation in some of Washington's most publicized disputes. His roles ranged from representing a member of the state legislature in pursuing libel charges against those who had wrongly accused the representative of being a Communist, to successfully representing King County in a case that forced baseball's American League to award a franchise that became the Seattle Mariners. Judge Dwyer was appointed to the court by President Reagan.[43]

Environmental groups in the Pacific Northwest were also working during the late 1980s to get the USFWS to list the northern spotted owl under the ESA but going slowly, as they were worried about losing in court if their arguments were not well grounded. Their hand was forced when, in early 1987, the USFWS received a citizen's petition to list the spotted owl as threatened or endangered under the ESA from Green World in Massachusetts, a tiny group of which few in the Northwest had ever heard. The USFWS then conducted a status review of the spotted owl and concluded in December 1987 that listing was not warranted despite expert advice to the contrary.

Shortly thereafter, environmental groups led by the Sierra Club Legal Defense Fund sued the USFWS, arguing that the agency's decision not to list the owl as threatened was arbitrary and capricious. Judge Thomas Zilly, in the federal district court in Seattle, agreed. In his opinion, issued in November 1988, he stated that the decision by the USFWS not to list "disregarded all of the expert testimony on population viability, including that of its own experts, that the owl is facing extinction and instead merely asserted its expertise in support of its conclusions. The Service has failed to provide its own or other expert analysis supporting its conclusions. . . . Accordingly, the [USFWS's] decision not to list at this time the northern spotted owl as endangered or threatened under the Endangered Species Act was arbitrary and capricious and contrary to law."[44] Listing of the owl as threatened was now inevitable.

While environmental groups were advancing their cause through administrative appeals and litigation during the 1980s, they had also redoubled their efforts on another front in their battle to stop old-growth logging: nationalizing the old-growth issue. They knew that their political power to influence the outcome would increase to the degree they could turn that issue into a national conservation cause.

Andy Kerr, a member of the Oregon Natural Resources Council (now Oregon Wild), was a key player in the nationalization effort. Until the mid-1980s, Kerr had found that the only way to effectively conserve natural federal forests was to persuade Congress to protect them in national parks and wilderness areas. And congressional conservation of public lands in a state just didn't happen without the support of that state's senior senator. In Oregon, that senator was Mark Hatfield, who served in the US Senate from 1967 to 1997 and who got Oregon wilderness increased in 1972, 1978, and 1984. Each successive bill also reserved more lower-elevation older forests, to the increasing consternation of the timber industry. After passage of the 1984 bill, Kerr and others sensed that Senator Hatfield had little interest in more wilderness bills in the future.

Kerr relates what happened next:

Time for a Plan B. Oregon Natural Resource Council had previously filed the first administrative appeal of a timber sale raising the issue of the northern spotted owl. Maybe we could do more on that. We soon decided to nationalize the plight of Pacific Northwest ancient forests by causing trouble in the congressional, judicial, and executive branches, as well as in the public arena.

After the 1989 injunctions, I was summoned to the office of Hatfield's Oregon director, where I heard out a last plea to keep this just among us Oregonians. Our Venn diagrams did not, would not, and could not overlap. I walked back to my office very frustrated and was handed a pile of phone messages that had come in during the last two hours. I needed to call back NBC News in New York, the *New York Times*, a writer for the *New Yorker*, among several others. Soon I was banned from every office in the Oregon Congressional delegation—save for Rep. (now Senator) Ron Wyden (D-OR). But that made more time for media and conferring with attorneys. The die was cast—we would create a national outcry over the continued clearcutting of old-growth forests to force change.[45]

Local activists were also taking matters into their own hands to stop old-growth logging in federal forests of the Pacific Northwest, which helped to further publicize the issue. Sporadic tree spiking and vandalism at sites of old-growth logging had gone on for years, especially in southwestern Oregon by the radical environmental group Earth First! However, protests were increasingly led by college students and other young people who largely followed the principles of nonviolent civil disobedience.

One such effort in particular galvanized public opinion—the North Roaring Devil timber sale up the North Santiam River on the Willamette National Forest. Logging began on a Friday in late March 1989 in the 33-acre old-growth harvest unit on a remote ridge about 60 miles southeast of Portland. While the timber sale was the focus of an environmental lawsuit challenging Forest Service timber management practices in the Willamette National Forest, the courts had not stopped the sale from proceeding.

A group of college students and other young people, who had been learning about the practice of civil disobedience in a workshop in Eugene, volunteered to go to the site, where a few Earth Firsters were already established.[46] The students camped on a bridge over the Breitenbush River and were there to greet Bugaboo Timber Company workers with a road blockage when they arrived for a third day of logging in the disputed area. The loggers gathered around the students' bonfire when they were unable to proceed into the cutting area, and the two sides spent several hours peacefully discussing their opposing views on old-growth logging. The protesters argued that the timber site contained old-growth trees that were hundreds of years old and was a nesting ground for spotted owls, which they said were endangered, while the loggers stressed their need for work to feed their families.[47] "I feel real sympathetic towards the loggers," said Leslie Hemstreet, a protester from Portland. "I know they have to make a living and they have to feed their kids, but we need to come up with an alternative for them instead of cutting these trees down."[48]

Sheriff deputies and Forest Service law enforcement moved in after the loggers left. The banner-carrying protesters obeyed an order to move off the bridge but then proceeded farther up the logging road and continued to block it. Authorities arrested them at a three-foot rock

Figure 5.1. Protesters arrested on March 31, 1989, during forest protests on the Willamette National Forest east of Salem, Oregon. (Gerry Lewin, © USA Today Network.)

wall, which Earth First! had previously built across the road, and took them to the county jail to be charged with trespass (fig. 5.1).[49]

In the end, the old growth was logged, but a new force in the spotted owl wars had come on the scene—college students putting themselves on the line to defend old-growth forests, such as Catia Juliana and Tim Ingalsbee, who met at the protest and went on to marry and organize protests in future years. Unlike the protests that had come before, this time the national media paid attention. The North Roaring Devil sale got coverage from the likes of *Good Morning America*, the *Today Show*, and National Public Radio. Old-growth logging was no longer just a regional issue.[50]

Also, people like Julie Norman, a whitewater river guide on the Rogue River in southwestern Oregon, got involved: "I was seeing the clear cuts across the heart of the Siskiyou National Forest. And I started wondering, well, this looks pretty dramatic." Like a lot of folks, Julie started with this concern and then began to learn about what she was losing by the clearcutting of old-growth forest. "And we began to quote Jerry Franklin because he was an esteemed agency person. And so people like him were fantastic and they would be our special guests every time we could get them to talk."[51]

Julie and others then used the old-growth criteria developed by
Franklin and his colleagues to map the remaining old growth and
overlay these maps on proposed timber sales. What they saw scared
them—the old growth was going fast. They appealed the sales, utilizing
NEPA to argue that the Forest Service had not divulged the environmental impacts of its action. Julie remembers that they lost many of
those battles, and then those committed to direct action would take
over: "The activists would just pour into an area and start protesting.
At one time there were maybe five or six people sitting in the top of
old-growth trees out in the Siskiyou National Forest."[52] People were
willing to give their lives to stop old-growth destruction.

Congress Intervenes: Section 318

The Oregon congressional delegation, a mixture of Democrats and
Republicans, was well aware of the legal activities in the courts of
Judges Dwyer and Zilly and was deeply worried that, as a result, federal
timber sales in owl habitat would be abruptly stopped. Therefore, the
entire delegation (Senate and House, Republicans and Democrats) and
the governor of Oregon met with representatives of the timber industry and environmental groups in Salem, Oregon, in June 1989. Their
concerns were further heightened by the USFWS's announcement in
the *Federal Register* the day before the meeting that it would list the
spotted owl as threatened under the ESA.

The delegation's goals for this Timber Summit were to ensure that
timber sales could proceed in the near term and to give impetus to
development of a long-term solution to the owl/timber crisis. At that
meeting, the delegation and the governor settled on an approach that
would insulate federal timber sales from judicial review for two years,
while also providing temporary protection for the most important
old-growth stands, staying out of the SOHAs identified in the Forest
Service's updated *Regional Guide*, and directing the Forest Service and
BLM to develop an owl strategy that would satisfy the courts. Such an
approach required congressional action.

Senator Hatfield and Congressman AuCoin of the Oregon delegation, each of whom was a senior member of his respective appropriation committee, then began negotiations with the forest industry and
environmental groups to gain enough support to pass the needed

legislation through Congress. Senator Hatfield was especially important in this process, as he chaired the Senate Appropriations Committee and was personally close to President George H. W. Bush, who had
recently succeeded President Reagan.

The forest industry wanted a commitment for a few years to a level
of timber sales close to the allowable cuts of the recent past. Endorsement by at least some environmental groups would be challenging but
was key to overcoming congressional resistance. The environmental-
group plaintiff (Seattle Audubon) in the spotted owl case in Judge
Dwyer's court, who helped precipitate this crisis, was not involved in
the negotiations. Neither were grassroots environmental groups across
Oregon and Washington. The national offices of the Audubon Society
and The Wilderness Society, though, were more willing to compromise.
The inclusion of important, even though temporary, protections for
old growth and northern spotted owl habitat, recognition of the ecological value of old growth, and a commitment to developing a credible
owl conservation plan enabled the national environmental groups to
withdraw their objections. Also, Senator Hatfield, whose staff led the
negotiations, said on the Senate floor that Section 318 was a one-time
emergency measure and promised to work through the relevant Senate
committees (especially Senator Leahy's Agriculture Committee) on a
permanent fix.[53] With all these commitments in place, Congress passed
the Fiscal Year 1990 Interior Appropriations Act in October 1989.[54]

Section 318 of that act directed the Forest Service and BLM to offer
a total of 7.7 billion board feet of timber within the range of the northern spotted owl—a level slightly less than twice the previous years'
average annual harvest. Most of this volume would come from spotted
owl habitat in old-growth forests. The timber sales would automatically
be viewed as complying with all environmental laws, insulating them
from legal challenge. This latter provision became known as *sufficiency
language*.

Section 318 also contained four telling provisions regarding the
future of federal forestry in the area of the northern spotted owl. First, it
called for minimizing "fragmentation of the most ecologically significant
old-growth forest stands" in meeting these timber targets in the Appropriations Act, the first reference to ecologically significant old growth
in congressional legislation and the first call (albeit temporary) for its

protection. Second, Section 318 called for old-growth forest stands to
be defined as those stands meeting the criteria specified in Forest Service Research Note PNW-RN-447—the definition of old growth that
had recently been developed by Franklin and his colleagues. Third, it
prohibited the placement of timber sales within the SOHAs identified
in the Forest Service's 1988 *Regional Guide*. And fourth, it directed the
Forest Service to revise the 1988 *Regional Guide* by September 30, 1990,
after considering "any new information gathered subsequent to the
issuance of the Record of Decision, including the interagency guidelines
for conservation of northern spotted owls that would be developed by
the Interagency Scientific Committee to address conservation of the
northern spotted owl." Congress passed Section 318 in October 1989
just as the Interagency Scientific Committee was formed. Thus, Congress asked the Forest Service to consider a forthcoming conservation
strategy from a group of scientists before that group had even begun to
develop it. Truly unusual times.

Citizen groups were formed to help select the forest to be logged to
try to avoid areas of special importance to people. In the end, though,
more than 100,000 acres of late-successional/old-growth forest were
logged as a result of Section 318, much of it owl habitat. The resulting
fragmentation also compromised the effectiveness of additional forest
that surrounded those cuts.

In some sense, Section 318 provided an economic bridge to a new
conservation plan for the northern spotted owl. However, it also made
the trade-off between owl protection and timber harvest in that plan
even sharper.

Perhaps most importantly, the "sufficiency" provision, which
(temporally) denied people access to the courts to challenge federal
forest policies, struck many as fundamentally inconsistent with the
right of citizens to petition the government to redress their grievances.
National environmental groups, and the broader conservation community, would not be party to such agreements in the future.

Scientists Answer the Call

The Interagency Scientific Committee mentioned by Congress in Section 318 was Chief Robertson's final attempt to regain control of the
spotted owl issue. In a bold and precedent-setting move, the chief,

along with the heads of the BLM, USFWS, and Park Service, convened an interagency group of scientists in October 1989 to create a scientifically credible conservation strategy for the northern spotted owl, independent of management staff and line officers (the regional forester, forest supervisors, and district rangers).[55] Robertson named Jack Ward Thomas, a Forest Service research scientist in La Grande, Oregon, to select and chair the committee.

Thomas was, in some ways, an unlikely leader for the northern spotted owl effort. He was best known nationally for his editing of *Elk of North America: Ecology and Management*,[56] an authoritative source for elk biology and management. He had never seen a spotted owl or studied owls in general. In another sense, though, Thomas was a natural choice to lead a potentially contentious committee of senior biologists. He had a knack for seeing the bigger picture and was a born leader. Smart, personable, and backwoodsy, Thomas was a master at dealing with people, whether they were scientists, agency leaders, politicians, reporters, or the public. Many of his projects had required synthesizing scientific information, which he thought was greatly needed, and he had the ability to express scientific concepts in plain language. Further, Thomas wasn't afraid to tackle large, complex problems, such as his effort to understand elk behavior by fencing 40 square miles in eastern Oregon and tagging all animals within that enclosure.[57] Forest Service leadership respected him as someone who would remember the agency's multiple-use mission. Perhaps most important for the tasks ahead, Thomas had real empathy for his fellow humans no matter where they stood on conservation issues.

Thomas picked five leading experts on wildlife in general and the northern spotted owl in particular for membership in the Interagency Scientific Committee (ISC): Jared Verner, Barry Noon, and Eric Forsman from the Forest Service research branch; Chuck Meslow from the Oregon Cooperative Fish and Wildlife Research Unit; and Joe Lint from the BLM. In addition, the ISC utilized the advice of designated agency managers and a group of knowledgeable observers representing the interests of the three states, National Park Service, USFWS, forest products industry, and environmental groups. He had brought together nearly every northern spotted owl expert in the world.

The ISC's mission was "to develop a scientifically credible conservation strategy for the northern spotted owl in the United States."[58]

The effort began with a series of public meetings at which the ISC and its knowledgeable observers discussed existing knowledge about the owl and its management. From these presentations and discussions, the ISC concluded that "the owl is imperiled over significant portions of its range because of continuing losses of habitat from logging and natural disturbances. Current management strategies are inadequate to ensure its viability. Moreover, in some portions of the owl's range, few options for managing its habitat remain open, and available alternatives are declining across its range. Delay in implementing a conservation strategy cannot be justified on the basis of inadequate knowledge."[59]

With this foundation, the six-member core group of the ISC began developing a conservation strategy that would ensure the owl's viability. The committee saw its charge as developing a reserve system sufficient to allow the northern spotted owl a high likelihood of persistence over 100 years in the context of a landscape that would be used largely for other purposes.[60]

How their effort would differ from those that came before them played on the minds of the ISC's members. How could they fulfill their charge to develop a scientifically credible strategy for the northern spotted owl? After some debate, they settled on "employing scientific methods to design a habitat reserve system"[61] in which they tested hypotheses about reserve design using a model that reflected both conservation theory and their understanding of the owl and its habitat needs. They knew that, in science, the process for reaching conclusions is as important, perhaps even more important, than the conclusions themselves.

The ISC used widely accepted concepts of reserve design from conservation biology in its quest, starting with the principle that species well distributed across their historical geographic range tend to be less prone to extinction than species confined to small portions of that range. That principle, combined with the NFMA viability standard to provide for species well distributed in the planning area, led the ISC to suggest a network of reserves (Habitat Conservation Areas, or HCAs) on public lands across the range of the northern spotted owl. The committee then utilized principles from conservation biology to suggest the size, content, and distribution of the HCAs. In particular, it recommended that the HCAs be large; covered by largely unfragmented, old forest; and close enough to each other to facilitate owl dispersal among them.

Within that conceptual framework, the ISC tested different hypotheses of potential networks of reserves across the landscape relative to their ability to provide a high likelihood of persistence. For that testing, it quickly constructed a mathematical model from theory, empirical information, and expert opinion about owl habitat needs and behavior and the size and location of needed habitat. Many alternative reserve sizes and maximum distances between reserves were simulated, including the Forest Service's proposed reserve system in the 1988 *Regional Guide*.[62] The ISC's analytical approach was, in concept, similar to what Lande had employed in his evaluation of the Forest Service's proposed reserve system a few years before, although this newer formulation was more comprehensive and elegant.

The ISC's modeling of alternative hypotheses enabled it to say with finality that the Forest Service's proposed reserve system would not work. As Thomas later summarized to Kathie Durbin for her book *Tree Huggers*, "The more we looked and the more we examined and the more we learned about the mathematical model, we recognized that [small reserves] wouldn't do it."[63] The agencies' existing plans, the committee declared, were "a prescription for extinction of spotted owls, at least in a large proportion of the owl's range."[64]

Having broken free from existing agency plans and having the ability to model and evaluate alternative conservation proposals, the ISC quickly developed the long-sought scientifically based conservation strategy for the northern spotted owl:

> Our strategy largely abandons the current and, we believe, flawed system of one- to three-pair spotted owl habitat areas (SOHAs), in favor of protecting larger blocks of habitat—which we term Habitat Conservation Areas, or HCAs.
>
> Large blocks of habitat capable of supporting multiple pairs of owls, and spaced closely enough to facilitate dispersal between blocks, are far more likely to ensure a viable population than the current SOHA system. Owls in an HCA containing multiple pairs will benefit from internal dispersal of juvenile owls as well as recruitment of dispersing birds from other HCAs. Owls in HCAs containing multiple pairs are less vulnerable to random fluctuations in birth and death rates. Large HCAs reduce the

impacts of habitat fragmentation and edges, and they are more resistant than SOHAs to small-scale natural disturbances.

The Committee has delineated and mapped a network of HCAs necessary to ensure a viable, well-distributed population of owls [fig. 5.2]. Wherever possible, each HCA contains habitat for a minimum of 20 pairs of owls. The maximum distance between these HCAs is 12 miles. Our 20-pair criterion is based on models of population persistence and empirical studies of bird populations. We have chosen 12 miles as the maximum distance between HCAs because this value is within the known dispersal distance of about two-thirds of all radio-marked juveniles studied.

The HCA concept applies primarily to BLM, USFS, and NPS lands, as delineated in the enclosed maps. The Committee strongly recommends that HCAs be established on State lands in certain key areas (as shown on the maps) to assure population connectivity.

We also recommend that resource managers of other State lands, tribal lands, other Federal lands, and private lands use forestry and silvicultural techniques and practices that maintain or enhance habitat characteristics associated with spotted owls.[65]

The ISC had achieved a major advance in regional conservation strategies. Its concept of a network of habitat reserves, each large enough to support a self-sustaining population of spotted owls and close enough to each other to allow dispersal, became the conceptual foundation for the scientific assessments that followed and was eventually embedded in the Northwest Forest Plan.

The HCAs proposed by the ISC were generally not covered by continuous northern spotted owl habitat. The Forest Service and BLM practice of dispersing clearcuts had highly fragmented federal forests, particularly at lower elevations in prime owl habitat. The landscape that resulted had the appearance of a moth-eaten blanket (fig. 5.3). Consequently, most HCAs were patchworks that contained residual older forest intermingled with much younger plantations. The ISC expected that the plantations would eventually grow into spotted owl habitat,

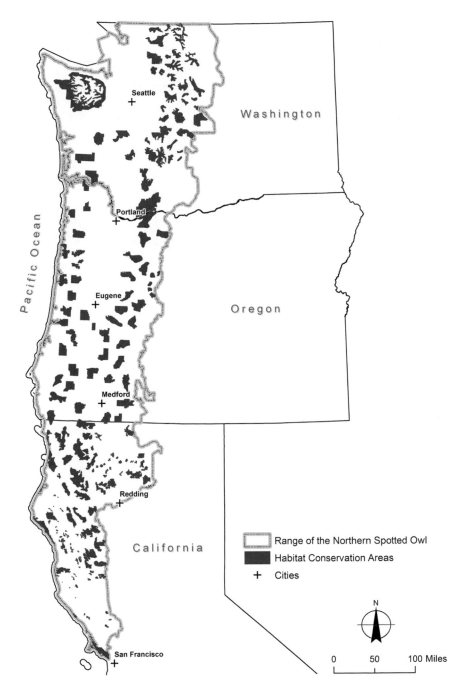

Figure 5.2. Habitat Conservation Areas (HCAs) in Oregon proposed by the Interagency Scientific Committee as part of its conservation strategy for the northern spotted owl. (Source: Thomas et al., *Conservation Strategy*.)

Figure 5.3. A typical "staggered setting" clearcut pattern on the Gifford Pinchot National Forest.

producing a large area of contiguous older forest and an increase in owl populations.

The ISC worked diligently to locate HCAs so as to minimize impacts on timber harvest, given the overriding goal of identifying a strategic network of large reserves that would conserve the northern spotted owl across its range. As an example, the ISC anchored HCAs in congressionally designated wilderness that contained large amounts of older forest to the degree feasible.

The ISC also recommended changes in the management of federal forest between the reserves, which they called Matrix, to ensure that these intervening areas were hospitable to successful movement of fledglings between HCAs: "We considered dedicating corridors of forests between HCAs to facilitate dispersal by juvenile owls, but decided corridors were unnecessary, provided at least 50% of the forest landbase outside of HCAs is maintained in stands of timber with an average d.b.h. of 11 inches or greater and at least 40% canopy closure."[66]

This provision became known as the *50-11-40 rule*. This approach— of modifying conditions between reserves to facilitate dispersal—was revolutionary in applied conservation science, which traditionally focused on corridors as the primary approach to connecting reserves.

The ISC concluded that its overall strategy would be sufficient to meet the viability standard in the 1982 Planning Rule for NFMA: "The Committee believes this conservation strategy, if faithfully implemented, has a high probability of retaining a viable, well-distributed population of northern spotted owls over the next 100 years. The HCAs on Federal lands contain 925 known pairs of owls, and we estimate the actual number to be about 1465 pairs. Regeneration of younger stands within HCAs on Federal lands should enable the spotted owl population to increase to about 1759 pairs." The committee also noted that a short-run decline in owl numbers was likely even with this conservation strategy, but they believed "the subspecies can withstand a reduction provided our strategy is followed."[67]

The ISC unveiled its report at a hearing conducted jointly by three congressional subcommittees in April 1990. At that hearing, Chief Dale Robertson said that implementing the ISC's proposal would reduce timber harvest levels on the national forests in the range of the spotted owl by 30–40 percent. An equal reduction was foreseen for BLM lands.[68]

One key element in the ISC's approach was initially missed by many in Congress: the ISC had not asked for the ultimate strategy for the owl. As Thomas said, "If we only wanted to do good for the owl, we could have gone home in thirty minutes. Stop cutting old growth and grow as much back as possible. . . . But we knew that wasn't practical. Anyone who thinks you can put forward a conservation strategy that ignores the needs of people is crazy. It will fail. The spotted owl would have ended up with nothing."[69]

Instead, the ISC compromised, realizing that the owl's present decline would take time to reverse. As Thomas explained, "We recognized we would give up 40–50 percent of the owls before the population came to equilibrium." Many members of Congress assumed that the report was a biological wish list and wanted to know how many acres the owls *really needed*. Thomas told them that the compromise had already been made. "We are already dooming up to half the owls . . . this is as fine a line as can be cut—there's no more room for a deal to be made."[70]

Release of the ISC's report in May 1990 unleashed a political thun-
derclap throughout the Pacific Northwest. Federal timber harvests
would decline by at least a third, costing thousands of good-paying
jobs, many of them in rural communities with few other opportunities
for employment. These coming impacts were on Thomas's mind as
he reflected in his journal soon after the ISC's report was released:
"The consequences of our action would descend upon us and follow
us all the rest of our days. We had done our job and we felt we had
done it well. Yet what we had done signified the end of an era, and
with the end of that era would come economic dislocation and human
suffering."[71] Further, Thomas painfully acknowledged, "The faces of
the thousands of people who will be hurt by this haunt my sleep and
my private thoughts."[72]

It was soon clear, though, that the ISC had provided a scientifically credible conservation strategy, as had been requested, and that
attempts to refute it would be unsuccessful. As William Dietrich summarized in his masterful book *The Final Forest*, "The wood industry
was deflated. For two decades, it had fought, fairly successfully, to
confine park and wilderness areas to the high alpine country of little
commercial value. Now it had been beaten by a bird. Almost overnight,
A Conservation Strategy for the Northern Spotted Owl had become a
textbook case of how to marshal scientific argument in a political and
sociological dispute."[73]

However, the ISC's report noted that development of a scientifically credible conservation strategy for the northern spotted owl had
dealt with only a small piece of the overall issue of how best to manage
these federal forests: "Adoption of the conservation strategy, however,
has significant ramifications for other natural resources, including
water quality, fisheries, soils, stream flows, wildlife, biodiversity, and
outdoor recreation. All of these aspects must be considered when
evaluating the conservation strategy. The issue is more complex than
spotted owls and timber supply—it always has been."[74]

Soon thereafter, the Forest Service, with the acquiescence of the
first Bush administration, sought to adopt the ISC's report as its conservation strategy for the owl despite its impact on timber sales. All
attempts to blow holes in that report had failed; embracing the report
might be sufficient to enable sales to resume in the owl habitat not

TRYING TO AVOID THE INEVITABLE
(Norm Johnson)

I worked for the Oregon governor's office in the late 1980s on federal forest planning, providing advice on how to minimize the reduction in timber harvest levels from those lands needed to accommodate other values. With release of the ISC's conservation strategy for the owl, my assignment shifted to helping the Oregon congressional delegation craft an alternative to the ISC's report that would conserve the owl at less cost to the federal harvest level.

The leader of the Northwest congressional delegation at that time was Senator Hatfield—a fierce defender of federal timber harvests and the rural communities those harvests helped support. He used his position on the Senate Appropriations Committee to direct and fund federal timber sales in the Pacific Northwest, and he sent word through his aides to me that if a lower-cost strategy could be found, he would consider legislation to mandate that strategy as the law of the land.

Toward that end, I obtained the latest maps of owl occupancy in Oregon's federal forests (the heart of the ISC's strategy and of greatest interest to the senator) and began to shift the HCAs such that they would capture more owls with less impact on harvest (but would also not be as well distributed). I then met with members of the ISC for a day and we went over my maps and logic. They made clear that they would not support such a shift in reserves.

I reported to Senator Hatfield at his Portland office a few days after that meeting. The conversation was short and to the point. He said, "Norm, did you find a modification of the ISC owl strategy that would allow a higher harvest level?" I said, "No, and I believe it can't be done in a way that the ISC will support." He looked at me with obvious disappointment and simply said, "Next."

Then the next person, a Bush administration staffer, had his moment with the senator. He asked the senator whether the White House could allow the Forest Service to endorse the ISC's report as its owl strategy. Senator Hatfield simply said, "No."

Thus, the Forest Service and Bush administration were left in limbo: they finally had a scientifically credible strategy for conservation of the northern spotted owl that they wanted to endorse through the NEPA process but could not do so because of the opposition of a powerful Oregon senator.

needed in the ISC's conservation strategy. However, the senior leader of the Oregon congressional delegation, Senator Hatfield, vetoed such an approach (see box on page 117).

Judge Dwyer Is Still Not Satisfied

The terms of Section 318 lasted only as long as the underlying appropriation legislation lasted: one fiscal year, during which the federal agencies offered almost two years of timber sales.[75] Thus, after the expiration of Section 318 on September 30, 1990, the *Seattle Audubon Society v. Robertson* litigation challenging the underlying owl plans in the 1988 *Regional Guide* could once again continue in Judge Dwyer's court. Also, on September 30, the Forest Service announced in the *Federal Register* that it would withdraw the *Regional Guide* and offer new timber sales in a manner "not inconsistent" with the recently completed ISC report on the conservation of the spotted owl. No advance public notice, comment, or environmental review accompanied the Forest Service's withdrawal of the *Regional Guide* and adoption of the new spotted owl policy. Environmental-group plaintiffs in the litigation before Judge Dwyer immediately challenged the withdrawal of the *Regional Guide*, arguing not only that it left relevant forests without any obligatory management direction for the owl, but also that the withdrawal was unlawfully effectuated without the environmental analysis and planning procedures required by NFMA and NEPA.

Judge Dwyer agreed with the plaintiffs that the Forest Service's withdrawal of the *Regional Guide* was illegal[76] and in May 1991, after an eight-day evidentiary hearing, enjoined all further timber sales in northern spotted owl habitat throughout the range of the species until the Forest Service could develop a legally adequate owl management plan.[77] Judge Dwyer gave the agency until March 1992 to develop such a plan, and his orders were upheld on appeal.[78]

Judge Dwyer's ruling in this case (by that time known as *Seattle Audubon Society v. Evans*) was truly groundbreaking. The judge's frustration with federal land management and regulatory agencies boiled over:

> The records of this case and of No. C88–573Z [compelling the listing of the northern spotted owl, and designation of its critical habitat] show a remarkable series of violations of the environmental laws.

The Forest Service defended its December 1988 ROD [Record of Decision] persistently for nearly two years. . . . But in the fall of 1990 the Forest Service admitted that the ROD was inadequate after all—that it would fail to preserve the northern spotted owl. In seeking a stay of the proceedings in this court in 1989, the Forest Service announced its intent to adopt temporary guidelines within thirty days. It did not do that within thirty days, or ever. When directed by Congress to have a revised ROD in place by September 30, 1990, the Forest Service did not even attempt to comply. The [US]FWS, in the meantime, acted contrary to law in refusing to list the spotted owl as endangered or threatened. After it finally listed the species as threatened following Judge Zilly's order, the FWS again violated the ESA by failing to designate critical habitat as required. Another order had to be issued setting a deadline for the FWS to comply with the law.[79]

Further, Judge Dwyer explained that "the problem here has not been any shortcoming in the laws, but a refusal of administrative agencies to comply with them." He further emphasized:

More is involved here than a simple failure by an agency to comply with its governing statute. The most recent violation of NFMA exemplifies a deliberate and systematic refusal by the Forest Service and the FWS to comply with the laws protecting wildlife. This is not the doing of the scientists, foresters, rangers, and others at the working levels of these agencies. It reflects decisions made by higher authorities in the executive branch of government.[80]

He concluded:

To bypass the environmental laws, either briefly or permanently, would not fend off the changes transforming the timber industry. The argument that the mightiest economy on earth cannot afford to preserve old-growth forests for a short time, while it reaches an overdue decision on how to manage them, is not convincing today. It would be even less so a year or a century from now.[81]

Yes, Judge Dwyer was on fire. However, he also included words in his opinion that suggested the way forward: "The ISC's report has been described by experts on both sides as the first scientifically respectable proposal regarding spotted owl conservation to come out of the executive branch."[82] With this statement the judge signaled that the ISC's strategy was a major step forward. That scientific effort had struck conservation gold, and no amount of political maneuvering and smoke screen could hide that fact. There was no turning back. Things would never be the same.

6

Congress Seeks Protection of Old Growth and Fish

Publication of the Interagency Scientific Committee's (ISC's) 1990 strategy for conservation of the northern spotted owl sent a shock wave through the Forest Service, BLM, local communities, and political leaders in the Pacific Northwest.[1] Obviously, debates over old-growth forests and conservation of species within them had not been resolved by the carefully managed, decade-long process of developing new national forest plans under the National Forest Management Act. Recently released national forest plans would need updating to accommodate this new owl strategy, with land allocations adjusted and projected timber harvest levels significantly reduced.

It quickly became clear, though, that the ISC's strategy for conservation of the spotted owl, even if accepted by the courts, would not completely settle the old-growth and species conservation issues in the Pacific Northwest's federal forests. Old-growth forests located between the ISC-proposed reserves would continue as the primary source of federal timber harvests, just as the conservation of these forests and their plants and animals was becoming a national issue. Many species besides owls lived in those forests, including the marbled murrelet (*Brachyramphus marmoratus*), which was in the process of being listed as threatened under the Endangered Species Act (ESA), and it was not clear that the ISC's owl strategy would sufficiently protect it.

Congress held hearings on the continuing old-growth issues in fall 1990, especially in the House of Representatives, and heard many different views. Two conclusions emerged from these hearings: the warring groups would not find common ground, and the Forest Service and BLM were incapable of leading the way through the thicket of protest and litigation. Thus, the committees sharing jurisdiction over the national forests in the House of Representatives (Interior and

Agriculture) pondered what to do, as did the Committee on Merchant Marine and Fisheries, whose subcommittee on Fisheries and Wildlife Conservation had responsibility for the Endangered Species Act.

The House Interior Committee, chaired by George Miller, Democrat of California, began fashioning legislation to address federal forest issues in the Pacific Northwest, with most of the effort occurring in an Interior subcommittee chaired by Bruce Vento of Minnesota. Toward that end, Vento's subcommittee worked with environmental groups from the Northwest to craft legislation that required a major expansion of old-growth reserves, to help protect those ecosystems and the species within them, and work with a scientific committee to select them.

The House Agriculture Committee took a different approach. Its forestry staff expert, Jim Lyons, urged that a panel of scientific experts be convened to look broadly at conservation of old-growth forests and dependent species. Lyons was committed to ensuring that any proposed solutions be based on the best available science, as he had done in the Society of American Foresters' Task Force on Scheduling the Harvest of Old-Growth Timber.[2]

Chair Kika de la Garza of Texas and subcommittee chair Harold Volkmer of Missouri, both Democrats, permitted Lyons to reach out to three respected senior scientists who had been key members of Lyons's old-growth task force: Jack Ward Thomas, who had also led the ISC effort; Jerry Franklin, who now was at the College of Forest Resources at the University of Washington; and John Gordon, who was Pinchot Professor and dean of the School of Forestry at Yale University. These three successfully urged the addition of Norm Johnson, a professor in the College of Forestry at Oregon State University and specialist in projecting timber harvest levels.

Volkmer invited the four members of this potential team to Washington, DC, in early 1991 to a hearing of his subcommittee, where they each presented assessments of the situation in the federal forests of the Pacific Northwest. After the hearing adjourned, the scientists were invited to lunch in de la Garza's office with Volkmer, Lyons, and a few others. As they talked and ate, Franklin observed Volkmer writing on the back of an envelope and asked the congressman what he was doing. Volkmer responded that he was writing down elements that needed to be included in a bill that would resolve the conflict over timber harvest, old growth, and owls. He returned briefly to his list, looked up, and said to Franklin, "You know, if you could just tell us where the good old-growth forest is, we could quickly resolve this issue." After a moment Franklin responded, "Well, if that is all you need, we could create that [map] for you," although he had no clear idea of how the four scientists would actually do that.

Discussion then turned to how the four scientists would form a team to develop an old-growth map and other materials that could help Congress address the old-growth/spotted owl/timber issue. The scientists' responses were indicative of the personal dynamics to come. Franklin continued to assert that it was possible to develop a map of old-growth forests quickly with the cooperation of the agencies, at one point suggesting a week or two. Thomas attempted to signal Franklin not to make such bold pronouncements by repeatedly trying to kick him under the table. However, he failed to get Franklin's attention, as Johnson was in the way and was the one who got kicked instead.

Volkmer led the discussion, holding a pen and a few sheets of paper before him. He questioned three of the scientists in turn: Jack—what do you need for owls? Jerry—what do you need for old growth? Norm—what would be the resulting harvest level associated with Jack's and Jerry's ideas? When challenged on his approach by other members of Congress sitting around the table, Volkmer said, "Well, it is better than what we have now." Volkmer apparently felt that he had the experts before him, and the right questioning might just elicit the answers he needed to craft legislation to resolve this thorny issue. The four scientists left the lunch feeling a bit shell-shocked, but definitely impressed with Volkmer's zeal to find a solution to the problems bedeviling federal forestry in the Pacific Northwest.

Through Lyons's leadership, Volkmer's back-of-the-envelope plan soon evolved into a formal request for the scientists to map the old-growth forests of the Pacific Northwest, evaluate their differing quality, suggest alternative conservation plans for protection of old-growth ecosystems and the species within them, and estimate the timber harvest level associated with each alternative. Moving from developing a plan to creating alternatives marked an important change from the approach taken by the ISC for the northern spotted owl: it shifted scientists from the role of making decisions to the more customary role of suggesting and evaluating alternatives. A letter, signed by the congressional leadership of the Agriculture Committee and the Merchant Marine Committee on May 22, 1991, reflected this approach.

There were also questions of funding. It was soon apparent that Congress had no money to provide for the scientists' work, although there was a vague promise that Congress might be persuaded to pay their out-of-pocket expenses sometime in the future. Johnson immediately went to Dean George Brown of the OSU College of Forestry for a line of credit. The dean advanced him $10,000 based on his promise to somehow pay it back, once he had received a commitment that the scientists' effort would truly be the development of choices rather than a single plan, as the ISC had done. Many timber companies and communities in western Oregon could be negatively impacted by the upcoming federal forest management decisions, and the dean knew there were storm clouds ahead. He wanted to be sure that the responsibility of selecting a forest plan was squarely and visibly placed with Congress—where it belonged.

With important details worked out, the authorizing letter was sent to the four scientists, and they returned to Washington, DC, for another hearing of Volkmer's forestry subcommittee. The four scientists had a pleasant exchange with the committee until they asked the chair how much time they would have to complete their assignment. Congressman Volkmer said that he hoped the scientists could complete the assignment in three weeks so that he would have time to review the results before meeting with constituents and lobbyists over the July 4 congressional recess.

Of course, the three-week deadline stunned the four scientists. After all, development of national forest plans under the National Forest Management Act in the range of the northern spotted owl had taken over 10 years, with a budget in the tens of millions of dollars and many hundreds of people involved. Even the ISC's analysis had taken six months. Fortunately, the authorizing letter included one provision that made this timetable seem somewhat plausible: the letter encouraged the four scientists to solicit the assistance of others who might be helpful in the endeavor, including resource personnel from the Forest Service, US Fish and Wildlife Service, and Bureau of Land Management.

Volkmer said he would personally write the heads of these agencies requesting their cooperation in this regard.

Soon the discussion was over, with many accolades and statements of support for the scientists' effort from Volkmer and others. As the scientists got up to leave and were literally halfway out the door, Volkmer hollered, "And don't forget the fish. We don't want some damn fish blowing up our solution after we are done!"

The scientists again were stunned. The authorizing letter did not mention fish, and there had been little discussion about fish in the scientists' meetings with Congress. Perhaps some little fish in Volkmer's congressional district in Missouri was giving him nightmares. At any rate, considering salmon and aquatic ecosystems on the federal forests of the Pacific Northwest would require a major expansion of their effort and the addition of scientists with that expertise. With that final admonition from Volkmer, they were on their way.

A key House committee, the Interior Committee, had not signed the authorizing letter. That committee, under George Miller and Bruce Vento, was working on its own with Pacific Northwest environmental groups to fashion legislation to protect special old-growth areas. Also, those groups were skeptical about the objectivity and conservation ethic of the four scientists, apparently believing that they would be too strongly influenced by the timber industry and its political allies in the Pacific Northwest. Thus, the Interior Committee held back and did not sign the authorizing letter.

However, it soon became obvious to members and staff of the Interior Committee that the four scientists might develop old-growth conservation strategies that would gain traction with Congress. Thus, they wanted to be part of the political oversight of the scientists' effort. Toward that end, Interior Committee staff contacted the scientists to say that the committee would endorse their effort if they would add one more scientist with strong environmental credentials to their group from a list of names the staff provided. The four scientists immediately rejected that offer as an attack on their independence but did file away the names they had been given for potential future alliances.

The four scientists knew they needed a name that would convey the importance and seriousness of their effort, comparable to that for the Interagency Scientific Committee. Given that their authorizing letter called for them to evaluate "ecologically significant old growth and late successional ecosystems," they settled on the Scientific Panel on Late-Successional Forest Ecosystems. That would help people understand they would take an inclusive approach to the forests they would consider. This was quite a mouthful and did not collapse into a very sprightly acronym (SPLSFE). Someone outside the group, perhaps with a little malice, coined the name Gang of Four (G-4), and that name stuck.

We Can Map Old-Growth Forests in One Week

As soon as Franklin and Thomas returned home, they took full advantage of their assumed power to engage agency professional staff in the effort. Franklin had decided that the best way to construct an accurate map of old-growth forests and rank the quality of those forests would be to rely heavily on the knowledge of technical staff from BLM districts and national forests, as well as scientists from Forest Service research stations. Hence, Franklin sent out requests for Forest Service and BLM technical personnel, who had local knowledge of the old growth and plants and animals in their districts, to bring their maps, aerial photos, and knowledge and join him for a mapping exercise in Portland in a few days. Also, he asked Tom Spies, a PNW Research Station expert on old-growth forests, to join him and sent out word to the area ecologists who advised federal foresters across the range of the northern spotted owl that he needed their assistance. Thomas, in turn, sent out the call for owl biologists who had helped him prepare the ISC's report, and for wildlife experts with knowledge of other old growth–associated species.

Without hesitation, the Forest Service and BLM quickly responded with teams of two to four resource specialists from each national forest and BLM district in the owl's range to assist in the mapping. They were joined by other qualified specialists from the PNW Research Station, the US Fish and Wildlife Service, the Oregon State Office of the BLM, and technical experts from state agencies. Most members of the ISC also participated. Word had gone out that something big was happening and people piled in to help and be part of the show. Without these resource professionals, the G-4 effort would have been only a shadow of what it accomplished.

The four scientists needed a large workspace in a hurry for an effort that they expected to grow to 100–200 people. Thomas undertook the task of finding such a space, with the help of Forest Service personnel, and rented a part of the Portland Memorial Coliseum. This original home of the NBA's Portland Trail Blazers would now house the science effort, providing approximately 10,000 square feet of space at $1,000 per day. Thomas left Johnson to pay for it.

Franklin and Thomas then went off to solve other problems for a few days. Johnson soon got a call from Lyons, who asked whether, while the scientists were at it, they would also design old-growth conservation strategies for California and eastern Oregon and Washington. Since Johnson had no idea how the G-4 would complete the assignment they had already been given, he declined that request.

When Franklin and his colleagues developed an interim definition of old-growth forests in the early 1980s (table 4.1), they defined classic old-growth stands based on their structural characteristics, including large, old trees; large standing dead trees (snags); and large logs on the forest floor. These early definitions of old growth were quite specific about numbers and sizes of such structural elements, and the need for all structural attributes to be present. The goal of that effort was to provide a rigorous definition of old growth so that marginally qualified stands would not be included in old-growth inventories. Stands that met these requirements were typically from 200 to 600 years old

However, research during the 1980s had increased knowledge of old-growth forests in terms of both structure and function. Much of this research had been carried out as part of a 10-year research program on Douglas-fir forests as wildlife habitat.[3] Utilizing this information as background, Franklin wanted to be sure to map and evaluate the quality of all forests in the owl's range that could potentially contribute to old-growth forest function, including wildlife habitat, even if they did not fully meet the definition of old growth that he had previously helped develop. Spies, who had been working on the structure and function of natural Douglas-fir forests throughout the 1980s, agreed with this more inclusive approach.

For guidance, Franklin and Spies looked to some younger forests that they knew had structural features that contributed to old-growth forest function. Both were familiar with many natural young forests that had developed following forest fires in the last half of the nine-teenth and early twentieth centuries. These forests had extensive biological legacies such as large, old, live trees, large snags, and large downed logs, as they had typically suffered only a single burn and had not been salvage logged. Franklin particularly focused on forests that he knew well in southwestern Washington near his boyhood home of Camas—forests that had developed after the 1902 wildfires in the Columbia Gorge region, including the parts of the famous Yacolt Burn that had not undergone reburns. These 80-year-old forests were already beginning to develop older forest characteristics and were on their way to becoming the old-growth forests of the future.

Ultimately, Franklin and Spies settled on 80 years of age as a rough threshold for identifying natural Douglas-fir forests that were begin-ning to provide significant habitat for species characteristic of older forests—for identifying the beginning age of late-successional forests. They would include natural forests at or beyond this age in the requested assessment of "ecologically significant old growth and late successional ecosystems," which they called *late-successional/old growth forests*.

Few ideas from this period have become more important to man-agement and conservation of the forests of the owl's range than the concept of late-successional/old-growth forests (LS/OG forests or just LS/OG). The concept recognizes that older-forest habitat function can begin relatively early (i.e., 80 years of age) in the natural developmental sequence and includes both mature forests (often portrayed as forests from 80 to 200 years of age) and old-growth forests (often portrayed as forests over 200 years of age) (plate 6.1). The concept is especially powerful because it dovetailed in 1990 with the dividing line between natural stands that had regenerated on their own following a natural disturbance such as wildfire or windstorm (generally more than 80 years of age in 1990), and managed stands that resulted from past clearcuts (generally less than 80 years of age). Perhaps not surprisingly, this age became a dividing line for describing where timber harvest could provide ecological benefits (stands less than 80 years of age) and where it could be ecologically harmful (more than 80 years of age).

The congressional committees wanted old-growth forests to be not only mapped but also graded as to quality. Franklin and Spies took the assignment to mean that they should map LS/OG forests—both

Table 6.1. Factors used in classifying the ecological significance of LS/OG Areas.

Factor	Characteristic contributing to higher ecological significance
Block size	Larger contiguous blocks of forest
Fragmentation	Little or no fragmentation
Location	Little other old forest in area
Stand attributes	Classic old growth[1]
Age	250–750 years
Productivity	Higher site productivity
Elevation	Lower elevation (relatively rare)
Spotted owl	Known/likely occurrence of spotted owl
Marbled murrelet	Known/likely occurrence of marbled murrelet
Other species	Known/likely occurrence of other LS/OG species

[1] As defined in J. Franklin et al., *Ecological Characteristics.*

mature and old—and develop a list of criteria for classifying LS/OG as highly ecologically significant (LS/OG1), ecologically significant (LS/OG2), and the remainder (LS/OG3) (table 6.1). Mapping would focus on larger aggregations of LS/OG stands that made suitable management units for old-growth reserves rather than on smaller individual patches. Given past cutting patterns, younger stands and cutover areas were often in areas mapped as LS/OG, which are called LS/OG Areas in this book.

The Forest Service and BLM, as well as the Audubon Society, The Wilderness Society, and the remote sensing laboratory at the College of Forestry at OSU, provided a variety of maps and satellite imagery that could help identify LS/OG Areas. Even with all these data sources, though, classifying the ecological significance of LS/OG Areas proved challenging.

The G-4 realized that they needed to learn quickly where LS/OG Areas were located on many millions of acres of federal land. They also knew that federal timber planners would have incorporated many of these stands into their five-year timber sale programs. How could they separate the mapping of old growth from concerns about whether such delineation would disrupt someone's timber sale program? Franklin and Thomas decided that they would try to accomplish this goal in two ways.

First, Franklin clearly described the purpose of the effort and asked that the mappers ignore the consequences of their LS/OG Area identification for timber sale programs, leaving the G-4 (and Congress) to take the responsibility for any plans developed from this information. Franklin emphasized to the mappers that he had great faith in their ability to help him identify and grade LS/OG Areas and that he had brought them to this large room, in part, to shield them from the outside world and enable them to concentrate on that task. To help put the mappers' areas of interest in perspective, improve consistency, and develop a sense of camaraderie, Franklin asked that the spatial arrangement of their worktables reflect the geographic distribution of the national forests and BLM districts across the range of the owl. Most had never been asked to do this before and they welcomed the chance to think and work in that way, as it would enable them to see what their neighbors were doing and converse with them.

Second, to further shield them from outside pressures, Franklin asked agency line officers (especially deputy regional foresters, forest supervisors, and deputy forest supervisors from the Forest Service, and state and district managers from the BLM) to meet with him before they went into the mapping room. When they arrived, they would be asked to wear orange hunting vests (purchased by Thomas for this purpose) to identify themselves as outsiders, and their presence would be announced. Early on, a supervisor and deputy supervisor entered the mapping room without Franklin's knowledge. When Franklin learned they were present, he used a microphone to request that they immediately come to the G-4 office. Upon learning what they would need to wear, they soon left. Subsequently, few line officers appeared during the LS/OG mapping, which helped to insulate the mappers from the managers.

With the aid of multiple databases and the expertise of resource personnel from the agencies, the universe of LS/OG Areas was identified on the 18 national forests and seven BLM districts in Oregon, Washington, and northern California within the range of the spotted owl. Rigid, detailed rules were quickly revealed as inadequate;

Figure 6.1. Jerry Franklin leading a discussion of the results of old-growth mapping on a national forest as part of the Gang of Four analysis, with John Gordon on his right and Eric Forsman with pen in hand on his left.

professional judgment was critical to the process and involved repeated interactions between Franklin and the resource professionals (fig. 6.1). Although the mapping teams from the national forests and BLM districts initially identified and ranked LS/OG Areas, Franklin, Spies, and others sometimes made significant modifications in developing the final maps, reflecting their analysis of the overall LS/OG systems, connectivity and consistency between administrative units, and the relation of federal forests to forests on adjacent state and private lands. As an example, the age and structure of adjacent nonfederal forests were significant factors in the analysis of potential LS/OG Areas along western boundaries of most Cascade Range national forests (especially from the Gifford Pinchot to Umpqua National Forests), as was the desire to include several large, relatively intact tracts of mature forest there as LS/OG1 Areas.

Almost 5.7 million federal acres were mapped as LS/OG1 Areas on federal lands in the owl's range (plate 6.2), and 2.3 million acres were mapped as LS/OG2 Areas. LS/OG Areas in national parks and wilderness were not considered, as they were already protected. Mapped areas included the LS/OG stands themselves and the intermixed younger stands, cutover forestlands, and meadows and lakes. Nonfederal forest that fell within a mapped LS/OG Area was not included in the totals.

In interpreting these rankings, it is critical to remember that total acreage was only one part of the picture in evaluating the quality of LS/OG Areas. The ecological condition of the stands within them had to be evaluated: some LS/OG forest was extremely fragmented, putting the interior conditions in these ecosystems and their associated species at risk and thus meriting a lower ranking.

Members of the ISC reviewed the network of LS/OG1 Areas to assess whether such a network would meet the standards and guidelines set forth in the ISC's strategy, under the assumption that the LS/OG1 Areas would be managed as prescribed in the ISC's report. ISC members working with Franklin concluded that the LS/OG1 Areas with some modest Owl Additions would fully meet the standards and guidelines of the ISC's strategy for the northern spotted owl. They also estimated that the resulting reserve network contained approximately 25 percent more known spotted owl pairs than the Habitat Conservation Areas proposed in the ISC's strategy.

Don't Forget the Damn Fish

With Congressman Volkmer's warning cry of not letting Congress be surprised by "some damn fish" ringing in their ears, the G-4 knew they needed expertise in fish and fish habitat even though those considerations were not explicitly part of the written charter the G-4 had received from Congress. One of the people the Interior Committee staff had requested be added to the scientific team was Gordon Reeves of the PNW Research Station. Reeves had recently been in the news because he disagreed with the Siskiyou National Forest's assessment of potential impacts of proposed logging on salmon habitat in southwestern Oregon, which made him sound like just the person needed to develop fish-oriented conservation strategies. Therefore, Johnson invited Reeves to join the scientists in Portland "for a few days" to suggest approaches that would better protect salmon habitat. In addition, James Sedell, an aquatic ecosystem scientist with the PNW Research Station and compatriot of Reeves, had also been alerted that he might be needed.

Sedell and Reeves met in Corvallis at 5:30 the next morning to drive to Portland, and an excited Sedell described this as an "incredible opportunity," saying that they needed to persuade Johnson and the other leaders that fish should be included in the effort and that they were the people to do it. Neither Reeves nor Sedell fully understood exactly what the effort was or what they might be expected to do, but Sedell told Reeves it was "a big deal and we need to be ready." Sedell drove as Reeves captured their ideas on paper.

They immediately agreed that they would most likely be asked whether and why native fish in the range of the northern spotted owl needed enhanced conservation efforts and how this related to the owl and old-growth forests. Reeves suggested that they start any conversation with a review of the recently published paper that documented how hundreds of salmon populations in the Pacific Northwest and California were declining in numbers and therefore warranted special attention.

As they headed north to Portland, the discussion turned to the components of a potential conservation strategy for salmon populations and other fish. They talked about what should be part of plan—some type of emphasis areas such as fifth- or sixth-field watersheds[4] where the management priority would be conserving or restoring aquatic and riparian habitat, an expansion of riparian management areas on federal lands to better protect the ecological processes that create and maintain aquatic habitats, and a way of specifying what management actions could occur within riparian areas.

Sedell and Reeves tossed ideas back and forth almost until the moment they turned into the Portland Coliseum parking lot. Reeves hurriedly completed a rough outline as they stepped out of the car. As Sedell was fond of saying, "They weren't peaking too soon!"

Sedell and Reeves met with the Gang of Four in a small room on the coliseum floor to present their case for inclusion of salmon conservation in the G-4 effort. They first reviewed with the G-4 the results of "Pacific Salmon at the Crossroads," a paper written by Willa Nehlsen, Jack Williams, and James A. Lichatowich and published by the American Fisheries Society just two months previously, in March 1991.[5] The authors compiled their report from field data gathered by fish biologists across the Pacific Northwest and identified over 200 populations

PACIFIC SALMON AT THE CROSSROADS
(Gordon Reeves)

The late 1980s and early 1990s were a particularly frustrating and stressful time for fish and aquatic biologists within the Forest Service, BLM, and state land management agencies across the West. One population of Chinook salmon, the winter-run Chinook on the Sacramento River in California, had already been listed under the Endangered Species Act, and biologists believed that many Pacific salmon populations (which include seagoing trout) were imperiled because of declining numbers, deterioration of freshwater habitats, and several other factors. However, federal land management agencies contended that populations and habitat were within acceptable limits, and special considerations were not necessary. Recently adopted and proposed forest plans reflected that view.[*]

Three members of the American Fisheries Society Endangered Species Committee—Willa Nehlsen (USFWS), Jack E. Williams (BLM), and Jim Lichatowich (Oregon Department of Fish and Wildlife)—undertook documentation of the status of Pacific salmon on their own time, outside their agency responsibilities and obligations. Over more than two years, they assembled data from federal and state management and regulatory agencies and discovered that 214 populations of Pacific salmon needed special consideration because of low or declining numbers. Only one was listed under the Endangered Species Act; all the others needed protection to prevent their demise. Of these, 101 had a high and 58 a moderate likelihood of extirpation (local extinction). The remaining 55 also needed special attention to prevent further declines. They summarized these results in "Pacific Salmon at the Crossroads: Stocks at Risk from California, Oregon, Washington, and Idaho."[†]

This paper had an immense impact on the management of these fish stocks despite the generally sour reception it received from federal and state agencies. Biologists were ecstatic over the recognition of their concerns. The credibility of the paper could not be seriously questioned: it has since been cited more than 1,300 times and became the springboard for formal assessments of these populations for ESA listing. It also provided the key argument for incorporating fish into conservation strategies for old-growth ecosystems during the G-4 analysis and the science assessments that came after it.

---

[*] Forest plans for the national forests of Oregon and Washington were adopted in the late 1980s, while those for California were still pending (completing the NEPA process), as were forest plans for BLM lands.

[†] Nehlsen et al., "Pacific Salmon at the Crossroads."

of naturally spawning Pacific salmon in decline and at risk of extinction. They identified habitat loss and timber operations as key causes of the declines, although other factors, such as dams and overfishing, were also important.

Sedell and Reeves next presented their preliminary framework for addressing conservation of salmon, other aquatic organisms, and aquatic ecosystems to the G-4. Their goal was to develop a watershed and fish emphasis for management of federal forests that would maintain and restore ecological functions and processes that create and maintain habitat in streams over time, with a special emphasis on the habitat of potentially threatened and endangered fish species. This included identifying watersheds where conservation of aquatic habitat would be the focus, expanding riparian management areas, and constructing a comprehensive restoration strategy.

At the end of their presentation, John Gordon said that having fish involved would be a "game changer" in the debate over old-growth forests. If people knew that salmon populations were declining as a result of old-growth harvest, the nature of the debate would shift away from owls versus jobs. Salmon harvest provided jobs, too, and many people had a favorite salmon population they pursued with rod and reel.

The G-4 unanimously agreed that fish would be part of the effort and asked Sedell and Reeves to assemble a team to help them. From then on, many rightly referred to the six scientists as the Gang of Four + 2 or G-4+2.

Sedell and Reeves immediately contacted fish biologists and hydrologists from the national forests and BLM districts in the owl's range and invited them to come to Portland with information on the status of Pacific salmon populations and habitat in their respective forests. Like their terrestrial counterparts, the aquatic group organized their tables geographically, which facilitated some of the first exchanges among the fish biologists and hydrologists working on adjacent forests as well as exchanges with wildlife biologists from their own and neighboring forests.

Staff from several national forests suggested that Sedell and Reeves look at recently adopted or proposed national forest plans to understand the potential impacts of these plans on fish and aquatic ecosystems. To their surprise, Sedell and Reeves found that many plans openly acknowledged that fish populations or habitat would decline as the plans were implemented. They also discovered that conclusions in some plans indicating that fish populations or habitat would improve were based on questionable modeling.

Sedell and Reeves worked with national forest and BLM district aquatic specialists to identify watersheds that contained habitat for potentially threatened populations of salmon or other fish or were sources of high-quality water. These Key Watersheds and related areas would form the nuclei of any broad-scale effort to recover potentially threatened fish species and populations. In total, they identified over 130 watersheds critical to the survival of Pacific salmon and seagoing trout within the range of the spotted owl—watersheds that provided habitat for 90 of the genetically distinct populations of salmon and trout that had been identified in the "Pacific Salmon at the Crossroads" article as being of special concern because of low or declining numbers. They placed a high priority on restoration of Key Watersheds and developed a package of special management directions for them.

Sedell and Reeves emphasized to their colleagues that wider riparian buffers across the federal landscape would provide additional protection from disturbance and help initiate recovery of degraded areas. However, they were unsure how far to push. If they proposed what they thought was needed, they were worried that it would be seen as so extreme that it would be rejected out of hand. So Sedell and Reeves went to John Gordon for advice—the member of the G-4 who had pushed hardest for their inclusion in the effort. Gordon was clear and direct: Congress should hear the whole story rather than be told later that more was needed. With that support, Sedell and Reeves plowed ahead to chart a new course for fish and riparian protection on federal forests.

Toward that end, their strategy for a watershed and fish emphasis called for a major expansion of buffers on fish-bearing streams. In addition, it called for buffers on seasonally flowing (intermittent) non-fish-bearing streams that remained largely unprotected in recently adopted or proposed national forest and BLM plans. These intermittent streams provided essential nutrients and structure (gravel and large wood) to the fish-bearing streams below. However, buffers on them represented a major new requirement, since most stream miles on federal forests were intermittent non-fish-bearing streams. Commercial timber harvest would generally be prohibited within the buffers.

They also identified the tens of thousands of miles of roads on federal lands within the owl's range as a major cause of watershed degradation. Landslides and debris torrents had been exacerbated by road drainage problems associated with culverts that were too small or infrequent, and by poor road design and maintenance. Their watershed and fish emphasis called for either improving these problem roads, such as by increasing culvert numbers and sizes, or removing them and restoring their rights-of-way to a natural condition.

Finally, Sedell and Reeves noted that timber harvest and roadbuilding activities had degraded many riparian buffers and recommended that the watershed and fish emphasis include ecologically sound programs to restore the buffers. These programs should include silvicultural treatments of riparian vegetation, erosion abatement, and in-channel engineering and planning in degraded areas to recover fish habitat.

All in all, the watershed and fish emphasis proposal by Sedell and Reeves, if fully adopted, would usher in a new era of federal forest conservation—one that would bring changes to management of these forests as monumental as the ISC spotted owl conservation strategy.

A Multitude of Choices

Franklin's team completed the mapping of LS/OG Areas and Owl Additions in six days, along with an initial delineation of Key Watersheds. On the seventh day, the G-4 and some senior advisers reviewed the results, made some final adjustments, rolled up the hand-drawn maps, put them in the trunks of their cars, and left the expensive coliseum space. However, they had no clear plan for next steps or where they would work. Fortunately, someone asked Johnson whether desk space at the Oregon State University Portland Center might be available to him as an OSU faculty member. The helpful people at the center welcomed the G-4 and provided two floors and a basement as workspace. The effort started modestly, but more and more scientists and resource professionals arrived as word got out that new conservation strategies for federal forests of the Northwest were being invented. Over the next two weeks, these efforts grew to fully occupy all three floors of the OSU Portland Center. Work went on for at least 16 hours a day in free-flowing continuous brainstorming sessions. Maps and graphics were all hand drawn (and redrawn) on Mylar, as there was neither time nor resources to do anything grander.

The G-4 (including Sedell and Reeves) went to work, under Johnson's leadership, developing a suite of alternative approaches to protecting LS/OG Areas, the species within them, and sensitive fish populations. Most alternatives were constructed from the modular pieces that represented the major components of the analysis:

- recently adopted or proposed Forest Plans (FP);
- the Interagency Scientific Committee's strategy for the northern spotted owl (ISC);
- highly significant LS/OG Areas (OG1) identified by the Gang of Four;
- significant LS/OG Areas (OG2);
- other LS/OG forests (OG3);
- added reserves to help conserve the spotted owl (Owl Additions, OA); and
- watershed and fish strategy (F) as drafted by Sedell and Reeves.

The alternatives that became a focus of later discussions with Congress are covered here (table 6.2). Alternative 2 represented the recently adopted or proposed forest plans. Alternative 4 modified the forest plans to incorporate the ISC's report. Alternatives 6 through 14 utilized LS/OG and watershed and fish components to sequentially add more protection of LS/OG species and ecosystems and fish populations. Such an approach enabled a marginal analysis of the benefits and costs of moving between alternatives.

Areas outside reserves under each alternative were called Matrix, as had been done in the ISC's report, and restrictions on timber harvest in Matrix were developed by the G-4 with assistance of agency specialists. Their goal was to enable timber harvest there, but in ways that would allow the persistence of many old-growth species in Matrix and assist in the movement of owls and other mobile old-growth species between reserved areas. Proposed restrictions on timber harvest in Matrix included the following:

- the ISC's 50-11-40 rule, which ensured that half of Matrix had an overstory canopy that would help owls successfully disperse;

Table 6.2. Components of key alternatives developed in the Gang of Four analysis.

Alternative	FP	ISC	OG1	OA	F	OG2	OG3
2	X						
4	X	X					
6	X		X	X			
8	X		X	X	X		
12	X		X	X	X	X	
14	X		X	X	X	X	X

- retention of green trees at final harvest to help reduce impacts on fish and wildlife and provide future sources of snags and downed logs;
- retention of some LS/OG forest in Matrix to help provide potential replacement stands for LS/OG forests that would be lost to natural disturbances; and
- longer rotations (120 or 180 years) to provide potential replacement stands for LS/OG forests.

Combinations of these requirements were bundled together (A, B, and C) for each alternative in table 6.2 beyond Alternative 2 (Forest Plans), with an increasing number of requirements included from combination A to combination C. Combining an alternative with a particular set of requirements created a choice for management of federal forests. As an example, Alternative 6 from table 6.2 became Choice 6A, 6B, or 6C depending on which set of forest management requirements were assigned to it.

Once the choices to be considered were developed, attention turned to assessing their implications for LS/OG species and ecosystems and for timber harvest levels. Four measures of protection of species and their habitat were chosen:

- ensure a viable population of northern spottted owls;
- provide adequate habitat for the marbled murrelet;
- provide adequate habitat for other LS/OG-associated species; and
- provide adquate habitat for sensitive fish species and populations.

An additional measure assessed the implications of each choice for retention of a functional LS/OG network in which populations of LS/OG-associated species could persist and spatially interact. This assessment tried to directly measure ecosystem effects of choices.

Wildlife and fish biologists assisting the effort suggested utilizing a qualitative risk analysis to assess the implications of any choice for LS/OG species and ecosystems. This ecological risk analysis used a seven-point likelihood scale ranging from a very high to very low probablity of persistence over 100 years. "Very high" indicated a probabililty of persistence of over 95 percent, while "very low" indicated a probability of less than 5 percent, and "moderate" a probability of 50 percent.

Thomas and Reeves led the ecological risk analysis with the assistance of experts on the respective species or ecosystems. The effort included a review of available knowledge on habitats of the species being rated and took less than a week to complete. It did not use sophisticated mathematical models because adequate databases and models were not available for most species. Rather, it relied on expert judgment. However, the G-4 was confident that further analysis would be unlikely to shift results by more than one risk level either way, such as from medium-low to low, or from medium-low to medium.

These risk ratings represented an aggregate assessment for all federal forests within the owl's range. The G-4 expected that more detailed analyses would show variation among national forests and BLM districts, and even among individual watersheds, with respect to the species or habitat risk under different choices.

Johnson worked with agency staff to estimate the likely timber harvest level on federal lands suitable for timber production under each choice for each national forest and BLM district in the owl's range. They first employed harvest scheduling simulators to project the maximum timber harvest over time that could occur on each national forest and BLM district under a given choice. Increasingly, though, national forest planners and specialists found these maximum values impossible to reach when all resources and uses were considered as required by forest plan standards and guidelines. After discussions with agency personnel, Johnson reduced the simulated timber harvest levels by 15 percent, believing that harvests would actually be 10–20 percent less than projected.

The G-4 developed 36 choices that varied in the type and amount of reserves, whether to apply the watershed and fish strategy, and restrictions on timber harvest between reserves. Choices of special interest to Congress (fig. 6.2) are described here, with the timber harvest level of the previous decade (1980–1 989) highlighted as the comparative basis for estimating expected declines in harvest under each choice.[6]

The G-4 noted that habitat on federal lands potentially represented only a portion of the life-history requirements for the marbled murrelet, other LS/OG-associated species (except the spotted owl), and sensitive fish species and populations. Management policies on federal lands would not necessarily ensure survival of these species and populations, as they spent considerable time on nonfederal lands. Thus, the risk analyses for these species focused on the adequacy of habitat for the portion of their life history on federal lands, rather than on overall species survival. Providing adequate habitat on federal lands in this analysis was a very similar standard to the requirement in the viability standard of the 1982 Planning Rule that "fish and wildlife habitat shall be managed to maintain viable populations of existing native and desired non-native vertebrate species . . . well distributed in the planning area."[7]

Key conclusions from the analysis were the following:

- Choice FP (the recently adopted or proposed national forest plans for national forests and BLM districts) provided low to very low probability of persistence for a century or longer for all five ecological resources—owls, murrelets, other LS/OG species, fish, and old-growth ecosystems—and a projected annual harvest that was 76 percent of the previous decade's average annual harvest. A decade of planning under a legal mandate to protect species and ecosystems had led to federal forest plans that, in the scientists' estimation, still left them at risk.[*]

---

[*]Duncan and Thompson, in "Forest Plans and Ad Hoc Scientist Groups," attribute the forest plans having that result to a complex combination of factors including agency expectations, the newness of viability analysis, the understanding of the level of protection needed, attitudes toward risk, and the place of these considerations among other goals. As Mike Kerrick, supervisor of the Willamette National Forest, told them: "At most, species viability was considered just one of a dozen or so attributes to be balanced."

Evaluation Category	\multicolumn{11}{c}{Probability of Retention for a Century or Longer}
	85-89 harvest	80-89 harvest	FP	4A	6A	8A	8C	12A	12C	14A	14C
Functional LS/ OG Network	--	--	VL	ML	M	M	H	MH	VH	H	VH
Viable Spotted Owl Population	--	--	L	H	H	H	VH	VH	VH	VH	VH
Marbled Murrelet Nest Habitat	--	--	L	ML	M	M	MH	MH	H	VH	VH
Other LS/ OG Species Habitats	--	--	VL	ML	M	MH	H	H	VH	VH	VH
Sensitive Fish Habitats	--	--	L/VL	L	ML	MH	H	MH	VH	H	VH

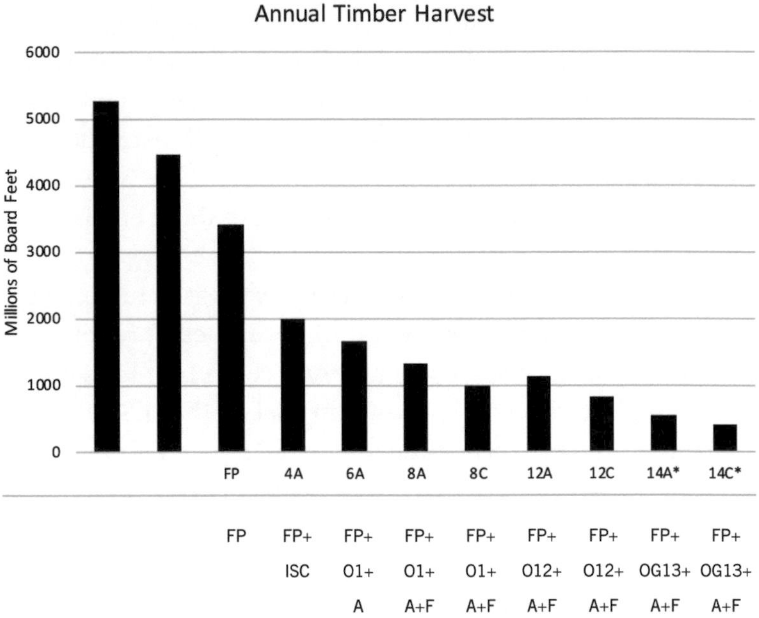

Annual Timber Harvest

	FP	4A	6A	8A	8C	12A	12C	14A*	14C*
	FP	FP+	FP+	FP+	FP+	FP+	FP+	FP+	FP+
		ISC	O1+	O1+	O1+	O12+	O12+	OG13+	OG13+
			A	A+F	A+F	A+F	A+F	A+F	A+F

Figure 6.2. Implication for protection of species and ecosystems and for projected timber harvest of some choices for federal forests in the range of the northern spotted owl in the Gang of Four analysis. Choices 8A and above were of most interest to congressional committees. Choice descriptors in the horizontal legend: FP = Forest Plans, ISC = Interagency Scientific Committee Plan, O1 = highly significant LS/OG Areas, O12 = highly significant and significant LS/OG Areas, OG13 = all LS/OG Areas, A = northern spotted owl additions, F = watershed and fish emphasis. Protection levels: VL (very low) to VH (very high); "A" and "C." = different timber management strategies for area outside reserves.
*First 20-year harvest, then harvest slowly rises.

- Choice 4A (Forest Plans + ISC + management requirement A) reflected the ISC's owl conservation strategy and had a projected annual harvest that was 45 percent of the previous decade's. The G-4 analysis confirmed that the ISC's strategy had a high probability of providing for viable populations of spotted owls. However, it found only medium-low or low levels of protection for other ecological resources. The predicted outcomes for this choice disappointed many people who had expected the ISC's owl strategy to also protect other LS/OG species. More protection would be needed.
- Choice 6A (Forest Plans + LS/OG1 + OA + management requirement A) used highly significant LS/OG Areas (LS/OG1) and Owl Additions as the building blocks in a reserve system for species and ecosystem conservation. These reserves were substitutes for the ISC's HCA reserve system. Choice 6A resulted in at least medium levels of protection for all resources being considered except habitat for sensitive fish populations, which received a lower rating. Projected annual harvest was 37 percent of the previous decade's.
- Choice 8A added the watershed and fish emphasis to Choice 6A, increasing the rating of habitat for sensitive fish populations to medium-high as well as increasing the habitat rating for other LS/OG species to that level, and providing the first choice with at least a medium rating for all ecological resources. Projected harvest was 30 percent of the previous decade's average annual harvest. The management requirement for lands outside reserves (i.e., A, B, or C) also influenced risks and harvest levels. Choice 8C, which included green tree retention at harvest and long rotations, as an example, boosted protection of all habitats to at least a high level, except for the marbled murrelet, with projected harvest at 22 percent of the previous decade's.
- Choice 12A added significant LS/OG Areas (LS/OG2) to the reserves of Choice 8A, which raised all protection levels to at least medium-high. Projected harvests were 25 percent of the previous decade's average annual harvest. Choice 12C, which included green tree retention at harvest and management on long rotations, boosted all ecological resources to at least the high level and

most to the very high protection level, with projected harvest at 18 percent of the previous decade's.

- Choice 14A built on Choice 12A by adding the remaining LS/OG forest (LS/OG3) to the reserves. It provided high to very high protection levels for all ecological resources. Projected harvest was 13 percent of the previous decade's. Choice 14C resulted in very high protection levels for all five resources, with projected harvest at 9 percent of the previous decade's.

These conclusions from the G-4 analysis foreshadowed the findings of later science assessments: achieving at least a medium-high level of protection for older forests and the species within them would require at least a 75 percent reduction in annual timber harvest from federal lands in the range of the northern spotted owl.

The Process of Democracy Must Go Forward from Here

The choices provided by the G-4 exercise were the most comprehensive and sophisticated set of conservation strategies for management of federal forest ecosystems and their constituent species ever presented to Congress. Perhaps it would have been more helpful to Congress if the G-4 had narrowed down the range of choices to fewer, more realistic choices or at least alerted Congress as to which might pass legal tests. Perhaps it would have been even better if the G-4 had recommended a single choice, as the ISC had done.

During the last few weeks of the analysis, the G-4 frequently tossed around the idea of signaling Congress as to which of the choices might meet legal requirements or be optimal. Ultimately, the G-4 decided not to do that for several reasons. First, the G-4 did not have the legal training that would enable them to credibly advise Congress on what choices would pass legal tests relative to the ESA and other laws. Congress could, in any case, change these laws if its members thought such action was in the public interest. Second, the group felt that once people settle on one choice as the best, they inevitably, and often unconsciously, begin to make it look better than the others. Finally, the G-4 understood that they were the analysts and Congress was the decision-maker, and they wanted to keep those roles clear. Thus, the G-4 never discussed which choices would meet the requirements of environmental law or polled themselves on which one was best. If Congress asked them for a recommendation, the G-4 had none.

The G-4 kept a news blackout on their findings from members of Congress and their staffs. Knowledge is power in policy and political processes, and policy makers expect to be given some warning if a study they commissioned contains politically difficult results. Occasionally congressional staffers would appear in Portland for a day or two and look over the shoulders of the G-4 to get a sense of things for their bosses, but the results shown in figure 6.2 were not made available, both because they were not entirely finished and because the G-4 wanted to be the ones to tell the story to Congress. Thus, staffers had to be satisfied with maps and overlays showing different alternatives. For once, members of Congress and their staffs would all see the results simultaneously at a scheduled briefing.

While the congressional committees that requested this work gave the G-4 three weeks to complete it, the G-4 asked for and received an extension of three additional weeks. That was all the time they would be given, however, and Lyons informed the G-4 that a briefing had been scheduled at the end of the sixth week of the effort—July 25, 1991. With that date set, analysis shifted into high gear and lights burned late at the OSU Portland Center.

In the final two days, Thomas led the effort to write a report explaining the approach taken and general conclusions, but without much detail, as it was all still being finalized. A draft copy was faxed to Gordon (one of the G-4) for his review, as he was in Europe. The G-4 also worked feverishly to finish the poster boards showing their maps and results that would be used in their presentation—much like college students scrambling to complete a term project before the due date.

Thomas invited Sedell and Reeves to dinner the last night before everyone was headed to DC. He thanked them for their involvement and expressed how much he and the other members of the G-4 appreciated the work they had done. Thomas also told them that if they got on the plane the next morning their careers would never be the same, and that they should be prepared for rough treatment from the management agencies and politicians. Everyone would understand if they decided not to go; their work would nevertheless be acknowledged.

However, Sedell and Reeves had no intention of being left behind; they were confident in the work they had done and were all in.

When Johnson got off the plane the next afternoon at Reagan Airport, Gordon was there to greet him. Without even saying hello, he waved a copy of the draft G-4 report at Johnson and bellowed, "I won't sign this crap." He claimed the draft report was rambling and inconclusive and did not do justice to the fine analysis of the last six weeks. That was a real setback, with the briefing scheduled for the next morning at 10:00. Knowing he had to have agreement from all G-4 members, Johnson then asked, "Well, what would you sign?" Gordon thought for a little while and committed to coming up with key points to emphasize by the next morning.

The morning meeting that Lyons had arranged was unusual in three ways. First, it was an in-house meeting restricted to members of Congress and their staffs, rather than a formal congressional hearing that would be open to the Bush administration, Forest Service, BLM, and the public. Second, many members of the three committees were present, rather than just their staffers. Third, everyone sat at the same level in a classroom setting, with chairs facing the front of the room where the Gang of Four (including Sedell and Reeves) stood and where tables held the poster boards summarizing their results.

Franklin, Thomas, and Johnson presented the maps and results. Charts tabulating both the ecological and economic consequences of a number of choices made the trade-offs clear: there would have to be major reductions in federal timber harvests, beyond what had been considered so far, to conserve the northern spotted owl, other species that inhabited LS/OG forests, and the fish that swam in the streams there (fig. 6.2).

Then Gordon summarized the G-4 findings:

We believe:
(1) A wide range of choices exists for managing late-successional forests in the Pacific Northwest. We have provided what we believe to be a full range of practical choices.
(2) Current Forest Plans do not provide a high level of assurance (low risk) for maintaining habitat for old-growth-dependent species.

(3) Projected harvest levels in the forest plans often overstate what can be achieved. Thus, our calculations started from a somewhat lower base than previous efforts.

(4) De-listing of the northern spotted owl over a significant portion of its range may be a realistic consideration under several of the choices presented.

(5) There is no "free lunch"—that is, no choice provides abundant timber harvest *and* high levels of habitat protection for species associated with late-successional forests.

(6) We have described the beginnings of a practical "ecosystem approach" to conserving biological diversity.

(7) We have provided a sound basis for decisions, given the time and information limits within which we operated. Science (at least as exemplified by the four of us and those that assisted us) has done what it can. The process of democracy must go forward from here.

There followed a period of stunned silence as members of Congress and their staffers took it all in. They had asked four scientists to develop strategies to conserve old-growth ecosystems and the species within them, and here were the scientists' findings. With the help of Sedell and Reeves, the G-4 had also addressed conservation of habitat for sensitive fish populations. The "process of democracy"—that is, Congress and the president—would need to lead from here. The scientists had done what they could.

The formal congressional hearing on the Gang of Four report occurred in the early afternoon. The Forest Service came out in force, led by Chief Dale Robertson and John Butruille, the regional forester for the Pacific Northwest Region and a tall, imposing ex-Marine. The chief and Butruille knew that changes to protect the northern spotted owl were needed, but they did not know that additional changes would be needed to protect other old-growth species. Most importantly, they did not realize that the forest plans that had just been adopted on their watch would not protect fish. In addition, they had been blindsided about this failing at a congressional hearing and (understandably) were mad as hell.

Chief Robertson stormed out once the scientists completed their formal presentations, but Butruille was soon standing red faced over

Sedell and Reeves, yelling at them for asserting that the national forest plans would not protect sensitive fish populations. The chief had just received awards for protecting fish habitat. How could their findings possibly be accurate? The Forest Service was not going to accept these findings without a fight.

As the scientists were leaving the hearing, one of Washington congressman Sid Morrison's staff asked Sedell and Johnson to come to dinner at Morrison's home. Congressman Morrison, a moderate Republican, was the informal leader of the Northwest congressional delegation at the time, a group almost equally divided between Democrats and Republicans. The congressman quizzed them gently at dinner to better understand the effort they had made and the results, and then let them go.

Thomas and Reeves got quite a different reception that night. They ended up at dinner with several middle-management people from the Forest Service who separated Thomas and Reeves, berated them individually about their assessment of the inadequacy of the new forest plans, and questioned their credibility and loyalty to the agency. Reeves then handed them a file folder with pages copied from some of the forest plans considered in their analysis; these approved plans clearly showed that fish habitat was projected to decline over the life of the plans, such as the habitat for young coho salmon on the Siuslaw National Forest. Despite this, the managers said they still could not believe that their plans would result in low levels of fish protection.

In between their meeting with Congress and dinner, the G-4 met the press. They explained that they had presented the congressional committees with three dozen choices for conserving old-growth ecosystems, species within them, and sensitive fish populations, which varied in the level of protection and associated timber harvest. They also repeated Gordon's summary comments. The members of the press who spoke up wanted none of this talk of choices and trade-offs; they said their newspapers and TV broadcasts could not describe all that—it was just too complicated. They asked repeatedly which choice the G-4 thought best, or, as they put it: "What do you want to happen?" They could not believe that scientists would do this work and come all the way to Washington, DC, without telling Congress what it should do.

The next morning the G-4 met with the members of the Northwest congressional delegation. They again went through the results. During a break, Congressman Bob Smith, a conservative from eastern Oregon, said: "You could have prepared a much shorter report—you could have left out any choice with an 'L' [low level of protection] in the ecological ratings [anything less than 8A]. Congress won't legislate anything with an 'L' in it." So a line had been drawn on what Congress would consider, but a congressman had made that decision, not the G-4.

After the scientists finished the briefing, Congressman Morrison rose and thanked them for their efforts. He said that they were not going to challenge the G-4 results—committee members believed that the G-4 had done the best they could and had portrayed the trade-offs as they understood them. However, Morrison did note that the "sticker price" of choices that gave a reasonable level of protection to species and ecosystems was extremely high.

The G-4 then headed back to Portland for a meeting with local press, citizen groups like the Yellow Ribbon Coalition (a community-based timber harvest support group), Portland Audubon, and others. The meeting went on and on, with most stunned and overwhelmed by the number of choices and the stark trade-offs portrayed in the G-4 analysis. The G-4 remained firm that they would not recommend a choice or even say which could be implemented without change in environmental laws. Mercifully, the G-4 were finally allowed to go home.

The G-4 had been given a crash course in the difficulties of holding to their story when so many hoped for another outcome, and in how different were the views of science held by scientists and policy makers (see box on page 150). They also realized that they would probably be identified with their effort in the Pacific Northwest for the rest of their professional lives.

The G-4 summarized their findings in a booklet with map overlays, such as plate 6.2, that enabled members of Congress to understand how the many choices had been built, and they delivered the booklet to the Agriculture and Interior Committees in midsummer 1991.[8] They had completed the task Congress had given them—a 60-page report with a score of map overlays and a multitude of choices and their implications.

The report included estimates of potential employment losses in the range of the northern spotted owl prepared by Brian Greber, a forest

MYTHS AND SCIENCE; POLITICIANS AND SCIENTISTS
(Jack Ward Thomas)

It has become the lot of the Gang of Four and associates to be the enforcers of reality. In that role, the Gang must constantly tell very powerful people, some of them Princes of Power, what they do not want to hear. There is extreme pressure to be reasonable, and this pressure is subtle, insidious, and largely self-imposed. I find myself wanting to be reasonable, wanting to please, wanting to help the Princes of Power maintain their illusions just a little bit longer—for these self-same illusions are illusions that natural resource managers helped create out of our cultural myths.

The new myth is that scientists and science will become the guiding light for natural resource management. Those who have such faith, strangely enough, do not know what science is. They view science as a product. They do not know that science is, instead, a process—a method in the never-ending search for truth. They do not understand that scientists are no better and no worse than others and are subject to all the vicissitudes of human behavior. The only difference is that when they think and analyze using the scientific method of formulating and testing hypotheses, and deduce and hypothesize yet again, they think in an uncommon manner.

> —Reflections by Jack Ward Thomas, on the way with members of the Gang of Four to meet with congressional staff about utilizing the Gang of Four report in legislation. From his journals, 1991. Used with permission of his wife, Kathy Thomas.

economist at the College of Forestry at Oregon State University. He included direct job losses in the forest industry and also those caused by indirect and induced effects. In addition to workers displaced, there would be indirect effects caused by changing business expenditures in the region, such as purchases of equipment, and induced effects caused by changing personal expenditures in the region, such as purchases of groceries and meals.[9]

It turned out that the magnitude of potential employment losses associated with the different choices in the G-4 analysis was highly sensitive to the starting point in the comparison. Therefore, Greber's analysis used three starting points—the recently adopted or proposed forest plans modified by the spotted owl strategy in the ISC's report, the average harvest from federal lands in the owl's range for the previous decade (1980–1989), and the average harvest for the previous five years from those lands that reflected the high harvests of recent experience.

Choice 8A, which achieved at least a medium level of protection for all ecological resources evaluated, can serve as an example of the importance of the point of comparison. Including direct, indirect, and induced employment effects, the G-4 report estimated that the number of jobs in the area associated with the harvest level in 8A would be 8,000 fewer than that provided by the level projected in the Forest Plans + ISC Strategy, 38,000 fewer than that provided by an average harvest equal to that of the previous decade, and 48,000 fewer than that provided by an average harvest equal to that of the previous five years. Including only direct effects (the timber industry jobs themselves), the impacts would be about half in each case. Thus, the estimated job losses associated with Choice 8A could range from 4,000 to 48,000. All employment estimates had their supporters among stakeholders and Congress.

While Congress was generally satisfied with the analysis and results, in the sense that members of the relevant committees did not challenge them, higher levels of the Forest Service were not. The agency was especially troubled by the findings of the low level of protection for sensitive fish populations associated with the recently adopted or proposed forest plans. Salmon were an iconic species for people of the Northwest and produced jobs in commercial and sport fishing that supported coastal communities. In addition, meeting treaty obligations with many tribes of the Pacific Northwest focused on maintaining salmon populations. Either the Forest Service had to refute the findings of Sedell and Reeves about the impact of the forest plans on salmon populations, or major changes in those plans were in the offing.

The Forest Service appointed a regional task force as well as a respected fish scientist (Jeff Kershner, national manager of Forest Service fish habitat programs) to separately review fish protection measures in its recently adopted or proposed forest plans. Each national forest had been given discretion on how much protection to provide for fish in those forest plans, and the plans varied widely in what would be provided. In August 1991, the regional task force reported that the new plans were not specific enough to determine whether they would protect the long-term viability of native fish runs. In October 1991,

Kershner reported on his investigation of Sedell and Reeves's protection measures, saying that they would provide "a strong basis for a conservation strategy for sensitive wild fish across the region."[10]

With these two reports, the questioning of Sedell and Reeves's findings and suggested policies lost steam and the agency began increasing protection of salmon and related species. Within two years, Forest Service scientists and professionals were using the watershed and fish conservation framework from the G-4 work as the foundation for protecting fish and aquatic ecosystems. Most strikingly, the Forest Service appointed a committee of scientists headed by Sedell and Reeves to develop a program for the management of riparian and aquatic ecosystems in national forests outside the owl's range. Sedell and Reeves patterned their effort on what they had done in the G-4 effort, and the results became known as PacFish and InFish. These two aquatic conservation strategies remain in use on many national forests in the Pacific Northwest.

In addition, Chief Robertson did not like Johnson's statement in the G-4 report that it was unlikely that the allowable cuts projected in the new plans could be attained. After all, the Forest Service had used Johnson's harvest scheduling model to estimate these harvest levels.

The fact that the Forest Service recognized the allowable cuts to be maximums and that Chief Robertson had recently sent out a directive that the many new environmental-protection standards in the plans should come first did not seem to carry much weight. Neither did the fact that forest supervisors and regional foresters were crying out to the chief that they could not meet their allowable cut levels while also meeting environmental laws and standards and resource goals.[11] Anyway, that train had left the station: the allowable cuts in the recently adopted or proposed forest plans in the owl's range were no longer meaningful because they lacked ecological protections as documented in the G-4 report.

By June 1992, both the House Interior and Agriculture Committees crafted legislation based on the choices provided in the G-4's analysis. The House Agriculture Committee, under the staff leadership of Lyons, settled on Choice 8A, including a glide path over the following decade down to the harvest estimate for 8A in the G-4's report. The House Interior Committee, under the direct leadership of chair

George Miller, who had become a champion of the Gang of Four's report, settled on something close to Choice 12C. Both bills were then approved by their respective committees with the idea that they would be voted on by the full House before the end of that congressional session in December 1992.

Congressman Tom Foley from eastern Washington was Speaker of the House, the highest standing in the House of Representatives that any congressional representative from the Pacific Northwest had ever achieved. As might be imagined, the timber industry of the Pacific Northwest focused its full attention on stopping consideration of legislation that would codify the findings in the G-4 report. In deference to the industry, Speaker Foley would not allow the G-4 bills from the Agriculture and Interior Committees to be considered on the floor of the House.

With that decision, there was no chance of congressional action based on the G-4 report. Federal timber harvest impacts from achieving acceptable levels of species and ecosystem protection (somewhere between Alternatives 8 and 12) were just too high. But there was also little congressional interest in weakening the environmental laws to enable lower levels of protection. Through its inaction, Congress pushed the spotted owl/timber problem back to the executive branch and the president—either the administration of President George Bush during the last year of his first term, or that of whoever would win the 1992 presidential election.

The G-4 Effort: What Difference Did It Make?

The Gang of Four completed their analysis in six frantic, intense weeks at minimal cost. Interest groups and agencies expressed skepticism about what could be accomplished in so short a time and with so little money, and how it could have much impact on the political and social maelstrom then consuming federal forest management in the Pacific Northwest. In addition, Congress did not pass legislation based on the G-4 analysis. Thus, it is logical to ask whether the G-4 effort made much difference. Some significant effects of the G-4 effort are noted here.

The results clearly showed that the Forest Plans + ISC Strategy, the Forest Service's proposal at the time, would not suffice as an overall conservation framework. It might give adequate protection for the spotted owl, but not for many other species that lived in old-growth forests and the streams that ran through them. This finding would help set in motion a major, unprecedented expansion in species conservation efforts on federal forests.

The G-4 report also raised questions about the usefulness of the National Forest Management Act (NFMA) forest planning process in general. According to the G-4 analysis, that mammoth planning process had resulted in plans that failed to protect species and ecosystems by all measures the G-4 utilized (fig. 6.2). Subsequently, the NFMA planning process would not be used in development of the Northwest Forest Plan (NWFP).

The terminology and concepts of LS/OG forests took hold. Out of necessity, Franklin and Spies had invented the concept of LS/OG forests to ensure that all forests that could make a significant contribution to older forest function were considered in analyses. They also settled on 80 years as the age at which younger natural forests began to attain some older forest characteristics. Both the concept of LS/OG forests and the 80-year threshold were enormously influential in development and implementation of the NWFP.

LS/OG Areas became an alternative foundation for conservation of the northern spotted owl and other species associated with these forests. The ISC had built a regional system of reserves, called Habitat Conservation Areas (HCAs), that were designed to help sustain viable populations of the spotted owl at minimal cost to timber harvest levels. To the degree possible, the ISC put the HCAs into wilderness and other reserved lands. As a result, the HCAs did not include some of the most ecologically important LS/OG Areas, which would still be available for timber harvest under its strategy. The mapping of LS/OG Areas provided the basis for an alternative reserve design, keyed to protecting the best remaining older forest.

The watershed and fish emphasis that Sedell and Reeves developed for the G-4 analysis became the starting point for the watershed and fish protection measures in the science assessments that followed up through development of the NWFP. Also, the components of this emphasis—Key Watersheds, expanded riparian buffers, and watershed restoration—became the foundation of similar strategies for the national forests in the rest of Region 6 (eastern Oregon and Washington) and other regions. What an amazing legacy for Congressman Volkmer's last-minute direction "don't forget the damn fish!"

The G-4 analysis demonstrated how ecosystem planning *could* be accomplished. With its focus both on ecosystems (LS/OG forests, aquatic ecosystems, watersheds) and on associated species, the approach taken by the G-4 demonstrated how an ecosystem-based plan might be formed and what it might be. A statement by Charles Meslow, a noted owl biologist, in an article titled "Spotted Owl Protection," said it best: the G-4 report provided a "glimpse of what a Pacific Northwest ecosystem plan might look like and cost." As such, it provided a guide to ecosystem planning in efforts that would come after it.

The G-4 analysis reframed the debate on the future of the federal forests in the owl's range from simply owls versus timber harvest to a multifaceted discussion of timber, wildlife, old growth, fish, and watersheds. It would take a while for the complexity of the problem to sink in, but it would not go away. In addition, the analysis provided a multitude of choices to consider. From Kiwanis clubs to middle schools to TV commentaries, the G-4 report provided fodder for countless debates and deeply serious discussions about the future of the federal forests and their meaning for the people of the Pacific Northwest. All knew that large changes were coming and that their form was still being designed.

Federal Agencies Try One More Time without Success

While committees in the House of Representatives debated how to legislate the G-4 report, three federal agencies (USFWS, BLM, and Forest Service) each made a last effort to escape the federal forest gridlock in the Pacific Northwest before the presidential election in November 1992. Now, though, they had the broader approach taken by the G-4 as the background for their efforts.

A USFWS team crafted a recovery plan for the northern spotted owl during 1991, utilizing a slight modification of the ISC's strategy that would become, if approved, a template for USFWS advice on how to recover the spotted owl. Ordinarily such a team would be composed of experts on the species in question. The membership of this recovery team was unusual in that it also included political appointees—a representative of the secretary of the interior, the deputy assistant secretary of agriculture, and representatives of the governors of the three involved states. The Bush administration was not leaving things just to scientists to decide.

The draft spotted owl recovery plan was published in the winter of 1992. It had two features often absent from such plans. It included an analysis of economic effects, when recovery plans at that time usually focused on what was necessary to recover a species and left estimating the cost of such efforts to others. Also, the recovery plan identified other species associated with old forests and briefly analyzed the benefits of the proposed owl plan to those species.[12] This identification continued the G-4 approach of considering effects on multiple species in conservation plans, while greatly expanding the number of species considered.

The Bush administration refused to sign the plan, citing unacceptable economic costs in terms of timber harvest reductions and associated employment impacts. Thus, it was issued as *Final Draft Recovery Plan*,[13] and there recovery planning stopped.[14] However, ideas from that effort, such as improvement in reserve design and the species considered, would carry over into the next science assessments.

At the same time, the USFWS, following a court order, designated critical habitat for the northern spotted owl, after public debate over criteria for delineating this habitat.[15] This 1992 designation was the agency's first attempt to describe the habitat across the owl's range that was critical to its survival.

The BLM did not operate under National Forest Management Act (NFMA) requirements to protect biodiversity and maintain habitat for viable populations of vertebrates. However, the northern spotted owl was now listed as threatened under the ESA, and that listing required a response by the BLM. Toward that end, the BLM developed an interim approach in 1990 to protect the owl and its habitat called the Jamison Strategy or Plan, named after Cy Jamison, then director of the BLM and leader in its development.

The Jamison Plan included a description of how the BLM would implement timber sales in owl habitat for fiscal years 1991 and 1992. In January 1991, the BLM submitted 174 timber sales to the USFWS for consultation. While the BLM did not submit the Jamison Plan itself, the sales ostensibly were based on it. The USFWS determined that 52 of these sales would, in fact, jeopardize the continued existence of the northern spotted owl. This outcome led the BLM to request the convening of the Endangered Species Committee to consider authorizing implementation of 44 of these sales regardless of deleterious effects on the owl.

When a federal agency concludes that complying with the ESA is too costly, it can request an exemption from the law's requirements. That request goes from the secretary of the interior to the Endangered Species Committee, or God Squad, which has the power to grant such an exemption. Congress added this process to the Endangered Species Act in 1988 at the request of Tennessee senator Howard Baker, majority leader of the Senate. Baker realized that the ESA might stop dam building in his state; unfortunately for Senator Baker, the God Squad's first task was to reject a proposal to exempt one of those dams (Tellico) from the law.

The God Squad consists of six officials of the executive branch (cabinet secretaries or department heads) and one representative from the affected states. It may grant an exemption to the ESA if it finds that

- there are no reasonable alternatives to the proposed action;
- the action is in the public interest;
- the benefits of such action clearly outweigh the benefits of alternative courses of action that are consistent with conserving the species or its critical habitat;
- the action is of regional or national significance; and
- the agency proposing the action has acted in good faith; that is, it has refrained from irreversible actions during the exemption process.

Five of the seven members of the God Squad must approve the exemption.[16]

Few exemptions from the ESA have been requested, and fewer still have been granted. The Bush administration turned to the God Squad as a last hope for the BLM to avoid complying with the conservation requirements for the northern spotted owl under the ESA. While 44 BLM timber sales may seem like a small issue over which to seek a waiver of the ESA, proponents on all sides viewed the decision as precedent setting.

An evidentiary hearing on the BLM request occurred in Portland, Oregon, in January 1992. Both expert and public testimony were given at the hearing, while competing rallies outside the hearing room vied for attention. As the days dragged on, some of the steam went out of the political theater as people realized how boring it all was.

All members of the God Squad had to be present to vote on the proposed exemption. The vote occurred in May 1992 in an Interior Department conference room in Washington, DC. In the days before the meeting, the secretary of the interior (who chaired the meeting) found it relatively easy to secure four votes in favor of granting the exemption, but the fifth and final necessary vote proved elusive. Securing the fifth vote (from John Knauss, administrator of the National Oceanic and Atmospheric Administration [NOAA][17]) for even a portion of the sales necessitated adding the requirement that the BLM would follow the *Final Draft Recovery Plan* in the future that had just been released by the USFWS (which was based largely on the ISC's recommendations)—the very plan the agency was trying to replace with the Jamison Plan. David Cottingham, a NOAA staffer, helped Knauss craft that requirement and gained much useful knowledge and experience that he would employ when he later worked for Katie McGinty on development of the Northwest Forest Plan. The requirement was added on the day of the vote, and then 13 BLM timber sales received an exemption from the ESA—truly a Pyrrhic victory.[18]

Next, litigants tackled whether the Jamison Plan was a legally adequate conservation plan. Environmental-group plaintiffs successfully argued in *Lane County Audubon Society v. Jamison* that the Jamison Plan was an action under the ESA because it authorized, funded, or carried out an agency decision that might affect the spotted owl. As such, they argued the Jamison Plan should have been submitted to the USFWS for consultation. The district court agreed but did not enter an injunction to remedy the consultation deficiency. However, on appeal, the Ninth Circuit Court of Appeals did so, halting all timber sales in owl habitat on BLM lands within the range of the owl until either the BLM submitted the Jamison Strategy for consultation or otherwise adopted an adequate owl conservation plan.[19]

Meanwhile, the Forest Service attempted to address the deficiencies Judge Dwyer had identified in his *Seattle Audubon Society v. Evans* decision where the agency had adopted a new conservation strategy for the owl without going through the NEPA process. The Forest Service adopted the draft recovery plan for the northern spotted owl proposed by the USFWS (based largely on the ISC's report) and then embedded it in an Environmental Impact Statement (EIS) and released it in winter 1992.[20] Environmental-group plaintiffs, again led by Seattle Audubon, challenged this decision in Judge Dwyer's court, arguing that it failed to adequately assess the environmental consequences of continued logging of owl habitat on other species associated with old-growth forests.[21] The EIS had briefly noted that the Gang of Four report had given a low to medium-low persistence rating for other terrestrial vertebrates under the ISC's owl strategy, and Seattle Audubon argued that the EIS needed to assess the effect of the ISC's strategy on other old-growth species. The judge agreed and stated that NFMA "requires planning for the entire biological community," and "whatever plan is adopted, it cannot be one which the agency knows or believes will probably cause the extirpation (local extinction) of other native vertebrate species from the planning areas."[22]

In addition, attorneys for Seattle Audubon argued to Judge Dwyer that the EIS failed to assess important uncertainties as required by NEPA.[23] The ISC had expressed some concern and uncertainty about its fundamental premise that owl populations would stabilize over the long term, even if 500,000 more acres of owl habitat outside the ISC's owl reserves were logged as allowed under the ISC's strategy. Also, the ISC's report and its conclusions assumed BLM participation in the strategy, which had not occurred as evidenced by the court's decision on the Jamison Plan. Judge Dwyer ruled that both these uncertainties needed further consideration.

Thus, Judge Dwyer once again struck down a Forest Service northern spotted owl conservation plan.[24] The judge subsequently issued another range-wide injunction on Forest Service timber sales in owl habitat pending compliance with the court's order, which was upheld on appeal.[25]

Federal timber sales in spotted owl habitat in the Pacific Northwest had ground to a halt. Never before had that happened to the mighty Northwest federal timber machine. Starting that machine again would take new ideas and considerable political will.

The largest shift in history in federal forest priorities and planning was underway. Scientific committees had become the mechanism by which the federal government tried to chart a path out of the legal quagmire over protection of species and ecosystems in federal forests. Soon a new president of the United States would be directly involved in finding a scientific solution to this problem.

7

The President and the Scientists

With no congressional solution in sight and the Bush administration unable to construct a plan that would survive judicial review, breaking the federal timber harvest gridlock in the Pacific Northwest was left to whoever won the White House in the next election. Democratic presidential candidate Bill Clinton promised to resolve the conflict, which led to a forest conference in the Pacific Northwest following his election, another science assessment, and a president choosing a federal forest plan for the owl's range.

My Administration Will Break the Gridlock

Early in his campaign, while the spotted owl/timber wars were at a high pitch, Bill Clinton met with friendly business and labor leaders at a hotel in Seattle. Katie McGinty, previously an environmental staffer to Senator (and now vice presidential candidate) Al Gore, was accompanying Clinton to advise him on environmental issues he might encounter on the campaign trail. As part of his briefing before the meeting, McGinty cautioned him not to engage on the spotted owl/timber conflict; the issue was intractable, and he should not get entangled in it. At best he should acknowledge its complexity and move on.[1]

As the meeting broke up, McGinty thought they were home free on the owl issue—it had not surfaced. However, as Clinton got up to leave, he ad-libbed that, if elected, his administration would break the gridlock (or words to that effect). McGinty was stunned both because the candidate had gone directly against her advice and because, if candidate Clinton were elected, delivering on that promise would end up on her plate.[2] He reiterated his commitment a few days later in Eugene, Oregon, at the backyard rally of a labor leader.

The candidate's commitment to resolve the federal forest crisis in the Pacific Northwest gained traction with the press and public in the final weeks before the election, and he added a proposal to immediately hold a Timber Summit to better understand the issues and possible solutions. Responsibility for planning that summit began in earnest soon after President Clinton was inaugurated in January 1993 and would rest primarily with five people, all of whom would later participate in development, adoption, and initial implementation of the promised forest plan:

- Kathleen (Katie) McGinty—deputy assistant to the president for environmental policy and director of the Office of Environmental Policy from January 1993 to December 1994, and then chair of the Council on Environmental Quality. She led the effort as President Clinton's direct representative—a lofty assignment for a twenty-nine-year-old former congressional aide.
- Tom Collier—chief of staff for Interior Secretary Babbitt and formerly a partner in a law firm with Secretary Babbitt.
- Jim Lyons—forestry staff, House of Representatives Agriculture Committee, on loan to the Clinton administration. Lyons would soon become undersecretary for natural resources and environment in the Department of Agriculture, where he oversaw the Forest Service and the Natural Resources Conservation Service.
- Tom Tuchmann—forestry staff, Senate Committee on Agriculture, Nutrition, and Forestry on loan to the Clinton administration. Tuchmann would later become head of the US Office of Forestry and Economic Development in Portland to assist in implementation of the Northwest Forest Plan.
- Jim Pipkin—soon to be appointed counselor to Secretary Babbitt and formerly a partner in a law firm with Secretary Babbitt. Pipkin would later lead development of interagency coordination in implementing the Northwest Forest Plan.

Two people who worked for Katie McGinty during crucial phases in development of the Northwest Plan also deserve mention: David Cottingham, who served with Katie in 1993, and Will Stelle, who served with Katie in 1994. They provided invaluable assistance to Katie in development and completion of the Northwest Forest Plan.

Early January 1993 saw a presidential inauguration with a multitude of inaugural balls and related celebrations. One of those balls was convened by a collection of environmental groups, delighted that a Democrat was back in the White House after 12 years of Republican presidents. Their expectations for the new administration were very high, and they placed resolving the federal forest crisis in their favor near the top of the list.

As with any presidential administration, the first few weeks involved appointing new staff, assigning offices, and moving in. Behind the scenes, though, work on resolving that crisis had already begun, with McGinty, Collier, and Lyons heavily engaged.

One key issue was the path that resolving the spotted owl issue might take, with two choices quickly surfacing: (1) a legislative political solution or (2) a regulatory administrative solution. Asking the Democratic-controlled Congress for a legislative solution that the president could support would certainly have been preferred, in part because it would be permanent and in part because it would take one more thorny problem off the president's plate. However, many bills addressing federal forest issues in the Pacific Northwest had been introduced in Congress over the previous few years and none had gotten very far. Also, the trade-offs outlined in the Gang of Four report to Congress were unacceptable to the Northwest congressional delegation, and without that delegation's support, Congress was unlikely to act. To gain that support, it appeared that Congress would need to weaken environmental laws like the Endangered Species Act, and the Clinton administration had no stomach for that.[3]

The stalemate on Capitol Hill made an administrative solution seem the most realistic possibility, which in turn meant that any proposed forest plan had to pass judicial review. Some in the administration still held out hope that Congress might endorse a reasonable, balanced forest plan from them, but it was clear that the ball was in the administration's court for the time being.

Collier immediately reviewed the three Dwyer decisions on the Forest Service's proposed conservation strategy for the northern spotted owl that had enjoined further timber sales in the owl's habitat. He quickly saw that Judge Dwyer's negative rulings stemmed largely from perceived and actual political interference in the science underlying development of the conservation strategy. He therefore concluded that the best approach to a legally defensible forest plan would be for scientific experts to develop a conservation plan clearly insulated from such interference.[4]

For his part, Lyons strongly believed that heavy scientific involvement in natural resource policy formulation was the best approach and right path to creating stable, well-grounded policies. He had taken such an approach in developing the Society of American Foresters' policy position on old growth and in helping create and coordinate the Gang of Four effort. He pushed it forward here.[5]

Thus, Collier for pragmatic reasons and Lyons for philosophical reasons had come to the same conclusion—turn the problem over to scientists. In addition, McGinty contacted Lyons almost immediately upon taking office and asked him for his assistance. It certainly looked as if they had found the path they would take.

McGinty and the others also quickly began work on the Timber Summit the president had promised he would hold. They first moved to lower expectations about what that meeting might accomplish: they changed the meeting's name to Forest Conference to send the message that they would hold a broad-based discussion of forest conservation and management issues, rather than a policy-making event. Further, they chose a town hall–style meeting in which interested parties would be invited to discuss the issues with the president and vice president. In such a setting, the team hoped to establish President Clinton's deep interest in the problem, provide the public a forum for stating positions and offering suggestions, and demonstrate the president's commitment to restoring the flow of federal timber to wood-product mills in the region. Also, they hoped the conference, scheduled for April 2 in Portland, Oregon, would provide a gateway to a science assessment that would aid the president in the search for a legally defensible forest plan.[6]

McGinty's planning group for the Forest Conference had arranged for a meeting on March 23 with President Clinton where they would seek his endorsement for their proposed approach and learn about any modifications he would like. To prepare for that meeting, McGinty sent a memo to Vice President Gore on March 17 that updated him on decisions they needed from the president about the conference agenda, its length, and related matters; preconference activities by staff; trips planned by cabinet officials to familiarize themselves with the issue; and recent developments.

Toward the end of the memo, McGinty briefly described a "deeply discouraging new report by Jack Ward Thomas and a team of Forest Service scientists that responded to Judge Dwyer's requests" for more information and analysis when he rejected the agency's 1992 spotted owl plan, including whether that plan would protect other old growth–associated species.[7]

McGinty summarized the findings of the Scientific Assessment Team (SAT):

> There are—in addition to the owl—387 old-growth related species which are at risk and mitigation measures to protect them will be needed to assure their long-term viability.
>
> The answer to . . . [this] question is grim. In short, the scientists have concluded that—if viable populations of the indicated species are to be maintained—harvest levels on Federal lands will have to be significantly reduced. (Note that the report does not specify a sustainable harvest level. BUT the trained eye will be able immediately to translate the science into forest management numbers.) BOTTOM LINE—Forest Service and Bureau of Land Management will have to reduce harvest levels on federal lands by 75% to 85% from historical levels. . . . Needless to say, this is an extremely powerful and sobering message.*
>
> The truth is that the President is not going to be able to make all well again in the timber industry and it is important to begin to bring some reality into expectations of what we are going to deliver.[8]

The Forest Service's spotted owl plan, which would substantially reduce federal harvests, was not enough to ensure persistence of hundreds of other species associated with old forests. More reductions

---

* The Gang of Four report had estimated that providing sufficient habitat for a high level of viability for old growth–dependent species and fish would require an approximately 80 percent reduction in harvest from historical levels. McGinty's staff were well aware of that report.

would be needed if they were to follow the scientists' advice. Grim news, indeed. McGinty's immediate worry was whether the SAT's findings would dominate and derail the Forest Conference three weeks hence. On that she would have to wait and see. McGinty and the vice president, though, could caution a naturally optimistic president to temper his promises at the conference.

Her longer-term worry, of course, was how to find the best solution she could in a very difficult situation. McGinty and the others briefed President Clinton on March 23 on their proposal to have a Forest Conference in which he, the vice president, and cabinet members would talk with stakeholders and scientists. They also urged that, toward the end of the conference, the president commit to convene a science assessment headed by Jack Ward Thomas to develop a forest plan that would resolve the crisis. They pitched this as the best way to get the injunctions lifted. The president agreed and said they should move forward with this approach. The stage was set for the Clinton administration to hear firsthand from stakeholders and scientists about federal forestry issues in the range of the northern spotted owl and discuss potential solutions, while increasingly realizing how difficult their task would be.[9]

The Forest Conference

On the morning of April 2, 1993, the president and vice president arrived at the Oregon Convention Center in Portland for the Forest Conference.[10] In addition, five cabinet members came with them: Interior Secretary Babbitt, Secretary of Agriculture Michael Espy, Secretary of Commerce Ron Brown, Secretary of Labor Robert Reich, and Carol Browner, head of the Environmental Protection Agency. The presence of cabinet members and other officials was intended to underline the importance of the problem to the administration and educate cabinet members about a controversy they would be expected to help resolve.

McGinty, though, had an additional reason for asking cabinet secretaries and department heads to attend. She wanted the Forest Conference to announce to the world that the Clinton administration was going to solve environmental problems in a different way than the Bush administration—federal agencies would work together on problems rather than publicly quarrel with each other as they had done. Fundamentally, McGinty believed that complex problems, such as the

Figure 7.1. President Clinton (far right), Vice President Gore (second from right), Labor Secretary Robert Reich (far left), and a panel of community members at the Forest Conference. (Source: Bureau of Land Management.)

one in the Pacific Northwest, required a coordinated effort by people and agencies with different backgrounds, knowledge, and perspectives to broaden and enrich their approach. Also, she wanted an ecosystem approach taken in planning the management of federal forests, which inherently would require crossing boundaries and jurisdictions and necessitate interagency collaboration. The conference would be a first step down that path, serving simultaneously as both a problem-solving and team-building effort.[11] All told, the meeting would represent the highest level of direct involvement of the White House in federal forestry policy since the era of Teddy Roosevelt.

The conference was organized into opening comments by political leaders followed by three panels: first were stakeholders who described how they might be affected by any proposed solutions (fig. 7.1), then scientists who described the scientific issues involved, and finally a mixed panel of scientist stakeholders who suggested the path forward. As Lyons noted, "In constructing the Forest Conference, we made a conscious decision to have the afternoon panels focus on the science associated with constructing a strategy to conserve the owl. This was in strong contrast to the morning presentations by locally elected officials, timber industry executives, environmentalists, and members of timber-dependent communities. It was the first time in dealing with the old growth/spotted owl issue that the products of Forest Service and academic research were given the attention they deserved since they were essential to developing a sound and legally acceptable solution."[12]

Who was not at the table was also telling as to the direction President Clinton and his staff planned to take: agency representatives and members of the Northwest congressional delegation who had been major historical players in the dispute were not there. While they were invited to be in the audience, the planning team had focused the event on Pacific Northwest citizens who would be personally affected by any outcome, and scientists who had expertise in the subjects being discussed. It was clear that the Clinton administration would not count on the agencies or delegation to break the gridlock but would take a different approach.

In opening the Forest Conference, President Clinton posed the fundamental challenge facing the gathering:

How can we achieve a balanced and comprehensive policy that recognizes the importance of the forests and timber to the economy and jobs in this region, and how can we preserve our precious old-growth forests, which are part of our national heritage and that, once destroyed, can never be replaced?

The most important thing we can do is to admit, all of us to each other, that there are no simple or easy answers. This is not about choosing between jobs and the environment, but about recognizing the importance of both and recognizing that virtually everyone here and everyone in this region cares about both.[13]

Each panel sat at a large, oblong table with the president, vice president, and cabinet secretaries, surrounded by an audience in specially built stands. The panelists were each allowed a three-minute opening statement and then the president and vice president asked them questions.

Stakeholders and scientists had wanted to show slides that would vividly illustrate the crisis at hand: plundered landscapes, working loggers, needy families, endearing creatures, and timber/owl trade-off charts. In the end, White House staff discouraged these types of images in pursuit of a congenial atmosphere to facilitate an upbeat discussion of possibilities with President Clinton and to encourage personal story-telling of the sort cherished by him and loved by the media.

The daylong conference succeeded in highlighting many different issues, perspectives, and values involved in the federal forest conflict in the Pacific Northwest for both Clinton administration officials and a national TV audience. In fact, it did more than that. The conference accomplished the notable goal of explicitly reframing the controversy as something bigger than the spotted owls. As Steven Yaffee noted in his seminal history of the conflict:

> The problem was defined as how to protect a broad range of environmental values within the old-growth ecosystem while dealing humanely within a regional economy that was undergoing a normal process of transformation. While the needs of some 480 old-growth dependent species were aired (and the threat of 480 more spotted owl controversies described), more important was a discussion of the impacts of timber activities on salmon and other fisheries. The Congressional delegation had tried in preconference negotiating sessions to keep the salmon issue off the table, but presentations at the conference made the very logical connection that one segment of the economy may well benefit from changes in other segments of the economy. Salmon populations could improve as logging declined and management practices changed.[14]

Economic panels at the conference addressed the many issues shaping the economy of the Pacific Northwest beyond impacts of owl lawsuits on timber supply. Panelists presented studies concluding that rising lumber prices were more a consequence of demands associated with recovery of the domestic economy than a result of owl-caused supply constraints. The key roles of automation and corporate invest-ment behavior in the decline in timber employment in the Northwest were highlighted. In closing one of the day's sessions, President Clin-ton reiterated the theme that had been central to his campaign: the unavoidable need to adjust in the face of a changing world. While he was sympathetic to the human costs of change, he noted that the average 18-year-old would need to shift the kind of work he or she did many times in a lifetime. In Clinton's words, "I cannot repeal the laws of change."[15]

At the end of the Forest Conference, the president stated that his administration would develop a plan to end the gridlock on federal forests in the Pacific Northwest in just two months. Further, President Clinton proclaimed that five principles should guide the effort:

> First, we must never forget the human and the economic dimensions of these problems. Where sound management policies can preserve the health of forest lands, sales should go forward. Where this requirement cannot be met, we need to do our best to offer . . . new year-round, high-wage, high-skill jobs.
>
> Second, as we craft a plan, we need to protect the long-term health of our forests, our wildlife, and our waterways. They are a . . . gift from God; and we hold them in trust for future generations.
>
> Third, our efforts must be, insofar as we are wise enough to know it, scientifically sound, ecologically credible, and legally responsible.
>
> Fourth, the plan should produce a predictable and sustainable level of timber sales and non-timber resources that will not degrade or destroy the environment.
>
> Fifth, to achieve these goals, we will do our best, as I said, to make the federal government work together and work for you. We may make mistakes, but we will try to end the gridlock within the federal government, and we will insist on collaboration not confrontation.[16]

McGinty accompanied President Clinton as he left the meeting. As the president got on the helicopter taking him to the airport, he turned to her and said, "We really have to help these people."[17]

McGinty had never seen a Douglas-fir old-growth forest, and she knew that here was her chance. Franklin, Thomas, Sedell, Johnson, Lyons, and a few others piled into a van with her and sped on to the

Bridge of the Gods across the Columbia River to the Wind River area of the Gifford Pinchot National Forest. She lacked warm clothes and it was raining, so Franklin loaned her a red timberman's shirt. Franklin then led the group into an old-growth grove and gave them a quick introduction to old-growth forest ecosystems. McGinty did not have much time to talk, but she certainly made clear that she was counting on this group to deliver a plan the president could stand behind.

The SAT Report: Taking Care of All God's Creatures

The Clinton administration did, in fact, get through the Forest Conference without it being derailed by the conclusions of the SAT report. Except for the statement of one scientist (Charles Meslow)—that a way needed to be found to avoid hundreds of additional issues like the northern spotted owl—SAT did not come up.

However, the SAT report fundamentally changed and expanded the framework for conservation of the species associated with old-growth forests. The pathway to creation of the Northwest Forest Plan cannot be understood without knowledge of it.

In the summer of 1992, Judge Dwyer invalidated the Forest Service's latest attempt to propose a legally acceptable strategy for conservation of the northern spotted owl, in part because the Environmental Impact Statement released with the strategy noted that the Gang of Four had found that the strategy would not protect other species living in old-growth forests.[18] In response, Chief Robertson formed yet another scientific team, the Scientific Assessment Team (SAT), to address this and others of Dwyer's concerns.[19] In addition, the chief directed the team to identify mitigation options to ensure that other species identified with old-growth forests would not be extirpated. He asked Jack Ward Thomas to lead SAT, which included scientists and species experts on relevant subject areas (appendix 1).

SAT's assessment of the protection of species associated with old-growth forests beyond the northern spotted owl is summarized here. A more detailed description of that assessment can be found in appendix 2.

The 39 species that had been identified by the Forest Service as associated with old-growth forests in the Dwyer litigation were terrestrial vertebrates (amphibians, reptiles, birds, and mammals). This selection was consistent with the regulatory requirement of the National Forest

Figure 7.2. A marbled murrelet on its nest on a large branch in an old-growth tree. (Source: US National Park Service.)

Management Act (NFMA) to provide habitat to ensure viable populations of native vertebrates.

One of those species, the marbled murrelet, had recently joined the northern spotted owl as threatened under the Endangered Species Act (ESA). The murrelet (fig. 7.2) is a small seabird that ranges from the Aleutian Islands in Alaska to Santa Cruz, California, and forages primarily on small fish and krill in the nearshore marine environment. Murrelets nest in coastal forests from Washington to northern California as well as on western Cascade Range slopes in Washington, at elevations up to 4,000 feet. Those forests are close enough to the ocean to allow murrelets to commute between their nests on land and foraging areas in marine waters. To be useful, forests must contain natural nesting platforms, such as large branches; trees with these structures typically occur in old-growth forests or in remnant old-growth trees in younger forests.[20]

Early on in SAT'S efforts, Jack Ward Thomas made a decision that would fundamentally reframe the conservation of species associated with old-growth forests: he decided to expand the analysis beyond vertebrate species associated with these forests, as directly required by the NFMA Planning Rule of 1982, to include all other species associated with those forests, including invertebrates, vascular plants, lichens, and fungi (plate 7.1). SAT gave three primary reasons for that expansion:

First, selecting and implementing a spotted owl habitat management plan is best conducted from a base of full disclosure and knowledge of potential effects of that plan on all species. Second, assessing effects on a broad variety of species groups better meets agency direction to provide for, and evaluate impacts on, the full range of biological diversity. Third, such a comprehensive approach lays the groundwork for a more complete approach to ecosystem management.[21]

Thomas, in a later interview, recognized the importance of the decision to assess all species and his role in making it:

We had to answer the question [on the effect of the 1992 Forest Service owl plan on other vertebrates] about the 39 species. I was assigned to head the [SAT] team to answer the Judge's questions. I went to Jim Overbay, Deputy Chief for the National Forest System who assigned the team, and said, "Look, why don't we quit evading the real issue and answer the appropriate question. The question is: There aren't 39 species associated with old growth, there are maybe 900 to 1,400 of them. Let's look at the whole spectrum of species."

Overbay said, "Okay, let's go." So I wrote the instructions that he would then give to us to follow. Now the [Bush] administration [in fall 1992] was not ready to talk ecosystems just yet. But when you consider 900 to 1,400 species and their interactions and interdependence, you are talking about ecosystems. The SAT report never made the headlines like ISC and FEMAT, but the SAT report was a truly crucial turning point. That is when we looked at *all* associated species.[22]

SAT evaluated the viability of species associated with old-growth forests in three steps: identifying species closely associated with those forests; evaluating their viability under the alternatives considered in the Forest Service's proposed 1992 northern spotted owl plan, about which Judge Dwyer had raised concerns; and identifying mitigations needed to ensure a high likelihood that none of these species would be extirpated from the national forests as a result of Forest Service actions.

Also, SAT planned to identify species that were so poorly known scientifically that their viability could not be judged.[23]

SAT identified 667 species as closely associated with old-growth forests, including terrestrial vertebrates, salmon populations, plants, fungi, lichens, and mollusks.[24] SAT then evaluated their viability under the alternatives in the Forest Service's 1992 EIS for a spotted owl conservation strategy that Judge Dwyer had rejected. SAT judged species receiving a high or medium-high viability rating were "likely to meet the population viability criteria presented in the regulations implementing the National Forest Management Act." SAT, in turn, judged species receiving a lower viability likelihood as being at viability risk.[25]

SAT next identified "mitigation options for habitat conditions conducive to providing for high viability" for species identified to be at viability risk.[26] These mitigations started with adoption of the standards and guidelines in the recently adopted or proposed forest plans and the ISC's strategy for the northern spotted owl, which provided sufficient protection for 280 species. That still left hundreds of species inadequately protected, as McGinty acknowledged in her memo to Vice President Gore before the Forest Conference. Therefore SAT proposed:

- additional protection of marbled murrelet nesting habitat;
- additional protection of habitat for at-risk fish populations by expanding the approach described in the Gang of Four report into a Riparian Habitat Conservation Strategy;
- additional standards and guidelines for birds and other species in upland forest; and
- additional protection for many mosses, fungi, and other invertebrates (especially mollusks), and surveys to identify sites occupied by some of these species prior to ground-disturbing activities.

The SAT effort and report changed the scope of old-growth conservation in several ways. First, SAT greatly expanded the number of species requiring viability analyses along with the diversity of scientific experts that would need to be engaged in this effort. Second, it set a medium-high likelihood of persistence as the minimum criterion for meeting the viability standard described in the 1982 Planning Rule

implementing NFMA and put a special focus on achieving a high likelihood of persistence in order to be confident that the necessary habitat
would be provided. Third, SAT linked protection of fish habitat to old-
growth conservation, thus bringing fish solidly into Judge Dwyer's pur-
view. Finally, and perhaps most importantly, the SAT report provided
elements of a federal forest plan that its respected team of experts
would support as fulfilling legal requirements for conserving species
closely associated with old-growth forests. Most of these experts would
later be on the science team created by the president, and they would
understandably bring the SAT conservation framework forward into
that effort, with profound implications for the development and implementation of the Northwest Forest Plan.

The SAT report and McGinty's grim interpretation of it also had
profound implications for the sense that the Clinton administration
could find a politically viable solution. As Rahm Emanuel, the president's
political director, stated in a memo to the president's chief of staff shortly
after the Forest Conference, "We need to address the gap between the
political reality and the scientific reality of this issue. The foundation for
decision-making on the timber issue has been placed in the scientific
community. However scientific analysis says we must reduce cutting by
70 percent. Clearly this would be economically destructive, not to mention the fact that it would be hard to sell politically. We are therefore in
a difficult and potentially perilous position."[27]

FEMAT: A Large Task and Little Time

After the Forest Conference, White House staff quickly created three
interagency working groups to help develop policies for Northwest
forests and timber-dependent workers and communities: the Forest
Ecosystem Management Assessment Team (FEMAT), headed by Jack
Ward Thomas; Labor and Community Assistance, headed by Peter
Yu (Department of Commerce); and Agency Coordination, headed by
James Pipkin.

By May 1993, the Forest Conference Executive Committee,
with representatives from 10 departments and offices of the Clinton
administration,[28] issued a letter to all three working groups describing
their mission. Several points in the mission statement proved especially
important for FEMAT—the centerpiece of the postconference effort.[29]

The mission statement first repeated the president's five principles that would guide FEMAT's work. Then it succinctly summarized the overall objective: "Our objectives based on the President's mandate [break the gridlock] and principles are to identify management alternatives that attain the greatest economic and social contribution from the forests and region and meet the requirements of the applicable laws and regulations."[30] The remainder of the instructions to FEMAT explained how to achieve that objective.[31]

Any management plan for federal forests had to meet current environmental laws including the requirements in the ESA, NFMA, and NEPA—the laws central to the litigation over the northern spotted owl and the injunctions on timber sales in its habitat. Regulatory relief from these laws and accompanying regulations, or from other environmental laws like the Clean Water Act, would not be considered as part of the solution to gridlocked federal timber harvest.

Planning needed to assume that the ISC's strategy for the northern spotted owl (as represented in the *Final Draft Recovery Plan*) had been adopted. Building the ISC's strategy into the recently adopted or proposed forest plans for federal forests significantly lowered the projected harvests in those plans, even though such a change had not successfully traversed the NEPA process and been formally approved. This direction would greatly reduce the magnitude of timber harvest and employment impacts attributable to management plans considered by the Clinton administration.

Meeting the "well distributed" viability standard in the 1982 NFMA Planning Rule would be the primary measure of ecological sustainability. Toward that end, FEMAT was directed to develop "alternatives that range from a medium to a very high probability of ensuring the viability of species"[32] associated with old-growth forests and at-risk fish populations. The implication was that such levels of viability would meet legal requirements, but this was not specifically stated, nor was it known at the time whether this would satisfy the courts. No direction was provided about how FEMAT should proceed if knowledge of a species was insufficient for a credible viability assessment to be made, although SAT had found this a significant issue.

The instructions to FEMAT also called for maintenance and/or creation of a connected or interactive old-growth forest ecosystem on the federal lands within the owl's range. However, most assessment of ecological sustainability in FEMAT's analysis would focus on individual species rather than directly on old-growth ecosystems, reflecting the emphasis on species protection in the ESA and NFMA.

In summary, FEMAT should suggest modifications to federal forest plans that would improve the viability of species associated with old-growth forests and at-risk fish populations sufficiently to meet federal law. Within that context, FEMAT should seek a plan that maximized economic and social contributions, which the Forest Conference Executive Committee later described to President Clinton thus: "Our instructions to the Ecosystem Management Assessment Team were to prepare a series of options that would, in essence, keep the timber harvest level from federal lands as high as possible consistent with environmental laws."[33]

In this analysis, FEMAT should treat the BLM's western Oregon lands similarly to national forests. The Federal Land Policy and Management Act, a multiple-use sustained-yield statute that provided overall governance to all BLM-managed lands in the United States, was cited in the mission statement given to FEMAT. However, the Oregon and California Revested Lands Sustained Yield Management Act (O&C Act)—the law specifically governing the BLM's management of most of its western Oregon forestlands—was not mentioned. This omission signaled that the Clinton administration would hold the BLM lands in western Oregon to the same environmental requirements as the national forests, whether or not the laws and regulations governing the BLM's western Oregon lands expressly contained those requirements. The statement in the instructions "To achieve similar treatment on all federal lands involved here, you should apply the 'viability standard' to the Bureau of Land Management lands"[34] affirmed this decision—one that would be a source of contention for years to come. In many ways, the Clinton administration's instructions fused the two agencies' missions in the quest to create a single ecosystem management plan over the range of the northern spotted owl.

In developing its plan, FEMAT was further directed to minimize the impact on nonfederal lands of protecting and recovering threatened and endangered species. Toward that end, FEMAT assumed that state and private land management would meet the requirements of state forest practice and land zoning regulations and any federal regulations directly applying to them, but no more. The instructions also called for FEMAT to note essential contributions from these lands to meet conservation objectives, which FEMAT did very generally by noting where federal policy might be insufficient to recover species.

Finally, the mission statement directed FEMAT to suggest how a coordinated interagency approach to planning and management might be advanced. With interagency conflict identified as a primary source of federal forest management gridlock, it is not surprising that FEMAT was tasked with suggesting strategies for overcoming this problem. As President Clinton said at the Forest Conference: "I was mortified when I began to review the legal documents surrounding this controversy to see how often the departments were at odds with each other, so that there was no voice of the United States."[35] The Clinton administration was determined to solve this problem.

With the president's short timeline of two months, FEMAT relied heavily on outcomes of the three major scientific efforts preceding it for its conceptual framework. Elements of those efforts that were of special usefulness to FEMAT are summarized here.

The ISC demonstrated that an expert team could develop a scientifically credible conservation strategy for a species at risk. Key elements of that strategy included a network of reserves across the federal landscape with each large enough to sustain a local population of northern spotted owls and close enough together that owls could successfully move between reserves; redundancy in the reserve network so that the strategy would be robust in the face of disturbance; and constraints on forest management in the areas between the reserves to improve the potential for successful owl dispersal. Also, the ISC utilized existing reserves on federal lands, such as wilderness and national parks, as much as possible in creating its reserve network in order to reduce impacts on timber harvest, while still meeting the distribution and shape criteria of the overall design.

The Gang of Four utilized the concept of late-successional/old-growth (LS/OG) forests, with 80 years as their minimum age; mapped and assessed the quality of LS/OG Areas on federal land; and aligned conservation reserves with concentrations of LS/OG forests. In the area between reserves, it shifted from clearcutting to harvesting with retention of trees and other biological legacies. Fish experts in the G-4 effort outlined elements of a scientifically credible conservation strategy for maintenance and restoration of fish habitat. In addition, the G-4 presented a suite of choices with different ecological risks and timber harvest levels rather than a single plan as the ISC had done.

SAT recognized the full suite of species closely associated with LS/OG forests, including nonvertebrates; utilized the viability standard in the 1982 Planning Rule to assess the adequacy of species protection; set a high likelihood of persistence as the level of protection to be achieved; and suggested mitigations for federal forest plans to enable them to meet this standard. In addition, SAT expanded the watershed and fish protection strategy of the G-4 report into a comprehensive strategy for conservation of aquatic ecosystems.

These three science assessments contained ideas and conceptual frameworks that could quickly be incorporated into alternative management designs for federal forest conservation and management. Perhaps most importantly, they also created a cadre of scientists and professionals experienced at organizing scientific information, developing well-grounded hypotheses, and applying expert judgment and intuition where solid information was lacking. It is also true, though, that using the scientific cadres that had conducted previous assessments could potentially limit imaginative consideration of alternative approaches.

FEMAT included most key leaders from these previous scientific efforts as well as many scientists and resource professionals who had assisted them. A social science contingent made an extremely valuable addition to the mix. Most people involved were federal employees, but some nonfederal scientists such as Franklin, Johnson, Margaret Shannon (University of Washington), and Brian Greber (Oregon State University) were also invited to participate.

Older scientists filled most of the lead scientist slots in FEMAT as they had in the ISC, Gang of Four, and SAT (appendix 1). However, many younger scientists made important contributions, including Shannon (social science), Kelly Burnett and Lisa Brown (aquatics), Robin Lesher (nonvertebrates), Sarah Crim (timber harvest scheduling), and Cindy Swanson (economics). In addition, Katie McGinty (then twenty-nine years of age) led the overall effort, representing the president.

The scientists and resource professionals worked in the US Bank Building, then the tallest building in Portland. The work location became known as the 14th floor of the Pink Tower, in recognition of the building's height and also its light pink shade in the afternoon sun. A guard stood just inside the door and admitted only team members.

The Clinton administration had made a blanket request for comments and suggestions from stakeholders following the Forest Conference, which were forwarded to FEMAT to use as it saw fit, but there was no systematic follow-up on ideas expressed at or after the conference. The administration had put its chips on scientists to find a solution to the impasse and did not want anyone or anything interfering with their work.

Agency line officers and timber management staff also were not asked to participate. Early on, Thomas and Johnson discussed putting some of these professionals in the different resource analysis groups. However, they soon recognized that such a move could create both the perception and reality that the work was subject to agency agendas, as most federal timber harvest in the owl's range came from LS/OG forests central to protection of the species and ecosystems that FEMAT would assess. Both the scientists and the Clinton administration were determined to avoid that outcome.

Thomas and Johnson also knew that excluding line officers and timber management staff from developing the plan for the lands they administered would lower chances for successful implementation. Thus, FEMAT attempted to reduce implementation difficulties by focusing on an ecosystem plan with minimal special rules for managing individual species; involving an implementation group to help evaluate and adjust proposed standards and the amount of timber that would be harvested; and employing other methods. Still, there is little doubt that this approach made acceptance and implementation of the resulting plan more difficult.

McGinty, Lyons, and others of the Clinton administration felt it was critical to let the scientists work unimpeded by political interference, so that this effort would not fall prey to the same troubles that sank those of the Bush administration. However, they also knew that the Clinton administration leadership would get nervous if the scientists were left on their own for long periods—particularly if rumors began circulating that the developing plan would significantly reduce timber harvest. To provide the administration with eyes and ears in Portland, Jim Pipkin, now counselor to the secretary of the interior, was given an office on the FEMAT floor. His task was to reassure McGinty and others in Washington that FEMAT was sticking to its assigned mission and to raise a red flag if it began running off the tracks. He never had to raise that flag. To make sure Pipkin knew the rules from the scientists' point of view, Jack Ward Thomas told Pipkin early on that he was welcome to observe, but that he would be out the door if he tried to interfere.[36] Thomas never had to take action.

Scientists and associated professionals organized themselves into analysis groups: terrestrial ecology, aquatic/watershed, resource (timber harvest) analysis, economic assessment, and social assessment. Each group had a dual assignment: suggesting components of alternative forest plans that addressed problems in their resource area and evaluating the alternative plans that emerged as viewed from their professional expertise.

This organizational description, though, does not adequately portray the dynamics of interaction among scientists or the emotional flash points. All knew they were working on a plan for the president of the United States—the most direct involvement of a president in federal forest policy and management in US history. In addition, Secretary of the Interior Babbitt had made it clear to the scientists early in the process that the administration was counting on them for its forest plan to resolve the spotted owl crisis. Finally, there was an overwhelming sense among the group that this effort was the end game after a multiyear series of science assessments and plans, and that this effort would produce the plan that would guide federal forests in the owl's range far into the future.

All these factors created an environment in which each different group or subgroup of scientists tried at times, often with some fire, to ensure that its view of the best approach would prevail. Biologists who had worked on the ISC saw the reserve design and Matrix restrictions (50-11-40 rule) from the ISC's strategy as the foundation of the landscape plan. Landscape ecologists who had worked on the G-4 effort saw LS/OG Areas as the basis of a reserve system. Aquatic and fish ecologists had their recently completed fish habitat strategy from the SAT report to offer. Those concerned about the wide variety of species associated with old-growth forests felt newly empowered by their SAT work and wanted to ensure that the needs of their species of interest were met. The resource analysis group wanted realistic assessments of the impacts on timber harvest of these different proposals and a systematic examination of trade-offs between species protection and timber production. The economists thought about economic and social effects at the regional and subregional level while the social science group thought about them at the community level. In addition, the sheer size of the undertaking, with hundreds of scientists and specialists involved to one degree or another, and the speed at which the work had to be finished made coordination and give-and-take difficult.

As a result, the relationships in the FEMAT endeavor were unlike those of the "band of brothers," who created the ISC's strategy for conservation of the spotted owl, or the camaraderie of the G-4, who expanded the conservation focus to old growth and fish, or the coordinated effort of SAT members, who expanded viability requirements to all species associated with old-growth forests. The FEMAT process was more chaotic, argumentative, frustrating, emotional, and exhausting than the earlier attempts. Yelling and table pounding at strategy meetings were not uncommon, and people occasionally threatened to give up and go home. No one did, though—the chance to take part in making history was too strong a draw and no one wanted to disappoint Thomas, who had been assigned to complete this arduous task. Officials representing the Clinton administration largely stayed out of the fray and left the scientists to work it out.

Designing Alternative Ecosystem Management Plans

FEMAT developed alternative management plans for the federal forests in the range of the northern spotted owl (called options) to find the option that would provide the "greatest economic and social contribution from the forests and region and meet the requirements of the applicable laws and regulations."[37] The scientists measured economic and social contributions mainly through the projected timber harvest level over time, and they measured meeting the requirements of laws and regulations mainly through the persistence likelihood of species associated with old forests and the streams that ran through them. However, it was not entirely clear what persistence likelihood would meet these requirements, or which species needed to meet that level, casting something of a conceptual haze over the scientists' work.

The scientists settled on developing options that answered three main questions: What system of reserves and management of the intervening forest was needed to protect LS/OG-associated species, particularly spotted owls and marbled murrelets? What actions were needed to protect and restore aquatic ecosystems, especially habitat for salmon populations? What level of timber production was possible under the management strategies proposed to address the first two questions? The options they developed reflected different scientists' hunches about the federal forest plan that would provide the most efficient answer: the highest timber harvest level associated with sufficient protection of species and ecosystems. Yes, scores of ecologists had been tasked with an economic analysis.

The options varied in four principal respects: size and placement of the reserve system; riparian buffer widths for streams; silviculture permitted within reserves and stream buffers; and limitations on silviculture outside reserves (table 7.1). Option 7 reflected the recently adopted or proposed forest plans modified by the ISC's strategy for the northern spotted owl and represented the status quo or existing situation in the eyes of the Clinton administration. Option 1, which contained the highest percentage of reserves, reflected placing all LS/OG Areas and all other LS/OG identified in the Gang of Four analysis into reserves. These two options were, in some sense, benchmarks on either end of the protection/production spectrum, with the other options falling in between. Among FEMAT scientists, Option 1 was known as the Green Dream and Option 7 as the Brown Bomb.

FEMAT grounded the options in a system of large reserves, called Late-Successional Reserves (LSRs), that were patterned after the ISC's strategy: large reserves, each with sufficient habitat to support a self-sustaining population of northern spotted owls, systematically spread across the range of the species and close enough to each other to allow adult and fledgling owls to move successfully between adjacent LSRs. In some options, Habitat Conservation Areas from the ISC's report (slightly modified by the spotted owl draft recovery plan) became the LSRs. In other options, configurations of LS/OG Areas from the Gang of Four report anchored the LSRs. However, all options except Option

Table 7.1. Characteristics of options fully developed by FEMAT.

Characteristics	Options						
	1	3	4	5	7	8	9
Late-Successional Reserves							
Source of reserves	LS	LS	HCA/LS	HCA	HCA	LS	LS/WA
Design	All LS	Owl	Owl	Owl	Owl	Owl	Owl
Silviculture permitted[1]	Non	TYM	TYM	TYM	TYM	TA	TYA
Added marbled murrelet reserves[2]	Already covered	Already covered	Yes	Yes	No	Yes	Yes
Riparian Reserves							
Scenario	1	2	1	2	FP	3	2
Silviculture permitted	AF	AF	AF	AF	FP	AF	AF
Limitations on silviculture outside reserves							
Rotation length	180	FP	FP	FP	FP	FP	180C
50-11-40 rule[3]	Yes	Yes	Yes	Yes	Yes	No	No
Green tree retention	6G	410	FP	FP	FP	FP	VAR
Marbled murrelet buffers[4]	Yes	Yes	Yes	Yes	No	No	Yes
Other species buffers	Yes	Yes	Yes	Yes	No	No	Yes

Source: Thomas, "Forest Management Assessment Team," 14.

Abbreviations: LS = LS/OG Areas from Johnson et al., *Alternatives for Management.*

HCA = Habitat Conservation Areas from the ISC strategy slightly modified by the USFWS proposed recovery plan.

WA = Key watersheds from the FEMAT effort.

Owl = Size and distribution of reserves designed to meet northern spotted owl conservation standards.

TYM = Treatment of young managed stands permitted.

TYA = Treatment of young natural and managed stands permitted.

TA = Treatment of stands up to 180 years of age permitted.

FP = Standards and guidelines from forest plans.

AF = Treatment allowed to advance goals of Aquatic Conservation Strategy after watershed assessment.

180 = 180-year rotation.

180C = 180-year rotation in California; otherwise based on forest plans.

6G = Leave six green trees of greater than average diameter.

410 = Leave four green trees plus 10 percent of area in tree aggregates.

VAR = Leave 15 percent of volume, at least half in aggregates, outside Coast Range.

[1] Silviculture permitted to accelerate the development of LS/OG characteristics and to prevent "large scale disturbances that would destroy the ability of the Reserves to sustain viable forest species populations" (FEMAT, *Forest Ecosystem Management*, 4-33). In addition to allowing treatment of young stands, Options 3 and 4 would manage a small portion of the late-successional reserves on long rotations.

[2] LS/OG1 and LS/OG2 Areas from the Gang of Four report reserved within marbled murrelet zone 1.

[3] Prescription calling for at least 50 percent of forest stands in Matrix to be at least 11" in diameter at breast height and have 40 percent canopy closure.

[4] Buffers would be applied around nest sites as they were found.

1, which reserved all LS/OG Areas, followed the large-reserve design laid down by the owl biologists in the ISC's report. Many options also designated smaller murrelet reserves along the coast, utilizing the highly significant and significant LS/OG Areas (LS/OG1 and LS/OG2) identified in the Gang of Four analysis.

Most LSRs included previously harvested areas occupied by young forests as a result of past harvest practices. Options generally allowed thinning of younger stands within LSRs to accelerate the development of LS/OG forest.

FEMAT developed an Aquatic Conservation Strategy (ACS) to protect and restore habitat for salmon and trout populations, broadening the approach developed by SAT and the Gang of Four analysis. The ACS contained four components:

- riparian buffers along streams, which were called Riparian Reserves;
- Key Watersheds crucial to recovery of at-risk fish populations (Tier 1) and protection of high-quality water (Tier 2);
- watershed analysis to tailor aquatic conservation to local conditions; and
- watershed restoration to help recover aquatic ecosystems.

These four components would operate together to protect the productivity and resilience of riparian and aquatic ecosystems through maintaining and restoring the geomorphic and ecological disturbance processes under which fish and other organisms had evolved.[38] All options utilized the ACS as their approach to conserving and restoring aquatic systems, except for Option 7, which utilized the recently adopted or proposed forest plans.

FEMAT's aquatics group, led by Sedell and Reeves, continued SAT's novel approach to determining the width of riparian buffers. SAT had employed the maximum average height that a tree can attain on a particular site, which it had called the *site-potential tree height*, as the metric in determining the width of the riparian buffer needed to protect aquatic ecosystems. Further, SAT had recommended riparian buffer widths equal to two site-potential tree heights on each side of fish-bearing streams and one site-potential tree height on each side of non-fish-bearing streams.

Sedell and Reeves, who had first utilized the site-potential tree height approach to setting buffer widths in SAT, believed that tree height distance away from the stream was a better indicator of potential wood recruitment or degree of shading than an arbitrary distance unrelated to ecosystem processes.[39] However, the SAT report had not justified this approach and recommendation for setting riparian buffers. To address that omission, Sedell requested that Kelly Burnett, a young PNW Research Station fish biologist who was assisting him, drift over a few yards from their cubicles in the Pink Tower to Fred Swanson's desk. Swanson was a longtime member of the Andrews science team that had done innovative work on forest and stream interactions. Sedell wanted them to characterize the effectiveness of riparian buffers as a function of distance from the stream expressed relative to the height of a site-potential tree. As Burnett explained it,

> Lacking Fred's depth of knowledge, I gathered the primary literature and headed to his cubicle. In addition to a narrative, we presented the information as graphs similar to those in McDade et al., "Source Distances for Coarse Woody Debris," which plotted the cumulative distribution of distances between the stream and the origin of large wood pieces found there. Fred relied heavily on his experience in the HJ Andrews Experimental Forest when drawing the curves for the various riparian functions in Figure 7-3 (above). I relied primarily on Chen, et al., "Contrasting Microclimates," for the curves in Figure 7-3 (below). Although Chen, et al. considered upland forests, we decided to use his paper because edge effects on microclimate had yet to be studied in riparian forests.[40]

In just an afternoon, Swanson and Burnett drew visualizations of relationships between forests and streams that made a lasting imprint on how scientists and managers thought about these relationships during the development of the Northwest Forest Plan and well beyond (fig. 7.3).

FEMAT options, except for Option 7, which utilized the buffer widths of the adopted or pending forest plans, measured riparian

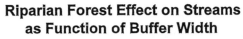

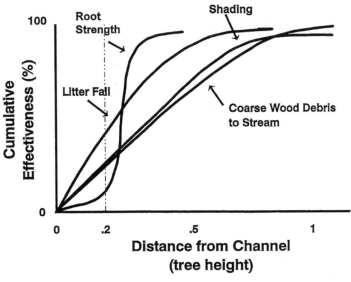

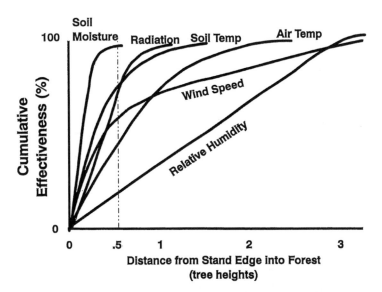

Figure 7.3. Generalized curves indicating (*above*) percentage of riparian ecological functions and processes occurring within varying distances from the stream channel, and (*below*) percentage of microclimatic attributes occurring within varying distances from the edge of a riparian forest stand. Distance expressed in terms of number of site-potential tree heights. (Source: FEMAT, *Forest Ecosystem Management*, 5-27.)

widths in terms of site-potential tree heights. They employed a buffer width of two site-potential tree heights for fish-bearing streams in the interest of maintaining the effects modeled in figure 7.3, reflecting what Sedell and Reeves had developed in SAT for providing a high level of protection of aquatic ecosystems. However, the aquatics group did not put the recommendation through the rigorous vetting that the ISC's spotted owl network received (chapter 5) in terms of examining the ecological benefits of alternative widths. Especially the second tree height for fish-bearing streams could be questioned since most riparian effects could be addressed within one tree height.

The aquatics group took a different approach to non-fish-bearing streams in which they considered three buffer scenarios, with each option (except Option 7) utilizing one of them:

1. A full site-potential tree height on all non-fish-bearing streams—the recommendation from the SAT analysis, often called *Full SAT*.
2. A full site-potential tree height on non-fish-bearing perennial streams in all watersheds and on intermittent streams in Tier 1 Key Watersheds; 1/2 site-potential tree height on intermittent streams outside Tier I Key Watersheds, often called *Half SAT*.
3. 1/2 site-potential tree height on non-fish-bearing perennial streams and 1/6 site-potential tree height on non-fish-bearing intermittent streams.[41]

Perhaps it is not surprising that different buffer choices were examined for non-fish-bearing streams since buffering those streams had not been done in the past.

Another crucial issue for the aquatics group was how it would define the stream system. Where does a stream begin? Should every gully that carried water during a storm be buffered? At one point some members of the group considered buffering all crinkles on the landscape as a non-fish-bearing stream. A quick walk on McDonald Forest near Oregon State University put an end to that idea: the entire landscape would be buffered. The group settled for a non-fish-bearing stream definition that required evidence of scour and other more conventional characteristics.

All SAT-based riparian scenarios represented a major expansion of riparian buffer systems compared to those utilized in the forest plans, particularly in the protection afforded non-fish-bearing streams. Major changes were afoot in riparian protection on federal forests.

Even though the stream buffers were called Riparian Reserves, timber harvest could occur within them after watershed assessment determined where such actions would advance the ecological goals of the ACS. For example, thinning of plantations would be permitted to accelerate the development of large trees that could eventually fall into streams. More generally, buffer widths in the Riparian Reserves were viewed by FEMAT's aquatic group as interim standards for riparian protection until watershed analyses were completed, at which time modifications in buffer widths and other considerations could be made.

The area outside reserves was called Matrix, as in previous efforts. Johnson worked with agency staff to project the likely harvest level over time from this area for each national forest and BLM district in the owl's range under each option. Silvicultural limitations, such as the 50-11-40 rule to aid owl dispersal and green tree retention at harvest, varied considerably across the options (table 7.1).

Not Good Enough

President Clinton had given FEMAT two months to complete its work. By the end of that period, FEMAT had developed and fully evaluated six options (1, 3, 4, 5, 7, 8). Initial review by Thomas, Pipkin, and others suggested that the highest harvest level of any option that might pass the species viability test was only one billion board feet, almost 80 percent less than federal harvests in the 1980s. A truly disappointing number for the Clinton administration, which had still hoped, even after SAT, that the scientists could craft a plan with less impact on harvest.

The administration requested a meeting in Washington, DC, to discuss the results. Thomas, Franklin, Raphael (terrestrial species), Sedell (aquatics), Johnson (harvest scheduling), Greber (economics), and Shannon (social and community) pulled together short presentations, flew to Washington, DC, and met with administration officials in the Indian Treaty Room of the Old Executive Office Building next to the White House in late May 1993.

Not surprisingly, the most spirited discussion concerned why the projected harvest levels associated with FEMAT's options, which Johnson had developed with agency staff, were so low. Clinton administration officials were especially upset that Johnson had reduced all harvest estimates to likely harvest levels, which integrated professional judgment with computer modeling, as opposed to the maximum harvest levels traditionally calculated and reported by the Forest Service and BLM. However, they made no convincing argument for using maximum versus likely harvest levels, especially given the president's direction to FEMAT that the "plan should produce a predictable and sustainable level of timber sales." Therefore, harvest estimates were not revised.

Toward the end of the meeting, the scientists acknowledged that the analysis groups in FEMAT had worked largely independently in constructing and evaluating the options; the short timeline made integration impossible. Most particularly, the selection of reserves had not considered locations of Key Watersheds. Several scientists thought that a more integrated approach might conserve species and ecosystems with less impact on timber harvests.

Lyons had previously asked Pipkin to check whether Thomas had the resources he needed to finish his task, including more time. Sure enough, he wanted another month, and Lyons alerted McGinty to that need.[42] Thus, the administration was ready when Thomas requested an additional month to develop a more integrated option. Although the additional time would violate the president's commitment to develop a plan in two months, the possibility of an option that would increase projected timber harvest while still protecting species and ecosystems was too good to pass up.

When Thomas returned to Portland on Friday, May 21, he gathered as many of the federal ecologists and biologists as possible and made an impassioned plea that they examine their maps to ensure that nothing was tied up in reserves beyond what was absolutely necessary: they needed to seek the most efficient option possible. Thomas explained in an oral history what happened next: "Jerry Franklin rather passionately declared that we had not yet done the job. We were all tired and completely exhausted. Franklin made a plea: 'Let's try one more option.' He took the lead in the development of Option 9. The team was pooped, but he got up and said, 'Come on guys, one more time.' Thus Jerry Franklin *fathered* Option 9."[43]

A small group of scientists led by Spies, Reeves, and Forsman immediately began brainstorming on how to better overlap LSRs and Key Watersheds, mainly by shifting LSRs into those watersheds. They worked late into that Friday night testing different ideas and developing draft strategies and maps for changes across the owl's range and learned much from that effort.

The next Monday, Franklin, somewhat independently, asked a number of scientists to join him in creating the new option, which they named Option 9, and set up a separate workspace for the effort. That included Forsman from the northern spotted owl group, who brought with him knowledge and insights from the previous effort; Sedell from the aquatics group; Johnson from the harvest scheduling group; and others. For starters, Franklin and Sedell noticed that shifting an LSR slightly so that it overlapped the Key Watersheds of the North Umpqua River would improve protection of species and ecosystems there, because it would reserve more high-quality old growth. Such a repositioning would also lessen impact on timber harvest because Key Watersheds had more harvest constraints than other watersheds. Utilizing this approach, Franklin and his team mapped an LSR system for the entire range of the northern spotted owl, in which the reserves captured LS/OG forest while also overlapping as much as possible with Key Watersheds.

Franklin's team also created an additional land allocation category, Adaptive Management Area (AMA), in response to urging from Tom Spies that they designate areas where experiments would occur on increasing the compatibility of timber harvest with conservation of species and ecosystems. Franklin selected ten AMAs containing over two million acres of federal forest that would otherwise be Matrix. Each AMA was given a specific mission, such as assessing alternative riparian system designs or assessing different ways to accelerate the development of LS/OG forest.

The AMAs were located where they could be buttressed by surrounding reserves, so that the experiments would not put the reserve network at risk. Many were near timber-dependent communities.

Once McGinty's staff learned that Option 9 would include a new category of land, they expressed concern to Johnson that this change would reduce projected timber harvest still further, since the planned AMAs contained significant forest that would otherwise be Matrix. Johnson told them he did not know what the effect would be, since the experiments that would be applied had not yet been designed. However, he also told them that if they wanted to ensure that no loss in timber production occurred, they could give these lands the same timber targets as if they were Matrix. McGinty's staff immediately adopted this solution. Thus, experimentation would be limited to approaches that would provide at least as much timber harvest as Matrix.

Option 9 reserved approximately four-fifths of the remaining LS/OG forest. It also specified a variety of standards for Matrix/AMA, which included a mixture of LS/OG and younger forest, where timber production was a key objective. These standards included requiring the retention of live trees during final harvests (fig. 7.4): at least

Figure 7.4. A few decades after a variable retention harvest in which significant elements of the preharvest forest were retained, mostly as aggregates, for the postharvest ecosystem. (Source: USFS, "Flat Country Project.")

15 percent of the trees in a unit were to be left standing after harvest on the site. Most retention would be in aggregates, as opposed to dispersed, in these variable retention harvests to increase both ecological and economic benefits.[44] Thinning stands up to 80 years of age was allowed in LSRs to accelerate the development of LS/OG forest, and thinning could be employed in Riparian Reserves to advance goals of the ACS. The 50-11-40 rule was not required under the argument that the reserves and other management standards were sufficient to enable successful spotted owl dispersal.

The effort Franklin led took a little more than a week to complete. Toward the end, he invited forest supervisors to visit, review, and comment on the plan proposed for their forest and talked to them about what had been done. Not much changed as a result of these visits, but they now had a better idea of what was likely to be carried forth to the Clinton administration.

Limited Choices with Most Forest in Reserves

FEMAT fully developed and evaluated seven options, including the new Option 9.[45] As previously mentioned, options varied primarily in size and placement of the LSR system, size of the riparian buffer system chosen, activities permitted within reserves, and restrictions on timber production in Matrix/AMA.

FEMAT recognized up to six land allocation categories on the landscape: Congressionally Reserved (mostly wilderness and national parks), Administratively Withdrawn (through forest plans and other administrative action), LSR, Riparian Reserve, Matrix, and AMA (fig. 7.5).

Wilderness and Administratively Withdrawn boundaries had been fixed before the FEMAT analysis began. However, LSRs were mapped over Administratively Withdrawn areas to ensure that the LSRs would continue in place if the withdrawals were redesigned in later planning processes. The areas in LSRs outside Administratively Withdrawn areas, which would otherwise be allocated to Matrix/AMA, varied from 17 to 35 percent of federal lands in the northern spotted owl's range (20 percent in Option 9)[46] and measured the additional reserves from LSR designation beyond those in the recently adopted or pending forest plans. Aggregating wilderness, withdrawals, and LSRs, reserved areas

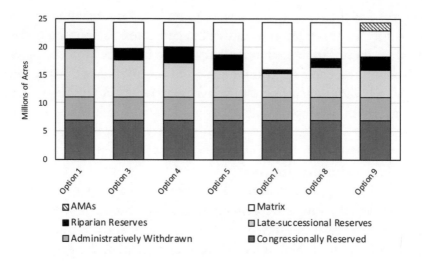

Figure 7.5. Distribution of federal lands among different allocation categories within the range of the northern spotted owl under different FEMAT options. Late-Successional Reserves shown as net of area that overlaps with Administratively Withdrawn areas. (Source: FEMAT, *Forest Ecosystem Management*, 2-24.)

totaled 63–81 percent of federal lands in the owl's range (66 percent in Option 9).

Riparian buffers were placed on streams in the allocations of LSR and Matrix/AMA. In Option 9, as an example, they covered approximately one-quarter of that landscape. However, they represented *additional* reserves in the accounting of figure 7.5 only in Matrix/AMA, where the buffer area varied from 3 to 12 percent of the federal forest (9 percent in Option 9).

The area outside reserves was classified as Matrix or AMA. Depending on the option, Matrix/AMA varied from 12 to 35 percent of the federal land in the owl's range (25 percent in Option 9). The reduction in Matrix/AMA from 35 percent in Option 7 (seen as the existing situation) to 25 percent under Option 9 came primarily from an expansion of riparian buffers and to a lesser degree from expansion of LSRs. The other options varied from 18 to 26 percent in Matrix/AMA—almost three-quarters of federal forests were in reserves of one kind or another in all options but 7.

The LSR, Riparian Reserve, and Matrix/AMA allocations were generally mixtures of LS/OG and younger forest, primarily because of prior dispersed cutting patterns that systematically spread small to moderate-sized clearcuts throughout federal forest landscapes. The LSRs contained higher percentages of LS/OG forest than did Riparian Reserves or Matrix/AMA, but they still contained significant areas of younger forests.

The portions of Matrix/AMA that were not too steep, too unstable, too unproductive, or otherwise unavailable in the underlying forest plans were considered suitable for timber production and were the source of the timber harvest levels reported below. In Option 9, as an example, approximately two-thirds of Matrix/AMA was suitable for timber production. FEMAT placed further limitations on silviculture in suitable forest. Option 9, for instance, designated minimum rotation lengths, retention of biological legacies at harvest, and buffers around locations of certain species (table 7.1).

All options (except Option 1) continued to rely on logging of older forest in Matrix/AMA, albeit at a reduced rate compared to the harvests of the 1980s. FEMAT estimated that Matrix/AMA in Option 9, for example, included approximately 21 percent old-growth forest and 24 percent mature forest, much of which was classified as suitable for timber production.[47] This older forest was expected to be the source of most projected harvest for the first 20 to 30 years of the plan's existence in all options except Option 1.

Species Viability as the Heart of Sustainability

A crucial part of FEMAT's effort was to estimate the level of protection provided by each option for terrestrial and aquatic species associated with LS/OG forests. FEMAT used viability analyses to make these estimates. The scientists keyed their evaluations to the "likelihood of maintaining sufficient habitat, well-distributed on federal lands to provide for the continued existence of viable populations of northern spotted owls and marbled murrelets . . . [with] similar assessments [conducted] for over 1000 plant and animal species closely associated with old-growth forests. The geographic bound was the range of the northern spotted owl; the time frame was 100 years."[48] This viability assessment became the central gauge of ecological sustainability and the foundation of overall sustainability analysis.

The regulations implementing the National Forest Management Act required that habitat be provided to ensure the viability of native

Table 7.2. Species (or species groups) considered by FEMAT to be closely associated with LS/OG forests.

Animals	Number
Vertebrates	
Amphibians	18
Fish	7
Birds	38
Bats	11
Other mammals	15
Nonvertebrates	
Arthropods (groups)	15
Mollusks	102
Plants	
Mosses/liverworts	106
Lichens	157
Vascular plants	124
Fungi	527
Total	1,120

Source: USDA and USDI, "Final Supplemental Environmental Impact Statement," appendix J2, 4.

vertebrates. FEMAT extended its viability assessment to all native species associated with LS/OG forests following the SAT analysis. Consequently, most species considered were not vertebrates (table 7.2).

With that expansion, FEMAT found itself in a whole new world where the population levels and distribution data for most species studied were somewhat or largely unknown. As Thomas and biologist Bruce Marcot later reflected, "The FEMAT work forced more clearly the issue of using new, uncertain, and disparate sources of information in evaluating large-area effects on ecosystems and species."[49] Further, Logan Norris, a respected forest scientist, noted the limitations that forced FEMAT to "lump geographic areas and species into large groups and utilize information sources that had not been the subject of independent peer review and referee evaluation," as well as the need to rely on expert opinion rather than quantitative analysis in viability assessments.[50] FEMAT worked at the cutting edge of ecology and wildlife and conservation biology, where irreducible uncertainty was inevitable.[51]

FEMAT set up expert panels, with large numbers of outside scientists, to conduct viability assessments on these species. The assessments first used a largely qualitative (very high to very low) outcome scale similar to that employed in the G-4 and SAT exercises. However, two concerns emerged after they were completed. First, the qualitative results from the viability assessment approach of these exercises did not lend themselves to a simple numerical aggregation across panelists to get a numerical score. Second, that approach made it difficult for panelists to register how certain (or uncertain) they were of their estimates. As the viability leaders struggled to use the results from the expert panels, they concluded that they needed an approach that overcame these problems. After much discussion and debate among FEMAT's leadership, Thomas decided to redo the viability assessment—a huge task.[52]

Dave Cleaves from the Southern Research Station, an economist and expert on decision-making under uncertainty, was called on to help revise the viability assessment process. He came for a few days and stayed a month. That is how it was then—you came when you were called and you stayed until the battle was over.

FEMAT then conducted a second round of panels using as many original panelists as were available. In that effort, the expert panels were asked to determine the likelihood of achieving four possible outcomes for habitat conditions on federal lands for each species or species group:

- Outcome A: Habitat is of sufficient quality, distribution, and abundance to allow the species population to stabilize, well distributed across federal lands (the viability standard in the Forest Service's Planning Rule).
- Outcome B: Habitat is of sufficient quality, distribution, and abundance to allow the species population to stabilize, but with sufficient gaps in the species distribution on federal land . . . to cause some limitation in interactions among local populations.
- Outcome C: Habitat only allows continued species existence in refugia, with strong limitations on interactions among local populations.
- Outcome D: Habitat conditions will result in species extirpation from federal land.[53]

FEMAT's viability leaders (mostly terrestrial biologists) asked panelists to assign 100 likelihood votes (or points) across the four outcomes for each species/option combination. Thus, a panelist could express complete certainty in a single outcome by allocating 100 points to it, express uncertainty by spreading votes across outcomes, or even refrain from expressing an opinion. The leaders then averaged the likelihood estimates of the panelists relative to each outcome.

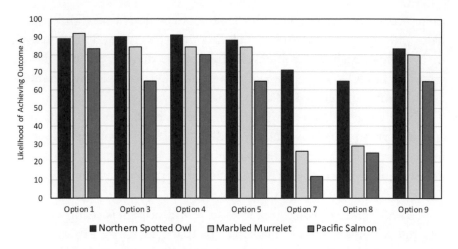

Figure 7.6. Likelihood of achieving Outcome A (sufficient habitat to maintain viable populations well distributed on federal land) for the northern spotted owl, marbled murrelet, and Pacific salmon under different FEMAT options. (Source: FEMAT, *Forest Ecosystem Management*, 2-31–2-41).

The viability leaders evaluated options in terms of whether a species (or group) attained an 80 percent or greater likelihood of achieving Outcome A. They felt that this likelihood and outcome combination represented a relatively secure level of habitat and that options attaining such a percentage should be viewed as meeting the viability standard. (See appendix 3 for more discussion of FEMAT's viability assessment process.)

Owl, murrelet, and salmon habitat that was the focus of FEMAT's landscape design did moderately well in the panel's evaluations. Most options (including Option 9) achieved at least an 80 percent likelihood of Outcome A for the spotted owl. FEMAT noted, though, that areas of special concern still existed for parts of the owl's range, so that contributions from nonfederal lands remained important. The marbled murrelet habitat evaluation focused on the adequacy of nesting habitat to support viable populations, and most options (including Option 9) also achieved at least an 80 percent likelihood of Outcome A (fig. 7.6).

The viability story for salmon is more complicated. Option 9 utilized Scenario 2 for Riparian Reserves, which halved the SAT-recommended buffer on non-fish-bearing streams outside Key Watersheds and enabled an increase in harvest over options that utilized the full tree height on those streams. Such an approach earned a 65 percent likelihood of

Outcome A (fig. 7.6) for federal salmon habitat from the expert review panel when it conducted its viability review of the new option in May 1993. In short order, news that Option 9 had achieved something less than the 80 percent likelihood level for salmon habitat got to McGinty's staff, who then called Reeves to understand why that had happened. Reeves told them that the expert panel had concluded that the time frame of 100 years in the assessment was too short to enable full recovery of the forest along streams. The 80 percent rating could not be reached even with wider buffers on non-fish-bearing streams. McGinty did not push any further and utilized Riparian Reserve Scenario 2 and the resulting 65 percent likelihood rating for Option 9 in her discussions with the president during June as he selected a forest plan. However, the FEMAT report documenting the relationship between salmon population viability and riparian scenario choice, published a few months later in July 1993, displayed a different conclusion:

> To increase the likelihood of achieving outcome A for fish habitat of all races/species/groups to 80 percent or greater in Options 2, 3, 5, 6, 9, and 10, we recommend two possible strategies. One strategy is to replace the Riparian Reserve 2 scenario used in these options with the Riparian Reserve 1 scenario [the scenario recommended in the SAT report that doubled the riparian width on non-fish-bearing streams compared to Scenario 2]. . . . A second mitigation strategy is to provide greater protection for Key Watersheds.[54]

Whether knowledge of the ability to raise the salmon habitat rating to the 80 percent likelihood level would have caused McGinty to shift to Riparian Scenario 1 for Option 9, and a lower timber harvest level, is unknown. However, this conclusion was utilized from then on in discussions about how to raise salmon habitat to the 80 percent level.

FEMAT scientists were generally confident in predicting outcomes for most other vertebrates since their life histories and habitat requirements were relatively well known. Their summary for nonfish vertebrate species indicated that most options provided at least an 80 percent likelihood of Outcome A for a significant majority. Birds were especially well protected. Not surprisingly, Option 1, which reserved all LS/OG forest,

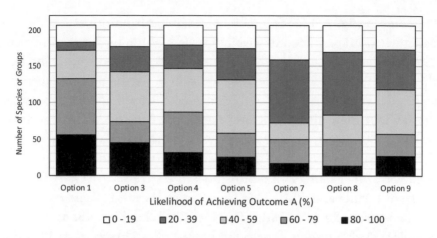

Figure 7.7. Number of invertebrates, nonvascular plants, and fungi expected to achieve various likelihoods of Outcome A (sufficient habitat to maintain viable populations well distributed on federal land) under different FEMAT options. Very few achieve a likelihood of 80 percent of Outcome A. (Source: FEMAT, *Forest Ecosystem Management*, 2-35.)

had the largest number of vertebrate species that achieved or exceeded the 80 percent likelihood threshold. FEMAT scientists also noted that mitigation measures beyond those contained in the options could be employed to improve the viability rating for vertebrates scoring less than 80 percent. Those additional measures would be especially important for species such as the red tree vole, about which little was known.

FEMAT considered six taxonomic groups in addition to vertebrates: vascular plants, lichens, fungi, mosses/liverworts, mollusks, and arthropods. Except for vascular plants, FEMAT scientists were much less confident in their knowledge regarding the distribution and habitat needs of these species (or species groups): only about one-quarter of them achieved an 80 percent likelihood of Outcome A, even when all LS/OG forest was reserved as in Option 1 (fig. 7.7).

A possible inference from figure 7.7 is that only 25 of 210 nonvertebrate species or species groups (12 percent) achieve an 80 percent likelihood of Outcome A under Option 9; that is, Option 9 does not sufficiently protect 88 percent of those species. However, a more accurate inference could be that panelists were uncertain about whether the species would have sufficient protection under Option 9 because so little was known about the species and their habitats. Lack of protection and lack of knowledge had been rolled up into a single rating. This subtle but important difference would haunt interpretation of, and

reaction to, these findings. Jack Ward Thomas later effectively summarized the difficulties associated with this approach:

> The people that wanted to force a zero cut of timber looked at
> the ratings of these species at risk. Well, maybe most of those
> species were at "high risk," not because we knew that they were
> at high risk, but because we simply didn't know enough to
> predict response to management alternatives. There was a risk
> associated with a lack of knowledge and not a risk associated
> with knowledge. Those are two different things. If I had to do it
> over, we would have made that differentiation.[55]

In addition to species viability analysis, FEMAT also evaluated the
likelihood of providing a functional, interacting, late-successional/old-
growth forest ecosystem. While such an evaluation might be seen to
subsume the test of viability for the system's component species and
groups of organisms, FEMAT scientists argued that

> an ecosystem will likely continue to function in some fashion,
> even in the absence of some component and perhaps even
> important species. Such a system is, however, no longer
> providing the same array of processes and functions once
> present. An impoverished ecosystem is not likely to be as
> productive and sustainable as one in which all the functions are
> provided. Clearly, the goal is to maintain functional interacting
> ecosystems and their complement of component species to
> maintain biodiversity.[56]

Tom Spies led a group of ecologists who assessed the likelihood
of maintaining a functional, interacting, late-successional/old-growth
forest ecosystem based on three components:

1. A relatively high abundance and diversity of old-growth communities and subregional ecosystem types well distributed
   across the region.
2. Occurrence of ecological processes and functions characteristic
   of old forests.

3. An interacting system in which the distribution of patches, and the landscapes in which they occur, provide for biotic flow to maintain viable species.[57]

They selected two major geographic areas based on the differing influence of fire—the dry eastern Cascades and Klamath provinces, and the moist northern and western provinces—which roughly correspond to the Moist/Dry geographic division used in this book (plate 1.3). They found that

the stability of a functional interacting old-growth forest ecosystem is less in the Eastern Cascades and Klamath Provinces than in the moister provinces due to the likelihood of large-scale disturbance (especially fire), current stand conditions, and the portent of global climate change within the 100-year evaluation period. The effects of human disturbance and land ownership patterns further weigh against maintenance of the old-growth forest ecosystems that were once present.[58]

Within that context, their evaluation of the moist provinces gave Options 1, 3, 4, 5, and 9 a greater than 70 percent likelihood of maintaining characteristics of late-successional/old-growth ecosystems, and their evaluation of the dry provinces gave these same options at least a 60 percent likelihood of maintaining ecosystem characteristics.[59]

This innovative analysis looked beyond the viability of individual species to the functioning of the ecosystems themselves. However, since the laws at issue (NFMA and ESA) focused on individual species, option evaluation did too. This analysis did draw attention to two very different types of ecosystems in the range of the northern spotted owl. That, in turn, influenced how standards and guidelines would be written for the plan that would be adopted, and it foreshadowed recognition of these differences in future policy debates.

Economic and Social Effects—Regional versus Local Perspectives

FEMAT was charged with estimating a broad range of economic effects of its options, including those on local and regional economies and national forest product markets. Most attention focused on the effect

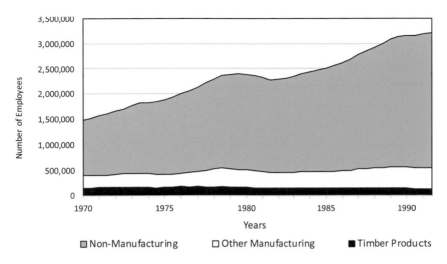

Figure 7.8. Wage and salary employment in owl-impact counties, 1970–1992. (Adapted from Greber, "Economic Assessment," 38.)

on employment in the range of the northern spotted owl, broken into timber industry losses and potentially offsetting employment gains. Also, the potential effect on federal payments to counties gained considerable attention.

FEMAT's economic group, led by Brian Greber, developed a useful economic history of the portions of Washington, Oregon, and California most influenced by timber harvests in the owl's range, calling them the *owl-impact counties*. The team's analysis told a story of a changing timber industry and a changing regional economy. In the 1970s, timber industry employment provided the economic foundation for many rural owl-impact counties, but by the late 1980s, that importance had waned (plate 7.2). Two primary reasons explain this change.

The industry had mechanized many mill jobs to reduce labor costs—the days were gone where a half-dozen men pulled boards off the green chain and sorted them into different lumber grades. Fewer jobs were needed to process each million board feet of lumber.[60]

Most importantly, overall regional employment increased significantly in the 1980s despite declining timber industry employment (fig. 7.8). This finding greatly weakened the argument that the region faced economic catastrophe without federal timber harvests.

It must be emphasized, though, that the regional growth was concentrated in metropolitan areas such as Portland and Seattle, where companies like Boeing, Microsoft, Nike, and Intel flourished.[61] Rural areas were largely left behind and timber industry jobs remained a significant part of the economic base in many rural communities within the owl's range. Thus, employment impacts from reductions in federal harvest changed from regional in nature to more local and focused on individual communities and areas.

FEMAT's economic group also documented that the timber harvest from nonfederal lands, which in the 1980s had made up about two-thirds of the total regional harvest, would continue.[62] That strengthened arguments that the region would not face economic catastrophe from reduced federal timber supplies. Furthermore, as log prices had increased with contractions in timber supply and international markets, the volume of logs exported overseas from private lands (20 percent of total harvest at its peak) had fallen. More timber was staying in the region for processing.[63]

Employment effects associated with the different options required estimates of likely timber harvest levels. Federal agencies usually estimate the timber *sale* level that can occur under a plan, rather than the actual timber harvest level in any year, in their allowable cut calculations, as their primary control of timber production is over the rate of sale. They offer timber for sale and the purchaser then generally has a number of years to harvest it. Thus, FEMAT projected timber sale levels.

The national forests and BLM districts in the range of the northern spotted owl contained approximately 21 million acres, of which 18 million were forested and 12 million were classified as tentatively suitable for timber production, which was the forest that remained after removing wilderness, other congressional and administrative withdrawals, and unstable and unproductive forest. The recently adopted or proposed forest plans removed approximately 4 more million acres, leaving a land base suitable for timber production of 8 million acres. That suitable land base was further reduced in each option by removal of LSRs, Riparian Reserves, buffers for individual species, and other considerations. As an example, FEMAT estimated that Matrix/AMA in Option 9 contained 4.1 million acres that were suitable for timber production, approximately 23 percent of the forested area. Importantly, this percentage was not evenly distributed across the owl's range. Coastal forests contained a much lower percentage of suitable forested area than did interior

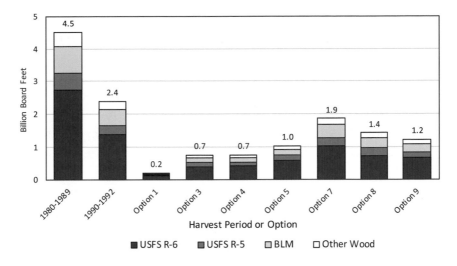

Figure 7.9. Average federal timber harvest per year over two historical periods and projected average federal timber sales (probable sale quantity) per year for the first decade under different FEMAT options. R-5 = California Region; R-6 = Pacific Northwest Region. (Source: FEMAT, *Forest Ecosystem Management*, 2-49).

forests because of the protections there for salmon, marbled murrelets, and spotted owls.[64]

Based on the area of forest suitable for timber production in each option, agency personnel projected timber sale levels under the rotation ages, yield restrictions, and management intensities prescribed in the underlying forest plans, constrained further by additional restrictions in each option (table 7.1). They consistently used a harvest flow policy of nondeclining yield (the rate of projected sales could not decline over time).

Johnson worked with agency staff to estimate the likely sale level under each option, which they termed *probable sale quantity* (PSQ), rather than estimating the maximum sale level, as had traditionally been done, in keeping with the approach he had started in the Gang of Four analysis. To accomplish this, Johnson worked with agency experts and made several downward adjustments in calculated levels, such as estimated losses from inoperable slivers of land in areas with dense stream networks and associated buffers.

The PSQ of individual options ranged from 0.2 to 1.9 billion board feet per year (fig. 7.9), with Option 9 having a PSQ of 1.2 billion board feet per year. All options included estimates of "other wood"—timber harvest volume obtained from partially rotted logs that were common in old-growth forests, which had averaged about 10 percent of total harvest volume in the past. This volume had generally not been included in allowable cut calculations, but it was real timber volume that would be processed into wood products, so it was included.

Projected sale levels under the different options were compared with actual federal harvests from two historical periods: 1980–1989 (the decade before sales in owl habitat were prohibited by the federal courts) and 1990–1992 (the period after sales in owl habitat were prohibited by the federal courts).[65] Projected timber sale levels were substantially lower than historical levels in all options. For example, timber sales under Option 9 were expected to equal approximately 25 percent of the average harvest in 1980–1989 and slightly less than half of the 1990–1992 harvest (fig. 7.9). Among the three states, Oregon would see the largest volume reduction, although Washington would experience the largest percentage reduction. Timber sales in coastal forests would be most affected.

PSQ estimates did not include volume that would be obtained from thinning in reserves to advance ecological objectives, or from salvage of postfire timber. The volume associated with such harvests was unknown, and allowable cut targets for them seemed inappropriate anyway. Still, that volume, as it occurred, would contribute to actual harvest levels.

The resource analysis group worked on refining their sale estimates up to the end of FEMAT. In a supplementary report submitted in August 1993, the group acknowledged considerable uncertainties surrounding the estimates, despite their effort to make them as realistic as possible in modeling the options. New planning processes such as watershed assessment would be employed; additional mitigations might be needed to bring the habitat for more species, such as salmon, up to the 80 percent likelihood level; and, most importantly, the timber harvest in the first few decades would still depend on the socially controversial practice of logging LS/OG forest.[66]

With the PSQ information for each option along with estimates of nonfederal future harvests, the economics group estimated timber industry employment in the owl-impact counties under each option (fig. 7.10). The group utilized two historical baselines for comparison: employment in 1990, which roughly represented timber industry

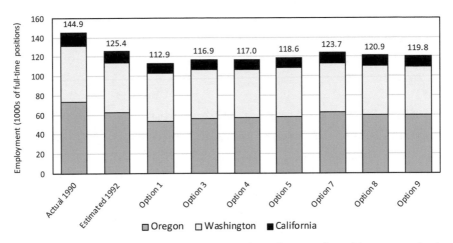

Figure 7.10. Average annual employment in the timber industry in the owl-impact counties in two historical years and projected average annual employment in the timber industry for the first decade under different FEMAT options. (Source: FEMAT, *Forest Ecosystem Management*, 2-55.)

employment after recovery from the recession of the early 1980s; and in 1992, after the owl injunctions on federal timber sales took hold. For example, the economics group estimated that industry employment under Option 9 would be 24,000 fewer jobs than it had been in 1990, and 6,000 fewer jobs than it had been in 1992. Under Option 1, which would end all harvest of LS/OG forest, timber industry employment was estimated at 32,000 fewer jobs than in 1990, and 12,500 fewer jobs than in 1992. Timber industry job losses were heavily concentrated in southwestern Oregon, the part of the owl's range most dependent on federal timber. Job loss estimates did not include forestry services, such as tree planting, which would also decline.

The economics group acknowledged that lower timber industry employment would impact local businesses, stores, restaurants, and other services—indirect and induced effects[67]—but did not include them in its estimates of employment effects:

> In a static view of the Pacific Northwest economy, every job in the forest sector supports approximately one job in other sectors. . . . Thus, in a static sense, job effects may be double the level suggested by direct jobs alone.
>
> In a dynamic view of the economy, other industries are growing and/or entering the region and may render many of

the indirect and induced effects equivalent to lost opportunities in the region rather than actual job losses. The proportions of indirect and induced effects that are actual job losses are hard to deduce.[68]

Thus, FEMAT's analysis arguably underestimated the extent and impact of timber industry job losses on timber-dependent communities by leaving out the effect of job losses in the timber industry on other employment. Those other job losses would be especially felt in rural communities that were not growing as were more urban areas. Leaving them out of the employment analysis would become a major source of criticism of FEMAT's results in the months and years that followed.

FEMAT's economic group noted that a large recreation and tourism industry existed in the region, with 50,000–80,000 jobs directly attributed to forest-based recreational opportunities. To the degree that retention of old-growth forest helped maintain or increase tourism, this industry would benefit. Also, they noted that thousands of individuals in the region were engaged in harvesting and marketing special forest products, especially mushrooms and floral products. However, they also acknowledged that many of these jobs were part time and seasonal and that it was not clear how adoption of FEMAT's options would affect that employment. Finally, they noted that restoration work associated with the different options, especially repairing watersheds and the roads within them, would create jobs. Their magnitude would depend primarily on federal appropriations.

Declining county payments were another economic issue of major concern. In fact, it was the first issue raised by county commissioners in spring 1993 when they reviewed FEMAT's results with the Oregon governor's office. County governments throughout the owl's range had historically received substantial federal payments generated from federal timber sales: 25 percent of gross receipts from Forest Service sales and 50 percent of gross receipts from BLM sales. These payments were designed to make up for lost property taxes since federal land cannot be taxed. Counties with extensive federal forestlands in the owl's range depended heavily on such payments, which provided them with hundreds of millions of dollars per year in the late 1980s. This dependence was especially high in southwestern Oregon, where most BLM forests were concentrated, and in-lieu payments enabled the counties there to keep taxes on private property in those counties exceptionally low. FEMAT's options would result in a major contraction in federal timber harvest and, consequently, substantial reductions in county payments, even if market forces brought a higher price per thousand board feet of federal timber sold. Some rural counties could face financial insolvency without special congressional appropriations.

FEMAT's social assessment group focused on identifying communities at risk. FEMAT social scientists first identified approximately 300 communities in the spotted owl's range that would be affected by federal forest management issues in the Pacific Northwest. They then conducted workshops, with participants drawn from a variety of governmental units (such as school districts and county and town governments), to explore the ability of these communities to cope with changes that would be associated with FEMAT's options.

FEMAT's social scientists utilized two central concepts to identify communities at risk from these changes: community consequences and community capacity.[69]

Community consequences involved the degree to which federal timber harvest levels influenced the ability of local residents to satisfy

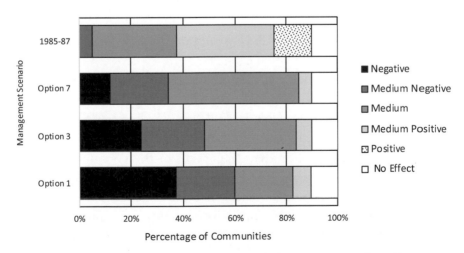

Figure 7.11. Predicted economic and social consequences for communities in the owl-impact counties of continuation of harvest levels of 1985–1987 or adoption of three FEMAT options. Option 3 had the projected timber sale level closest to that of Option 9. (Source: FEMAT, *Forest Ecosystem Management,* 7-58.)

their overall needs and expectations, affected employment and income generation opportunities, and other considerations. Workshop panelists were asked to qualitatively rate the likely effects of FEMAT options within one to three years of implementation, with ratings ranging from very positive to very negative. Consequences shifted from mostly positive under the 1985–1987 harvest level to increasingly negative as harvest levels declined across the options presented (fig. 7.11). Panelists estimated that the lower Columbia River, Olympic Peninsula, southwestern Oregon, west-central Oregon , and interior northern California would be especially hard hit by federal harvest reductions.

Community capacity involved the ability of residents and community institutions, organizations, and leadership—formal and informal—to meet local needs and expectations. As FEMAT's social scientists wrote: "Communities with lower capacity have reduced ability to maintain community relationships and improve well-being. These same communities are less resilient and have reduced ability to contend with changes of any sort. A community's capacity is only as high as its physical infrastructure, human capital, and most importantly the manner in which residents and groups devote energy to community issues."[70] Workshop panelists rated the capacity of the communities in their area to respond to impacts of federal timber harvest reductions on a qualitative scale from low to high. Panelists rated 45 percent of the communities as having low to medium-low capacity to respond to lower harvests, 25 percent as having medium capacity, and 30 percent as having medium-high to high capacity.

FEMAT's social scientists then integrated the two sets of ratings to identify communities that were expected to experience negative to moderately negative consequences and that also had low to medium-low capacity to respond. They identified such communities as *most at risk*. This analysis found that 27 percent of the communities they studied fit that description.

FEMAT's social scientists characterized these at-risk communities: "Communities that are small, isolated, lack economic diversity, and have low leadership capacity are also more likely to be classified as 'most at risk' than others. Residents of these communities may find it difficult to mobilize and respond to changing conditions. They are likely to suffer unemployment, increased poverty, and social disruption in the absence of assistance."[71]

These findings provided analytical support to the idea that some communities would have great difficulty in responding on their own to reductions in federal timber harvest, while others would adjust more easily. In addition, these findings were about real people and communities rather than somewhat abstract employment numbers.

Since FEMAT's social science group evaluated communities at risk one by one, they could name the specific 27 percent of communities identified as most at risk and most in need of assistance to survive and prosper. Toward that end, they created a map identifying these communities and sent it to Clinton administration staff in Washington, DC. Clinton's economic assistance team could then use it, if they wished, to focus recovery efforts on those communities that would be hardest hit by reductions in federal timber harvests.[72]

Ultimately, the map was not made public or included in FEMAT's report. FEMAT's social scientists offered reflections on the decision to withhold the map and the general approach to identifying at-risk communities:

> Risk labels can be a double-edged sword. Among the many problems associated with determining risk is the question of how to predict social and individual resilience. The presence of risk in a community may lead to increased survival strategies of individuals. For example, woods workers as an occupational group have shown themselves to be resilient and innovative, capable of subsistence and survival strategies during economic downturns. But at some point, persistent stress will overcome personal, cultural, and social reserves. Labeling communities *most at risk* can also paralyze and demoralize community members, increase social disruption, and from the labeling itself, create indirect impacts on communities (for example red lining of communities by banks). It is for these latter reasons and because of the need to involve locals in the self-assessment process that we choose not to report individual community ratings.[73]

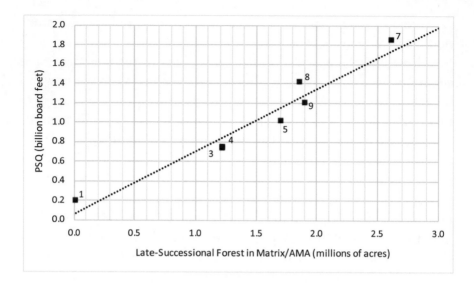

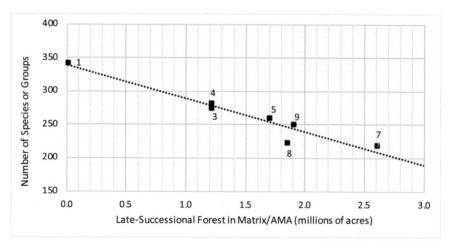

Figure 7.12. (*above*) Projected probable sale quantity (PSQ) as a function of the LS/OG forest in Matrix/AMA under different FEMAT options. (*below*) Number of species or groups of species rated as having a greater than 60 percent likelihood of outcome A (sufficient habitat to maintain viable populations well distributed on federal land) as a function of LS/OG forest in Matrix/AMA under different FEMAT options. (Source: FEMAT, *Forest Ecosystem Management*, 2-29.)

However, the location of many of the small, isolated communities lacking economic diversification and likely to be at risk was not a secret. Community leaders in places like Forks, Washington; Sweet Home, Oregon; and Hayfork, California, were quite vocal that their very futures were on the line.

To enable policy makers and the public to visualize overall trade-offs, Marty Raphael (FEMAT's deputy team leader) prepared two summary charts portraying the relationship between species protection and harvest of LS/OG forest—the core issue of FEMAT's analysis. Truly, no free lunch had been found (fig. 7.12).

Option 9 Becomes the President's Plan

While the scientists were developing forest management options, the Forest Conference Executive Committee, with representatives from the Departments of Agriculture, Interior, Commerce, Labor, Treasury, and Justice and relevant White House Offices, held a series of meetings to hammer out its recommendations to the president. Under McGinty's guidance as chair, they worked on developing cohesive strategies for accommodating the scientists' findings; making policy changes, such as a log export tax policy, that might increase wood supply in the Pacific Northwest; providing labor and community assistance; and improving interagency coordination.[74]

The completed FEMAT analysis was immediately made available to the Clinton administration in early June 1993 so that the Executive Committee could evaluate the options and their projected outcomes and effects. Toward that end, McGinty, with the assistance of Jim Pipkin and others, prepared an 18-page memo for the Executive Committee summarizing its proposals, as well as a three-page attachment on the political situation.[75]

The memo first reminded the president that he had directed cabinet secretaries to develop a balanced plan within 60 days to break the stalemate in the Pacific Northwest and had formed three interagency working groups, including FEMAT, to complete that task. It next stated the "bottom line":

Excessive timber cutting in the previous decade has left us very little decision space. The Administration is under a court order to submit by July 16 a draft environmental impact statement on a proposed land management program that complies with the law. . . . Compliance with conservation and land management provisions [of environmental laws] will necessarily result in timber harvests that are far below the levels of the 1980s. Our primary opportunities to be creative and "put people first" come under the community/labor assistance and institutional reform programs and a new forest management option [Adaptive Management Areas] we recommend that emphasizes community involvement.

It then asked President Clinton to approve Option 9 as his forest management plan, approve proposals for labor and community assistance, make policy changes to increase nonfederal wood supply, and approve a plan for agency coordination to implement the plan.[76]

Most of the memo addressed the options developed by the scientists, noting that "our instructions to the Ecosystem Management Assessment Team were to prepare a series of options that would, in essence, keep the timber harvest level from federal lands as high as possible consistent with environmental laws." The memo further identified the relevant laws as the Endangered Species Act and the National Forest Management Act.[77] It keyed on the viability ratings for the northern spotted owl, marbled murrelet, and salmon populations—the species that had been the focus of the landscape designs of the options—in judging which options would meet legal requirements. Further, it stated that "we believe that any plan that achieves 'high' or 'medium high' viability ratings for key species will likely be deemed to be in compliance with the ESA and NFMA as they relate to federal land," while acknowledging that the "issue is not free from doubt." [78] The low viability ratings for many species, especially nonvertebrates, were not mentioned.

Options 7 and 8 were immediately dropped from further consideration (fig. 7.6) due to insufficient viability ratings for key species. Of those that remained, "the one that produces the largest amount of timber (1.2 bbf) is an option newly developed by the scientific team, referred to as the 'unified reserve' option (Option 9)." [79] Thus, the best the scientists could do resulted in a very disappointing outcome for the Clinton administration, well short of the level that might address "political reality" relative to demands of labor and industry.

The memo also noted that Option 9 "protects only some (not all) old growth; indeed, about 50 percent of the timber harvested will be at least 150 years old."[80] Surely a flash point with environmentalists.

In addition, the memo highlighted the Adaptive Management Areas of Option 9 where intensive ecological experimentation and social innovation would occur. "It thus incorporates some of the cooperative, community-based principles that you laid out at the Forest Conference."[81] It noted that these areas would be very popular with local communities and labor and stridently opposed by environmentalists, as the approach would seem to give local communities more control over federal forest management.

The memo estimated that adoption of Option 9 would result in 6,000 fewer timber industry jobs than current (1992) levels. It also argued that economic assistance would be needed for an additional 5,000 to 10,000 people already displaced by contraction of industry employment from 1990 to 1992. [82] To help workers and communities transition to a new future, the authors recommended an integrated program to improve economic conditions in timber-dependent communities and a decade of federal payments to counties to make up for the loss of revenue sharing associated with timber sales."[83] They were under no illusions, though, that these commitments would satisfy industry, labor, and the counties.

The memo emphasized that "the new plan will be based on ecosystem management" and recognized that "ecosystems are not generally contained within the geographic boundaries of a single ownership," and that their management often involves federal and state agencies with regulatory responsibilities. "As a result, ecosystem management of federal lands will require vastly increased agency coordination." To implement this new vision, the memo called for a new approach to planning where the "the beginning place for analysis is the individual watershed and the most logical planning unit is a physiographic province." Most of the planning work would be done by province teams that would include the relevant federal agencies, states, tribes, and others affected by the decisions made. Also, consultation under the ESA would be revised to engage the USFWS and NMFS early in the process. [84]

The authors urged that the president move quickly, as an early draft of this memo had been leaked to environmental groups who then attacked the administration's plan (as they understood it) in vivid language in the *Washington Post*, calling Option 9 "political science, not biological science." As a result, "anxiety over this issue on the Hill and among key interest groups has increased tremendously."[85]

However, the Executive Committee's proposed endorsement of Option 9 set off a last-minute battle within the administration, especially once staff and advisers saw the projected timber harvest. Key advisers, such as Rahm Emanuel, vehemently argued that the economic and political cost of Option 9 was just too high. Endorsement of it would greatly damage the president's standing with labor groups, who were important political allies in future elections. President Clinton had run for president the previous year as a friend of working families, and yet he was set to endorse a plan that would provide significantly lower timber industry employment than in the recent past, with a potentially devastating impact on many rural communities. How could he endorse Option 9 and still tell labor groups, with a straight face, that he was with them?

Yet President Clinton had chosen the path on which he now found himself: he had publicly directed a group of scientists to craft a plan that would provide the highest economic and social contribution possible while meeting legal requirements. In response, the scientists had tried their best. There really was no Plan B. Yes, he could refuse to endorse Option 9 and leave the Dwyer injunction in place until a better plan was developed or environmental laws were changed, while putting all the administration's energy into helping with economic transition. However, after all the buildup that had occurred and the promises he had made, that outcome would be a major defeat for the new administration.

The president was known to put off tough decisions. Yet the administration was under a court order to submit by July 16 a draft environmental impact statement on a proposed land management program that complied with the law.[86] To help move him along, the White House had previously announced that a decision would be forthcoming by July 1, 1993, and that day had arrived. Delaying the decision to endorse Option 9 until the last moment left no time to brief key congressional committee chairs and allies who might comment favorably on the plan. Consequently, all members of Congress would hear about the president's decision without forewarning, making them even less willing than before to take any responsibility for it.

President Clinton held a live announcement to release the forest plan and an explanation of it, called *The Forest Plan: For a Sustainable Economy and a Sustainable Environment*, which he coauthored with Vice President Gore.[87] In President Clinton's announcement and the associated document, the plan was described as the "President's Plan." For the first time in the history of the United States, a president had put his imprimatur on a forest plan as his own.

The direct connection between the plan and the president is surprising for at least three reasons: the Record of Decision for the plan would be signed by the secretary of agriculture (since the Forest Service was in the Department of Agriculture) and the secretary of the interior (since the BLM and the Park Service were in the Department of the Interior), legally making it *their* plan, not the president's; future political blame for any plan failings would fall squarely on the president; and the close association of President Clinton with the plan could raise questions about the plan's status once his term ended. Presidential ownership, though, sent a clear message to Clinton's administration and the Pacific Northwest about how important successful implementation of this plan would be to him.

The Clinton-Gore report described the forest plan in terms of the three parts: forest management, economic assistance, and interagency coordination. It generally followed the structure and content of the Executive Committee's memo, while also putting the best, most upbeat spin on what had been accomplished.

The report highlighted that the forest plan would protect watersheds and the most ecologically valuable old-growth forests, ensuring sufficient habitat for the spotted owl, marbled murrelet, and threatened fish populations along with scores of other species. Adaptative Management Areas were cited as an innovation that would enable community participation in developing novel forest management approaches. The president's commitment to achieving a forest plan based on sound science and within existing law rang clearly through in both his speech and the associated white paper. Despite the disappointing timber harvest level, there would be no request by the Clinton administration to loosen federal environmental laws.

The administration committed to providing $270 million in new funding for job assistance in fiscal year 1994 and $1.2 billion over five years, estimating that the resulting assistance would add employment opportunities exceeding the 6,000 jobs directly impacted by the decision.[88]

The white paper highlighted that interagency cooperation in the development of the President's Plan had created a new model for federal forest planning, citing both the joint Forest Service and BLM development of a forest plan and integration of regulatory agencies (US Fish and Wildlife Service, National Marine Fisheries Service, and Environmental Protection Agency)[89] into the process. The white paper called for a continuation of an interagency approach to forest planning into the future, based on watersheds and physiographic provinces, that keyed management to the unique ecology of each region.[90] If implemented, this approach would revolutionize federal forest planning, which had traditionally been done by individual national forests and BLM districts.

A Stormy Reception

Upon its release, the President's Plan was castigated in apocalyptic terms by all sides.

The AFL-CIO—the umbrella organization of most labor unions in the country, with over 14 million members—sponsored a press conference immediately after the plan announcement, at which representatives of home builders, woods workers, and sawmills dependent on federal timber spoke. Jay Power of the AFL-CIO led off and expressed the "extreme disappointment of the AFL-CIO with the President's Plan." He said it would "hit the Pacific Northwest like a battery of misguided cruise missiles . . . and force tens of thousands of forest products workers into unemployment." Also, he charged that the plan would force many workers in related fields, such as transportation and construction, out of their jobs and would deeply harm counties dependent on in-lieu payments from federal timber sales for many of

their services such as roads, schools, public health, police, and libraries. Other speakers at the conference echoed Power's denunciation of
the plan.[91]

Many people in timber-dependent communities were devastated,
fully realizing that their livelihoods and communities would bear the
costs of protecting species and ecosystems. Previously, the Forest Service had assured them that the federal harvests were sustainable. Yes,
there would be ups and downs as the economy fluctuated, but people in
timber-dependent communities could count on the agency offering the
same harvest volume for sale over time. They had settled and bought
homes in these communities with the promise of sustained yield as
the bedrock of their economic future. Almost overnight, that security
was gone. In its place was the fragile promise of new jobs, provided
that displaced workers were willing and able to retrain and perhaps
relocate, and provided that the Clinton administration and Congress
supplied funding.[92]

Northwest tribal nations were deeply unhappy with FEMAT's process and results. FEMAT's social science group had documented that,
as specified in the treaties they signed to give up their land, tribes retain
sovereignty rights to fishing, hunting, and gathering over large areas of
the owl's range, because such lands were part of their ancestral lands.
Further, they noted that "these rights have been interpreted through
case law to have precedence over subsequent resource uses and must
be accommodated by agencies. . . . Tribes must be consulted on a
government-to-government basis regarding policy development, . . .
planning design, and project formulation."[93] Yet tribal nations within
the range of the northern spotted owl were not involved in development of the President's Plan. Further, Option 9 did not provide a high
level of protection for salmon, a resource of vital interest to them.

Shortly after Jack Ward Thomas submitted FEMAT's report to
McGinty, representatives of Northwest tribal nations met with Pipkin,
Thomas, and Johnson to hear about the FEMAT effort and an outline
of the results. Joe DeLaCruz, chairman of the Quinault Tribe, immediately raised his hand at the start of the meeting and asked how many
Bureau of Indian Affairs or tribal professionals and scientists were part
of FEMAT. The three FEMAT representatives looked at each other
and realized that none had been asked to take part. Upon hearing that

response, Chairman DeLaCruz queried, "Then why should we believe you?" and the meeting went downhill from there.

Ted Strong, executive director of the Columbia Intertribal Fish Commission and beloved tribal leader, somewhat later provided a biting critique of Option 9 that was broadly representative of tribal perspectives:

> The FEMAT dealt with the issue of anadromous fish protection entirely in the context of the likelihood of a given option providing viable populations over the next 100 years. The FEMAT admitted that it did not design an option to achieve the most biologically sound viability levels. At no point did the FEMAT address how much harvest of fish, if any, would be possible under any of the options. At no point did FEMAT's report describe how or whether any of the options would be consistent with the treaties of Indian tribes.[94]

After their initial blast at the plan just before its release, environmental groups took a few months to read the 1,000-page FEMAT report in depth. Then they came on like thunder. The response of Mike Anderson for The Wilderness Society reflected their views: "The result of this politically compromised process was the ecologically disastrous Option 9. The environmental weaknesses of the plan were glaring. Forest reserves omitted more than a million acres of old growth. Most riparian reserves were only half the size recommended by FEMAT scientists. And, most disturbing, the scientists estimated that another 400-plus species, ranging from salmon and steelhead trout to lichens and salamanders, would be at risk of extinction."[95]

Congress grumbled but took no action except to hold a hearing in the House of Representatives by all three committees that had jurisdiction (Agriculture, Interior, and the Merchant Marine Subcommittee of the Commerce Committee). Committee members from the rural portions of the owl's range dominated that hearing and directed most of their anger at FEMAT's economic analysis. They were outraged that the estimated 1992 timber industry employment, which reflected legal injunctions on federal sales, had been used as one of the bases for estimating employment loss, allowing the Clinton administration to claim that the forest plan would result in a loss of only 6,000 jobs. They were also upset that the loss of indirect and induced employment in communities attributable to the plan was not included in FEMAT's analysis. Congressman Peter DeFazio, whose southwestern Oregon district would suffer the largest employment losses, was especially upset. He declared, "We asked for credible assumptions and instead you gave us incredible assumptions."

Much of the hearing was devoted to questioning Greber, leader of FEMAT's economics group. The congressional representatives wanted him to project the expected declines in employment associated with implementing Option 9 when he used the average harvest level of the 1980s as the baseline and the estimate from a Forest Service study (cited in the FEMAT report) of 16.5 total jobs per million board feet of harvest, which included direct, indirect, and induced economic effects.[96] Greber, after a few false starts, estimated the loss at over 60,000 jobs, which provoked further shouts of anguish and dismay.

The congressional outrage directed at the job loss estimate that the Clinton administration had brought forth might be dismissed as political grandstanding, but it also revealed why it was so hard to reach agreement on the job losses attributable to the President's Plan. A job loss estimate was highly sensitive to the reference period or employment level for estimating the loss and the breadth of employment effects included, as discussed in the last chapter. Employment effects associated with the reduction in federal harvests could range from about 60,000 jobs lost (using the harvest of the previous decade as the reference and including direct, indirect, and induced employment effects) to at least 10,000 jobs gained (using the likely harvest if the Dwyer injunction on timber sales in owl habitat had remained in place as the reference and including direct, indirect, and induced employment). The job loss estimate of 6,000, on the other hand, used the employment provided by the harvest in 1992 as the starting point and considered direct job loss only. And none of these estimates included possible employment opportunities in recreation services or forest and watershed restoration. The differences in approach to estimating employment impacts made reasoned discussion of the employment implications of the President's Plan largely impossible in Congress and almost everywhere else.

Perhaps oddly, while the congressional representatives challenged the plan's economic conclusions, no one challenged the approach taken to estimating the projected timber harvest level itself—that of estimating likely versus maximum levels—that had so riled the Clinton administration earlier in FEMAT's process. Nor did they challenge much else in the President's Plan, and the hearing ended without further pyrotechnics.

8

The President's Plan and Judge Dwyer

Development of the President's Plan was one very important step in changing the management of Northwest federal forests. Others would be needed: working through the National Environmental Policy Act (NEPA) process for federal actions, and a final decision by President Clinton. Then it would be back to Judge Dwyer's court. This chapter tells that story.

The next step in adopting the President's Plan for management of federal forests within the northern spotted owl's range was to fulfill NEPA's requirement for "major federal actions significantly affecting the environment." A Draft Environmental Impact Statement (DEIS) would be needed that described a preferred alternative and its potential environmental effects, as well as an analysis of other alternatives. The completed DEIS would then undergo public review and comment, and the preferred alternative would be revised and published in a Final Environmental Impact Statement and Record of Decision.

The Clinton administration's top priority in this effort was to craft a plan that would meet Judge Dwyer's legal requirements and avoid the repeated rejections that the Bush administration had incurred. While the NEPA documents displayed the options developed by FEMAT as the alternatives, Clinton had selected Option 9 and endorsed it as *his* plan, so there was never much doubt as to which one would be chosen as the preferred alternative in the NEPA process.

Producing a DEIS for a major action like a plan for federal forests in the owl's range could take years, but this one was released only a short time after announcement of the president's choice of Option 9 on July 1, 1993. This speed was possible largely because the team charged with preparing the DEIS drew directly on FEMAT's report for technical details and alternatives and began writing the DEIS before the FEMAT report was released.[1]

The public responded to the DEIS with over 100,000 comments. Most were extremely critical of the President's Plan and generally followed familiar story lines. Some critics predicted that the plan would result in ecological disaster for species and ecosystems, while others countered that it would result in an economic and social catastrophe for workers and rural communities.

More Species Protections Needed

While the Clinton administration did not back away from Option 9 based on the public review, the harsh criticisms highlighted the need to carefully review it for signs of legal vulnerability.[2] An interagency legal team, which included representatives from the Departments of Agriculture, Interior, and Justice and the Environmental Protection Agency, was created by the Clinton administration to assist the administration with legal interpretations and positions. Intense criticism over the apparent lack of protection for several hundred species quickly captured the legal team's attention. They were particularly concerned about comments echoing Mike Anderson's statement that under Option 9, "the scientists estimated that another 400-plus species, ranging from salmon and steelhead trout to lichens and salamanders, would be at risk of extinction."[3]

Lois Schiffer, assistant secretary of natural resources in the Department of Justice, had become an increasingly important advocate for a high species-viability standard to increase the probability that Judge Dwyer would find the outcome acceptable. She had been with the Clinton administration since it came into office. At each step of the way, she had argued for the highest standard feasible, and she did so here again with great effect.[4]

The Forest Service created a new science team, the Species Assessment Team (SAT II), to assist the legal team on species viability concerns and related issues (appendix 1).[5] Dick Holthausen of the Forest Service, who had been involved in past science assessments and spotted owl recovery planning, led SAT II. The team included scientific experts on the species and taxonomic groups that had been given relatively low viability ratings under Option 9—salmon, red tree voles, salamanders, bats, invertebrates, fungi, lichens, and other plants and animals. In addition, it included experts on habitat needs of the northern spotted owl and marbled murrelet, because the US Fish and Wildlife Service (USFWS) had raised concerns in its review of Option 9 about their level of protection. SAT II was limited to federal employees, unlike FEMAT, which included professors from regional universities, among others.[6]

In FEMAT's analysis, viability panels had been asked to determine the likelihood of achieving four different outcomes on federal land for species associated with late-successional/old-growth (LS/OG) forests:

- Outcome A—Habitat managed to maintain viable populations well distributed across the planning area;
- Outcome B—Habitat managed to maintain viable populations with gaps in distribution;
- Outcome C—Populations relegated to refugia; and
- Outcome D—Extirpation (local extinction) likely.

The viability standard, as laid out in the 1982 Planning Rule to implement the National Forest Management Act, required Outcome A for vertebrate species.[7] FEMAT scientists had argued that providing habitat for species sufficient to achieve an 80 percent likelihood of Outcome A would meet that requirement.

The interagency legal team asked SAT II to suggest mitigations that would provide sufficient habitat on federal lands for vertebrates to achieve two goals: bring all vertebrates to at least an 80 percent likelihood of Outcome A, and eliminate any chance of extirpation (Outcome D).[8] In other words, up to a 20 percent likelihood of Outcome B or C was allowed but no likelihood of Outcome D, adding an additional criterion for adequacy of protection beyond that employed by FEMAT. These became known as the viability screens for vertebrates. Meeting the two requests meant reexamining protective measures in Option 9 for salmon populations, several species of salamanders and bats, the red tree vole (an important prey species of the northern spotted owl), and numerous bird species, including the northern spotted owl and marbled murrelet.[9]

Viability screens were also developed for nonvertebrate species, even though the 1982 Planning Rule for the National Forest

Management Act (NFMA) contained no specific requirement regarding these species. The statutory language in NFMA, however, did call for "maintaining plant and animal community diversity to meet multiple use-objectives,"[10] and potential viability problems for many nonvertebrate species had been identified in FEMAT's report under Option 9. To satisfy the intent of the diversity clause, the legal team asked SAT II to suggest mitigations that would provide sufficient habitat on federal lands for nonvertebrates to achieve two goals: increase the projected viability of nonvertebrate species to at least 80 percent of Outcomes A + B (habitat managed to maintain viable populations with gaps in distribution permitted), and eliminate any chance of extirpation (Outcome D).[11] These became known as the viability screens for nonvertebrates. SAT II also addressed mitigation measures for species whose habitats were largely unknown and identified species whose extirpation risk was strongly influenced by actions on nonfederal lands.

The task before SAT II was challenging, with so many species, habitats, and situations to evaluate. In total, SAT II identified 490 species, species groups, or populations that were inadequately protected under Option 9, most of which were not vertebrates (table 8.1). Almost 45 percent of all the species identified in FEMAT as being closely associated with old-growth forests were represented in this list. In other words, SAT II quickly concluded that Option 9 provided sufficient protection for approximately 55 percent of the species associated with LS/OG forests, and more analysis of their protection was not needed.

This list included the three species that were the focus of the landscape design in Option 9—the spotted owl, marbled murrelet, and salmon. In the case of the spotted owl and marbled murrelet, the USFWS's draft biological opinion expressed doubts about whether Option 9 would avoid jeopardizing their survival and assist in their recovery.[12] In the case of salmon populations, FEMAT had estimated that Option 9 would provide habitat with a 65 percent likelihood of achieving Outcome A—less than the desired 80 percent likelihood.

SAT II developed an extensive list of potential mitigation measures (table 8.2) to help decide how Option 9 should be augmented. These mitigation measures were counted on, either individually or in different combinations, to enable each of the 490 species (or species groups) to pass the screens previously described if it were possible to do so

Table 8.1. Number of species (or species groups) reviewed in FEMAT that were given additional analysis in the refinement of Option 9 to create the Northwest Forest Plan.

Species/species group	Number assessed in FEMAT	Number needing additional analysis
Vertebrates		
Amphibians	18	12
Fish (races, species, groups)	7	7
Northern spotted owls	1	1
Marbled murrelets	1	1
Other birds	36	2
Bats	11	7
Other mammals	15	3
Nonvertebrates		
Bryophytes	106	9
Fungi	527	255
Lichens	157	75
Vascular plants	124	17
Arthropods (groups or ranges)	15	4
Mollusks	102	97
Total	**1,120**	**490**

Source: USDA and USDI, "Final Supplemental Environmental Impact Statement," appendix J2, 4.

through federal forest management. Some potential mitigation measures increased the area in reserves, some proposed landscape controls on forest management activities, others proposed modifications in standards for forest management in Matrix/Adaptive Management Area (AMA), and a number of measures fell under the general category of "Survey and Manage" requirements.

SAT II scientists faced a complicated, multifaceted viability problem that had to be addressed by a collective effort, since changes benefiting one species could also potentially benefit another species. As an example, widening stream buffers or entirely reserving Key Watersheds would not only increase salmon viability but also benefit many other species. SAT II developed a mitigation table, with the help of numerous outside experts, to identify which mitigation measures contributed most fully to moving the species of concern across the viability thresholds.

Table 8.2. Additional mitigation measures considered by SAT II.

Survey and Manage
1. *Survey to acquire additional information on species particularly poorly known and difficult to detect*
2. *Survey and protect sites of species where found*
3. *Protect known locations of the species*
Riparian Reserve[1]
4. Apply Riparian Reserve Scenario 1 where particular species found
5. *Apply Riparian Reserve Scenario 1 throughout the range of the northern spotted owl*
6. Apply Riparian Reserve Scenario 1 throughout the range of the coho salmon
7. *Ensure that riparian management in Adaptive Management Areas equals that in Riparian Reserve Scenario 1*
8. Provide additional buffers around small wetlands
Watershed Protection
9. Remove all lands within Tier I Key Watersheds from the suitable timber base
10. Build no new roads in Tier 1 Key Watersheds
Matrix/AMA Management
11. *Provide well-distributed coarse woody debris*
12. *Emphasize clumped green tree retention*
13. *Provide additional no-harvest buffers around cave entrances*
14. *Modify site treatment practices to minimize soil and litter disturbance*
15. Provide dispersal corridors for red tree voles
16. Apply the 50-11-40 rule to assist spotted owl dispersal
Other Measures
17. Remove inventoried roadless areas from the suitable timber base
18. Retain all old-growth forests in Marbled Murrelet Zone
19. *Retain old-growth fragments where little old growth remains*
20. *Retain 100 acres of habitat around each known spotted owl activity center (in Matrix/AMA)*
21. *Protect specific sites from grazing*
22. *Limit impacts in recreation areas on certain fungi and lichen*
23. *Identify species-specific measures*

Source: USDA and USDI, "Final Supplemental Environmental Impact Statement," appendix J2.

Note: Measures incorporated into Option 9 to create the preferred alternative in the Final Environmental Impact Statement for the NWFP are in *italics*.

[1] See explanation of Riparian Reserve scenarios in chapter 7.

Filling out those tables was much more than a technical exercise, though—it involved discussion and negotiation among SAT II members and associates as they attempted to find the changes that had the most bang (in terms of species assisted) for the buck (in terms of reduced timber harvest). Once again, ecologists were engaged in an efficiency analysis as they consciously searched for mitigations with multiple benefits that would not break the timber bank. As an example, they quickly realized that the widening of Riparian Reserves on non-fish-bearing streams was one such mitigation (measure 5 in table 8.2). First, it raised the likelihood of providing sufficient habitat on federal land for salmon populations up to at least 80 percent of Outcome A, countering criticism of the lower level in Option 9. Second, it increased dispersal habitat for the northern spotted owl, which pleased owl scientists on SAT II who had argued strongly for reinstating the 50-11-40 rule in Matrix/AMA for that purpose. Third, it bolstered protection for a wide variety of other species. Utilizing the harvest scheduling analysis done by Johnson and colleagues in *Sustainable Harvest Levels and Short-Term Timber Sales*, a supporting document of FEMAT, they could see that the cost of the mitigation was relatively modest for all it accomplished—just what the team was looking for.[13]

SAT II provided its mitigation tables to the people developing the Final Environmental Impact Statement (FEIS), along with lengthy explanations of how combinations of mitigations would benefit different species. Ultimately, the FEIS team incorporated 14 of the 23 potential mitigation measures devised by the SAT II scientists into the preferred alternative (Option 9) of the FEIS (table 8.2). These measures included expanding buffers on non-fish-bearing streams, protecting habitat around northern spotted owl nests in Matrix/AMA, and protecting old-growth fragments in watersheds where little old growth remained.

The mitigations adopted in the FEIS shifted approximately 700,000 acres from Matrix/AMA allocations to various reserve allocations (3 percent of federal lands and 16 percent of Matix/AMA in the range of the spotted owl). Most of this shift came from expanding riparian stream buffers on non-fish-bearing streams. Most of the rest came from two other adjustments: locating small (100-acre) habitat reserves around each known owl activity center on Matrix/AMA lands, and moving 100,000 acres from Matrix/AMA allocations to a category called Managed Late-Successional Reserves (LSR) in eastern Washington and northern California to fill in gaps in the LSR network for better owl protection (a mitigation not listed in table 8.2). With these mitigations in place, the long (180-year) rotations originally prescribed for the California national forests in Option 9 were eliminated as unnecessary, and rotations in existing individual forest plans were used instead.[14]

Also, many species were assigned to one or more Survey and Manage categories:

- protect known sites;
- survey project areas for the presence of certain species before undertaking action and protect them if found;
- conduct extensive surveys for species of concern and protect them where found; and
- conduct general regional surveys for species of concern to increase knowledge about their location and habitats.[15]

Altogether, almost 400 species (mostly nonvertebrates) were assigned Survey and Manage requirements.

The second Survey and Manage category—survey project sites before undertaking actions and apply protective measures—was particularly concerning, as its implementation would directly affect the timber sale program. This category of Survey and Manage was required for 74 species: 6 vertebrates (red tree vole, Del Norte salamander, Larch Mountain salamander, Shasta salamander, Siskiyou Mountains salamander, Van Dyke salamander), and 68 nonvertebrates (50 mollusks, 2 fungi, 3 lichens, 5 bryophytes, and 8 vascular plants).

In a last crucial step, SAT II and the FEIS team met in March 1994 with administration officials to describe these additional mitigations, which had been incorporated into the FEIS, and receive final approval to move ahead. Katie McGinty, Lois Schiffer, and Jack Ward Thomas (now chief of the Forest Service) were among those at the briefing. Thomas strenuously objected to including the Survey and Manage mitigations, arguing that they were not needed and would wreck the timber program. However, McGinty and Schiffer were concerned that a plan that resulted in poor viability ratings for hundreds of species would fail to receive Judge

CHANGING AN ECOSYSTEM PLAN TO A SPECIES-BASED PLAN
(Jerry Franklin)

The decision to adopt the Survey and Manage mitigations in Matrix/AMA land allocations for a wide variety of species greatly troubled Jack and me. We were not involved in developing or evaluating the mitigation options: I had returned to the university and Jack had become chief of the Forest Service.

The species at issue were not listed as threatened or endangered under the Endangered Species Act; rather, they were primarily species about which little was known, such as their geographic distribution and habitats, except that they had been observed in LS/OG forests. In our eyes, reserving 80 percent of the remaining LS/OG forest on federal forests had a high probability of adequately protecting these species; special treatment should await evidence that they were at risk. If inadequate information about an array of species was the real problem, the correct course would be for agencies to obtain such knowledge with surveys starting in the reserved lands, such as LSRs and Riparian Reserves. If the reserves were shown to be inadequate after such studies, further mitigation measures could be placed on Matrix/AMA lands.

Most fundamentally, we objected that the robust coarse filter approach of Option 9, which emphasized specifying macrohabitat conditions such as vegetation cover types and their successional stages, was being rendered ineffective by an avalanche of species-based fine filters.[*] As Thomas famously declared, "Option Nine was manipulated and morphed into Option One" (Option 1 being the FEMAT option reserving all LS/OG forest).[†]

[*] J. Franklin, "Preserving Biodiversity."
[†] Steen, "Interview with Jack Ward Thomas," 18.

Dwyer's approval. In addition, the FEIS would have to be rewritten, which would delay the process for many months. In the end, McGinty decided to include the Survey and Manage mitigations in the plan that she would submit for presidential approval.[16] Franklin also disagreed with including the Survey and Manage mitigations when he later heard about it. Both he and Thomas believed that this change fundamentally altered the basic nature of Option 9.

The interagency NEPA team developing the FEIS projected that the additional mitigation measures would reduce the timber harvest level under Option 9 by approximately 10 percent (100 million board feet) to 1.1 billion board feet, including "other wood." Thus, the FEIS substantially improved the viability of nearly 400 species with only a 10 percent reduction in harvest—quite an accomplishment.[17]

However, a primary cost for improving viability scores was left out of the timber harvest calculations: SAT II could not estimate harvest impacts of the surveys and protections for species under the Survey and Manage provisions. The second category of mitigations (survey project sites before undertaking actions and apply protective measures) had the greatest potential for reducing the timber harvest level estimated in the FEIS, as buffers would probably be needed if surveys found a category 2 species in a project area. SAT II scientists, though, did not know how many individuals of these species would be found, and the amount of habitat protection that would be needed for species (if found) had not yet been determined. Thus, they could not estimate the harvest impact. The illusion that the estimated timber sale level of 1.1 billion board feet included Survey and Manage effects would haunt plan implementation.

With the mitigation issues resolved, the NEPA team worked with the Clinton administration to write the Record of Decision (ROD), which would describe, explain, and justify the decision. NEPA specialists and legal staff from the Forest Service and BLM wrote the first draft of the ROD. However, Forest Service staff had difficulty reconciling the ecosystem approach taken by the President's Plan with the Forest Service's 1897 Organic Act and the Multiple-Use Sustained-Yield Act— two foundational laws for directing Forest Service management of the national forests. Therefore, Katie McGinty and Tom Collier asked Jim Pipkin to reframe this foundational portion of the ROD, which led to the addition of the following statement:

> The President charged us to use an ecosystem management approach. In this vein, we defined the planning area for this strategy as the federally administered lands within the range of the northern spotted owl. We asked our scientists to assess not only effects on individual species of each of the alternatives, but also the likelihood that the alternatives would provide for a functional and interconnected old-growth forest ecosystem. We involved all the relevant federal agencies at an early point in the

planning process and asked them to coordinate their efforts to the extent possible.

Such an approach proceeds from our statutory authority. . . . These statutes invest in our Departments broad discretion to rely upon our expertise to manage the lands under our administrative authority in a manner deemed to best meet the purposes Congress has delineated. One such purpose is to provide for the long-term sustainability of all of the forests' many natural resources, including the species that inhabit them. Through its utilization of ecosystem management principles, our decision is designed to meet this purpose more effectively and efficiently than previous planning efforts. . . . Thus, our decision, while emergent in some important respects, is nevertheless firmly grounded in the authority reflected in our statutory schemes as well as the best federal land management agency tradition of crafting approaches that meld dynamic concepts with the legal duties under which we are charged to carry out our stewardship responsibilities.[18]

When that statement was challenged in subsequent litigation, the courts affirmed the agency's broad discretion to adopt an ecosystem approach.

Framing the Final Presidential Decision

The two cabinet secretaries responsible for implementing the President's Plan formally authorized it by signing the completed ROD. While President Clinton had endorsed Option 9 as his plan on July 1, 1993, the responsible officials were actually the secretary of agriculture (since the Forest Service was in the Department of Agriculture) and the secretary of the interior (since the BLM and the Park Service were in the Department of the Interior). In this case, though, the president retained the ultimate decision of whether to accept the revised Option 9 and seek legal affirmation from the courts. Toward that end, McGinty and her staff prepared a decision memo for President Clinton. While the memo was finished on April 7, 1994, the president did not see it until April 14, just in time to meet the schedule his administration had set to file the plan with Judge Dwyer's court.

McGinty's memo makes several important points. The President's Plan would now be named the Northwest Forest Plan (NWFP).[19] This change might help it endure beyond the president's term. Also, McGinty expected immediate litigation over the NWFP in Judge Dwyer's court and was uncertain how the judge would rule. Further, McGinty realized that the NWFP satisfied none of the warring parties (environmentalists, timber industry, and labor), Congress was not interested in stepping in, and the governors were keeping their distance. Except for a few supportive congressional representatives, the Clinton administration was alone in advocating adoption of the new plan.[20]

McGinty described the major changes in the plan and the resulting timber sale impacts, noting that these changes "also included a 'survey and manage' commitment for those species and habitats for which little information is known. The effect of these changes [as estimated by the interagency team in the FEIS] was to reduce the average annual timber sale quantity from 1.2 billion board feet to 1.1 bbf." With the strong adverse reaction from Jack Ward Thomas over including Survey and Manage requirements undoubtedly in the back of her mind, she added: "Arguments are now occurring over whether the 1.1 figure is accurate or too optimistic. At this juncture, we are maintaining that the estimate is the best possible given the information we now have."[21]

McGinty's memo also reported that less harvest would likely occur on nonfederal land in the NWFP area in the next decade than had been estimated by FEMAT, which increased expected job loss in the forest industry by 4,000 workers.[22] However, McGinty argued that the NWFP itself would not *cost* jobs and a gave a number of reasons for this conclusion. Most timber industry job declines were the result of events that preceded the Clinton administration, while the NWFP would allow modest levels of timber sales to resume by enabling injunctions to be lifted on harvest in owl habitat. Moreover, the "unprecedented $1.2 billion in community assistance and worker retraining" would create thousands of employment and training opportunities and new jobs.[23]

McGinty described the NWFP as "a genuine watershed event in the evolution of land management policies in this country . . . [and] when the dust settles it will serve as a positive and constructive force for change in the region."[24] She also expressed a major concern with "the complexity of the plan itself and our ability to implement it promptly and thoroughly." Therefore, McGinty "intended to oversee the implementation of the plan closely."[25]

Finally, McGinty noted that "the key to our success from a political and substantive standpoint is to proceed in a steady and predictable way with plan implementation, all the while continuing our significant efforts at assisting those communities who are very much in the throes of major transformations."[26] Clearly, President Clinton and McGinty knew that change was coming to the timber-dependent communities and timber workers within the area of the NWFP. While they would try to ease the economic and social burden of the changes, they knew they could not stop them.

In her memo, McGinty noted that agency estimates about how much timber they could deliver over the first five years were discouraging. The Forest Service projected that the 1995 volume would be half of the projected 1.1, and BLM made no projections at all.[27] The agencies were well aware that they would be harvesting timber under a vastly different set of rules than in the past, including Survey and Manage, and that they had to be cautious in what they promised the president.

Seeing these agency estimates made the president wonder which numbers he should believe. Intensely frustrated at this outcome of their long journey, he wrote a number of questions and comments on the memo such as: "I find it amazing that we cannot assure the amount of harvesting at 1.1 bbf after all this work."[28] He seemed to sense that achieving his goal of a stable (if modest) timber supply of about one billion board feet might be slipping away.

McGinty's memo to President Clinton contained a one-page foreword, written a week after the memo, from R. Paul Richard, who served as deputy staff secretary. The deputy staff secretary is the conduit to the president, ensuring that presidential documents reflect a consensus of opinion of the president's advisers. Richard first provided more detail on the Forest Service's estimate of short-term harvest levels. He then reported: "The memo has been reviewed by the Office of Management and Budget, National Economic Council, Domestic Policy, and other relevant members of senior staff. There was consensus that while the overall plan is sound, we should do all that we can to minimize the fallout from possible heightened expectations [of short-term harvest levels]." With such overwhelming staff endorsement, the president did approve the NWFP a few days later after more discussions with McGinty. It could now be sent to Judge Dwyer for his approval.[29]

The Northwest Forest Plan

The NWFP recognized six major land allocation categories in the 24 million acres of federal lands in the range of the northern spotted owl (fig. 8.1, plate 8.1). Late-Successional Reserves (LSRs) partially overlapped with Administratively Withdrawn areas and Riparian Reserves. Here, overlap with Administratively Withdrawn areas is shown as part of the LSR allocation to indicate that LSR standards take precedence where they are more restrictive.[30] Also, overlap with Riparian Reserves is shown as LSR; in those areas both sets of standards apply. Compared to Option 9, LSR area increased a few percentage points and Riparian Reserve area increased almost 20 percent.

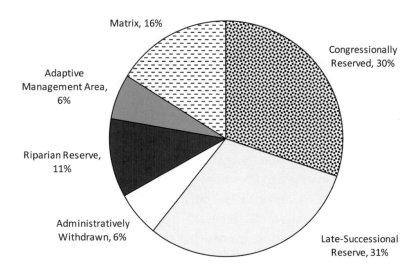

Figure 8.1. Land allocations of the Northwest Forest Plan. Congressionally Reserved includes wilderness, national parks, and other reserves. LSR (gross area) includes overlap with Administratively Withdrawn areas and Riparian Reserve. Riparian Reserve area within Matrix and AMA is displayed. Thus, the Matrix and AMA area is the net of embedded Riparian Reserve. (Source: USDA and USDI, "Record of Decision.")

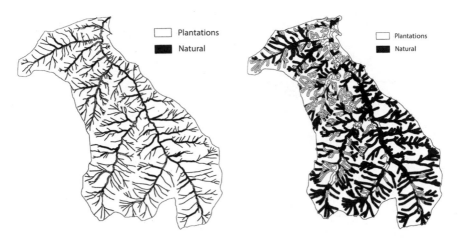

Figure 8.2. (*left*) Riparian buffers in the Willamette National Forest Plan and (*right*) Riparian Reserves in the Northwest Forest Plan in the Augusta Creek watershed of the McKenzie River drainage in western Oregon. Plantations (open canopy) that developed after past clearcutting covered much of the area designated as Riparian Reserves, especially in the northern portion of the watershed. (Source: FEMAT, *Forest Ecosystem Management*, 5-40, 5-43.)

The areas unclaimed by the other land allocations were designated Matrix or AMA, and these areas were the basis for calculation of the regularly scheduled timber harvest (the probable sale quantity). The ROD emphasized that such harvest would be done "only in compliance with standards and guidelines designed to achieve conservation objectives."[31]

Doubling the Riparian Reserves on non-fish-bearing streams, as recommended in SAT II, completed the transformation of the stream buffer system on federal lands from a minor portion of the landscape to one of its defining features (fig. 8.2)—quite an accomplishment for a strategy that had begun in the Gang of Four effort only three short years before.

The NWFP recognized two types of Key Watersheds (fig. 8.3). Tier 1 Key Watersheds, covering 8.1 million acres, were to serve as refugia for aquatic organisms or had high potential for restoration. They had a primary objective of aiding the recovery of fish that had been listed or would likely soon be listed as threatened or endangered under the Endangered Species Act (ESA). Tier 2 Key Watersheds, covering 1.1 million acres, were to provide sources of high-quality water. All other watersheds were designated as nonkey watersheds. Watershed assessments were required in Key Watersheds before activities could be undertaken, and these watersheds had priority for restoration.

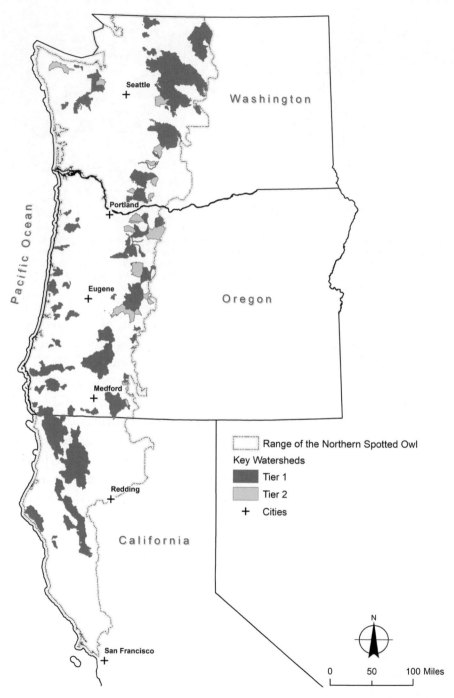

Figure 8.3. Key Watersheds designated in the Northwest Forest Plan. (Source: US Forest Service Regional Ecosystem Office, Northwest Forest Plan Document Library.)

The ROD for the NWFP included a separate document of standards and guidelines, which contained extensive requirements governing timber harvest and other activities. Some key provisions are covered here. See appendix 4 for details.

Late-Successional Reserves and Riparian Reserves were not reserves in the classic sense of prohibiting all silvicultural activities and associated timber harvest. Both LSRs and Riparian Reserves were a mixture of LS/OG forest and younger forest. Most of the younger forest had resulted from clearcutting of LS/OG forest followed by replanting of commercial tree species. The resulting stands were primarily plantation-like tree farms with little of the compositional and structural diversity found in the natural forest. The ROD allowed thinning of young forests in both categories of reserves where these actions would advance reserve objectives, because "current stocking and structure of some of these stands were established to produce high yields of timber, not to provide for old-growth-like forests. Consequently, silviculture can accelerate the development of young stands into multilayered stands with large trees and diverse plant species, and structures that may, in turn, maintain or enhance species diversity."[32]

The ROD also allowed limited salvage logging in LSRs and Riparian Reserves after wildfire and other stand-replacing disturbances, as long as negative effects on the protection and enhancement of LS/OG ecosystems could be minimized and the objectives of these land-use allocations still achieved. This allowance of salvage logging proved to be one of the most controversial provisions of the NWFP and was subject to repeated litigation.

One innovation in Option 9 was the creation of Adaptive Management Areas (AMAs), where, as described in the ROD, landscape units were

designed to develop and test new management approaches to integrate and achieve ecological, economic, and other social and community objectives. The Forest Service and BLM will work with other organizations, government entities and private landowners in accomplishing those objectives. Each area has a different emphasis to its prescription, such as maximizing the number of late-successional forests, improving riparian conditions through silvicultural treatments, and maintaining a predictable flow of harvestable timber and other forest products.[33]

Most AMAs were "associated with sub regions impacted socially and economically by reduced timber harvest from the federal lands."[34] FEMAT had emphasized that it was important for projects in the AMAs to start quickly for the areas to meet their promise and warned that "it is absolutely critical that initiation of activities not be delayed by requirements for comprehensive plans or consensus documents beyond those required to meet existing legal requirements."[35] The ROD also acknowledged the need to get started quickly but added the goal of encouraging local community involvement in deciding what to do and how to do it.[36] This approach reflected President Clinton's interest, expressed at the Forest Conference, in collaborative planning as a mechanism to reach agreement on the projects to undertake on federal forests. The Applegate Partnership and Watershed Council on Forest Service and BLM lands in southwestern Oregon, which became the Applegate AMA, provided a shining example of how this goal might be achieved.

Further, the ROD and subsequent clarifications included statements to ensure that AMAs and the Riparian Reserves within them would be given protection equivalent to that of the rest of the federal forests.[37] The hope that looser rules would be allowed in AMAs to enable experimentation was negated, in part, by these instructions.

Judge Dwyer Rules the Plan Meets the Law

Given the litigious environment that led to the development of the NWFP, it is not surprising that the litigants who had forced its promulgation would test its validity. Both environmental groups (led by Seattle Audubon) and timber industry groups (led by the Northwest Forest Resource Council [NFRC], a trade association of firms dependent on federal timber) immediately challenged the nascent NWFP upon its release in early 1994 under the auspices of an ongoing case, *Seattle Audubon Society v. Lyons*, and under two new cases, *Northwest Forest Resource Council v. Thomas* and *Northwest Forest Resource Council v. Dombeck*. After integrating the claims in the new cases into *Lyons* and considering their merits, Judge William Dwyer finally provided the legal certainty that the agencies and people of the Pacific Northwest had been seeking since the late 1980s.[38]

Judge Dwyer evaluated dozens of claims separately brought against the NWFP in these cases. In an opinion that is quite brief given the number and complexity of the claims presented, he dismissed all challenges to the NWFP and upheld the plan as consistent with the ESA, NEPA, and NFMA,[39] while prefacing his ruling with laudatory observations as well as cautions:

> For the reasons given below, the court finds that the federal defendants have acted within the lawful scope of their discretion in adopting the 1994 forest plan. The question is not whether the court would write the same plan, but whether the agencies have acted within the bounds of the law. On the present record, the answer to that question is yes.
>
> The order now entered, if upheld on appeal, will mark the first time in several years that the owl-habitat forests will be managed by the responsible agencies under a plan found lawful by the courts. It will also mark the first time that the Forest Service and BLM have worked together to preserve ecosystems common to their jurisdictions.
>
> The Secretaries have noted, however, that the plan "will provide the highest sustainable timber levels from Forest Service and BLM lands of all action alternatives that are likely to satisfy the requirements of existing statutes and policies." ROD at 61. In other words, any more logging sales than the plan contemplates would probably violate the laws. Whether the plan and its implementation will remain legal will depend on future events and conditions.[40]

Four categories of claims by the litigants[41] have special relevance to the approach taken in the NWFP to conservation and management of federal forests: lack of authority to adopt an ecosystem plan; misuse of the viability standard; violation of NEPA; and lack of recognition of the special mandate for BLM lands.

The NFRC claimed that the agencies lacked the authority to adopt an interagency plan on an ecosystem basis, in that such an approach would violate the multiple use and sustained yield principles of the Multiple-Use Sustained-Yield Act and NFMA. Judge Dwyer noted that "conservation on this [ecosystem] basis is entirely consistent with those principles. The plan designates millions of acres for programmed logging. . . . Congress's mandate for multiple uses, including both logging and wildlife preservation, can be fulfilled if the remaining old growth is left standing, but not if logged to the point where native vertebrate species are extirpated."[42] Further, he explained, "Given the current condition of the forests, there is no way the agencies could comply with the environmental laws *without* planning on an ecosystem basis."[43] Finally, he noted that "the courts have repeatedly encouraged the Forest Service, the BLM, and FWS to turn from disparate strategies for managing LS/OG forests to a cooperative approach" and that "the Forest Service will doubtless have to coordinate its planning efforts with other agencies such as the Fish and Wildlife Service and the Bureau of Land Management."[44] Thus, Judge Dwyer confirmed that interagency planning on an ecosystem basis was not only consistent with the law but absolutely necessary. For the first time, a court had passed judgment on the ecosystem approach, and it had passed with flying colors.

Given the central role of the viability standard in development of the NWFP, it is not surprising that Judge Dwyer had to address numerous claims contesting the validity and use of that rule. He rejected them all.

The NFRC argued that the viability standard in the 1982 Planning Rule interpreting NFMA was itself fundamentally inconsistent with the law. Judge Dwyer noted that NFMA directed the secretary of agriculture to issue regulations to provide for diversity of plant and animal communities and that the secretary then promulgated the viability standard, which was within his legitimate authority to interpret as to the meaning of the statute. Thus, Dwyer ruled that the viability standard was consistent with NFMA and with multiple-use management.[45]

Seattle Audubon challenged the adequacy of the level of protection given to a number of species. The judge noted that the ROD for the NWFP argued that the plan satisfied the requirements of the statute and its implementing regulations because it would provide for habitat to support the continued persistence of vertebrate species, with Option 9 providing an 80 percent or better likelihood of persistence for all vertebrates where it was possible to do so. Dwyer concluded that the secretaries did not act arbitrarily or capriciously in making these findings.[46]

The NFRC argued that the viability standard had been unlawfully applied to the conservation of nonvertebrates. Judge Dwyer noted that the ROD stated: "By its own terms, the regulation applies only to vertebrate species. Nevertheless, consistent with the statutory goals of providing for diversity of plant and animal communities and the long-term health of federal forests . . . [the plan] satisfies a similar standard with respect to non-vertebrate species to the extent possible."

He concluded that "the federal defendants were bound by law, and by the obvious fact of species interdependence, to consider the survival prospects of species other than vertebrates. Their chosen method of doing so was within the legitimate scope of their discretion."[47]

The NFRC also argued that the viability standard could not be applied to BLM lands. In response, the judge quoted the NWFP ROD: "Use of the regulation's goals in developing alternatives applicable to BLM lands served the important policy goals of protecting the long-term health and sustainability of all federal forests" in accordance with the BLM's governing statutes, including the Oregon and California Revested Lands Sustained Yield Management Act (O&C Act) and the Endangered Species Act. Thus, the judge held that "the Secretary of Interior had not acted contrary to law in deciding to manage the BLM forest so as to preserve viable populations of vertebrate species."[48] In other words, Judge Dwyer upheld the agency's use of its discretion to protect species and the ecosystems on which they depended.

Judge Dwyer then turned to the NEPA claims, quickly confirming that the agencies had considered a reasonable range of alternatives and potential cumulative effects from nonfederal actions. He then spent considerable time on NEPA's requirements for addressing scientific uncertainty and controversy, as well as monitoring and adaptive management, which were central to the litigants' claims.

The judge acknowledged the issue, raised by both sides, of scientific uncertainty surrounding two key aspects of the NWFP: northern spotted owl viability and the new Aquatic Conservation Strategy (ACS). He found that while there was uncertainty around whether the NWFP would provide for persistence of the owl over the long term, the agencies had acknowledged the uncertainty and addressed it in the

EIS for the plan, which was what NEPA required.[49] Similarly, while he recognized that there was uncertainty about whether the ACS would conserve aquatic species, he found that "the effectiveness of the ACS is still subject to debate among scientists. If the plan as implemented is to remain lawful, the monitoring, watershed analysis, and mitigating steps called for by the ROD will have to be faithfully carried out, and adjustments made if necessary."[50] Dwyer's determination of legality, therefore, was premised on faithful implementation and reasonable adaptation by the Forest Service and BLM.

Regarding the claim of inadequate consideration of opposing scientific views raised by both sides over protection of the northern spotted owl, Dwyer noted that reputable scientific opinion supported the secretaries' views. He then stated:

> The plan must be designed by the agencies, not the courts. The question for judicial review is whether NEPA's requirements have been met. A disagreement among scientists does not itself make agency action arbitrary or capricious, nor is the government held to a "degree of certainty that is ultimately illusory." . . . The FSEIS [Final EIS for the NWFP] does take a "hard look" at the available data and opposing opinions. There is a reasoned discussion of a myriad of factors. . . . Careful monitoring will be needed. . . . But on the present record the FSEIS adequately discloses the risks and confronts the criticisms, as required by NEPA.[51]

Seattle Audubon challenged the adequacy of the monitoring provisions. Dwyer expressed some skepticism that the agencies would be able to implement such a far-reaching plan and its attendant monitoring. However, he ultimately upheld the monitoring measures prescribed in the plan: "The plan includes monitoring for implementation, verification as to results, and validation as to the underlying assumptions. The monitoring program is described above. . . . As written it is legally sufficient. It remains, of course, to be carried out. Monitoring is central to the plan's validity. If it is not funded, or not done for any reason, the plan will have to be reconsidered."[52]

The NFRC further claimed that the NWFP's creation of extensive reserves would violate the principle of sustained yield required by the

O&C Act. Therefore, the group claimed that such an approach was prohibited on the BLM O&C lands that made up more than 80 percent of the BLM's western Oregon forests. Judge Dwyer rejected these claims, emphasizing the secretary's duty to comply at the same time with other applicable statutes such as the ESA.[53] Citing other cases, Dwyer explained that these legal decisions confirmed that the BLM must fulfill conservation duties imposed by other statutes in managing the O&C lands.

In finding the NFRC's claim unpersuasive, Judge Dwyer further explained that "the Secretary of the Interior has, and for many years has exercised, broad authority to manage the [O&C Act] lands; the BLM is steward of these lands, not merely a regulator. Management under [the O&C Act] must look not only to annual timber production but also to protecting watersheds, contributing to economic stability, and providing recreational facilities."[54] It was this discretion that the Department of the Interior exercised in creating reserves on O&C lands, and Dwyer cautioned: "If this ruling were to be reversed on appeal, the ROD would have to be reconsidered because of the loss of important [old growth] and riparian reserves."[55]

The Ninth Circuit Court of Appeals affirmed Judge Dwyer's opinion in all respects.[56] Over the course of his opinion, Judge Dwyer repeatedly noted that the newly adopted Northwest Forest Plan was legal as written, but that future events—implementation—would dictate whether the plan remained legal and viable. Significant ecological reserves, surveys, monitoring, mitigation, and adaptive management were all required by the NWFP, and because each aspect of the plan was inextricably linked to its other provisions and concepts, tinkering with one provision could necessitate an equal—and in some cases, opposite—reaction to maintain the integrity of the plan.

Interagency Coordination for Ecosystem Management

As the NWFP wound its way through the courts, another effort was nearing completion: the creation of a new interagency structure to guide plan implementation. President Clinton had mandated interagency coordination in the development of the NWFP, and that coordination had created a new model for federal forest planning. The Forest Service and BLM jointly developing a forest plan was unprecedented, as was integration of regulatory agencies (US Fish and Wildlife Service and National Marine Fisheries Service) into the forest planning process. Further, the ROD acknowledged that "these standards and guidelines call for a high level of coordination and cooperation among agencies during implementation,"[57] and McGinty put a priority on making that happen. She saw that applying the ecosystem approach to agency planning in a way that could withstand review by the court was key to the success of the NWFP. To do so would require change in the insular way that agencies carried out their planning efforts, adoption of a broader vision of implementation, and a willingness to work toward a set of common objectives across the landscape.

Toward that end, McGinty asked Jim Pipkin, who had been the White House representative in the FEMAT office in Portland helping to smooth plan development, to chair the agency coordination work group that would develop an institutional framework for how the federal agencies would work together to implement the NWFP. A new approach to forest planning based on provinces and watersheds, announced by the president when he first endorsed Option 9 on July 1, 1993, was at the heart of this effort, as was the continued interagency cooperation in all aspects of plan implementation. At first, that proposal was met with resistance: when Pipkin first met with Forest Service leaders, they said in essence, "Tell us what you want, and we'll do it." By that they meant doing the work in-house and not through an interagency framework. The BLM, though, was more receptive (because of helpful advice from BLM Oregon state director Elaine Zielinski), and by the time the interagency plan was announced, agency leaders had generally become more accepting of it.[58] Still, it remained to be seen whether an interagency approach to ecosystem planning, based on provinces and watersheds, would occur.

A linked set of interagency groups was formed to ensure coordination and implementation of the NWFP from the White House down to the local project level (fig. 8.4). The Interagency Steering Committee in Washington, DC, and the Regional Interagency Executive Committee in Portland dealt with major policy issues, although their task was somewhat eased because the conflict between the land management agencies (Forest Service and BLM) and the regulatory agencies that protected endangered species (USFWS and NMFS) had been

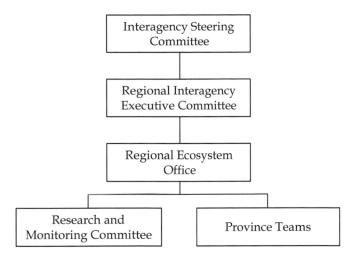

Figure 8.4. Structure of interagency groups designed to oversee and coordinate NWFP implementation. All were in Portland except for the Interagency Steering Committee, which was in Washington, DC. (Source: Adapted from USDA and USDI, "Record of Decision.")

substantially resolved by the NWFP, which had the fierce support of the Clinton administration.

Project-level consultation with the USFWS and NMFS would still be needed to ensure that the provisions of the ESA were being met. Before the NWFP was adopted, that consultation had taken many months, with much dispute and debate about the adequacy of protection provided by the Forest Service and BLM. To announce and ensure that those days of conflict were over, the regional leaders of the land management and regulatory agencies in the Pacific Northwest issued a joint memorandum to all their respective regional and field staff in 1995 directing them to adopt streamlined procedures for all ESA consultations, noting that projects that followed the NWFP were "not likely to jeopardize" listed species, and consultation for them could be "rapidly concluded." As a result, review and approval time frames were greatly reduced in Oregon and Washington.[59] Battles between the BLM and the regulatory agencies would remerge during the second Bush administration, but peace and friendly relations between the land management and regulatory agencies were the order of the day for the remaining years of the twentieth century.

Implementation guidance at the project level on a day-to-day basis came mainly from the Regional Ecosystem Office in Portland, which was responsible for working with the Forest Service and BLM in evaluating and resolving consistency and implementation issues in project development. Proposed actions in the LSRs, Riparian Reserves, AMAs, and Matrix were subject to its review, as was development of guidance for Survey and Manage implementation.[60] While the Regional Ecosystem Office did not have formal decision-making authority, it could and did signal when proposed projects were headed for administrative and legal trouble, which was usually enough to get project leaders in the field to revise their proposed sales. It also oversaw the development of monitoring protocols by the Research and Monitoring Committee, an important element of commitments made in the NWFP.

Province teams were set up to coordinate analyses and planning for each of the 12 provinces (plate 1.3) to help tailor the NWFP to the ecological and social setting of particular areas. Province teams had representatives of federal agencies, states, counties, tribes, private landowners, and environmental groups. However, the teams had great difficulty fulfilling their intended role for many reasons, including resistance of the Forest Service and BLM to giving up planning to such groups, and concerns by private landowners that the effort was a veiled attempt to apply the NWFP to private lands.[61] The province-planning effort never really took hold.

Harvest Level Adjustments: Real and Imagined

During 1994 and 1995, the Forest Service and BLM began merging the NWFP with their own plans. The FEIS had projected a timber sale level of 1.1 billion board feet per year under the NWFP. However, 10 percent of that came from the Other Wood category, which was not historically included in allowable cut calculations. Removing that category from the FEIS timber sale estimate lowered the projected sale level to 1.0 billion board feet per year. As individual national forests and BLM districts amended their forest plans to incorporate NWFP land allocations, standards, and guidelines, further reductions dropped it to about 875 million board feet.[62] That level, almost 30 percent below FEMAT's estimate, became the starting timber target for implementation of the NWFP.

Approximately 675 million board feet would come from the national forests and 200 million from BLM lands, with about 75 percent from Moist Forests and 25 percent from Dry Forests. In all cases, the volume included in these estimates came from Matrix and AMA. Thinning volume from reserves or future salvage was not included, as the future amounts were not known.

The quick descent of the projected timber sales under the NWFP to 875 million board feet was undoubtedly a bitter disappointment to Jack Ward Thomas, who was now chief of the Forest Service, and to many who had worked to help the Clinton administration find an equitable solution to the federal forest crisis in the Pacific Northwest. Also, it certainly did not make it any easier for Thomas to deal with the new Republican Congress.

However, Thomas and others held out hope that one mandated effort under the NWFP might reverse that decline. Watershed analysis, a component of the Aquatic Conservation Strategy, was one of the first tasks the Forest Service and BLM undertook in plan implementation, in which agency professionals fit the specifications of the NWFP to the site-specific conditions in each watershed and provided the basis for the watershed planning approach desired by the Clinton administration. As part of that effort, the interim Riparian Reserves could be adjusted.

The main architects of the Aquatic Conservation Strategy (Sedell and Reeves) saw Riparian Reserve widths as a starting point in the negotiations to follow as part of watershed analysis: "Interim widths are designed to provide a high level of fish habitat and riparian protection until watershed and project analysis can be completed."[63] They fully expected them to be narrowed as the reserves were tailored to site-specific conditions, which in turn would increase the amount of Matrix/AMA and projected harvest.[64]

Thomas would call Reeves every few weeks to find out how watershed analysis was going and whether it was producing the expected increase in the area for timber production. However, the Riparian Reserves had been integrated into the mitigations prescribed in SAT II for many species, and any modification of these reserves would have to consider effects on these species, as noted in the ROD's standards and guidelines:

> Any analysis of Riparian Reserve widths must also consider the contribution of these reserves to other, including terrestrial, species. Watershed analysis should take into account all species

that were intended to be benefited by the prescribed Riparian Reserve widths. Those species include fish, mollusks, amphibians, lichens, fungi, bryophytes, vascular plants, American marten, red tree voles, bats, marbled murrelets, and northern spotted owls. The specific issue for spotted owls is retention of adequate habitat conditions for dispersal.[65]

The Forest Service and BLM lacked the expertise, time, and resources for this kind of analysis. Thus, watershed analysis left the so-called interim Riparian Reserves largely as originally specified, and increases in timber production acreage and projected harvest did not materialize.

PART 4

Best-Laid Plans

9

Traumatic Change and Creative Adaptation (1995–2010)

Judge Dwyer's decision in December 1994 upholding the Northwest Forest Plan (NWFP) set the stage for a new approach to managing the 24 million acres of federal forests in the NWFP area. But the controversies were not over, and implementation would not proceed entirely according to plan during the first 15 years of the NWFP.[1]

A newly elected Republican Congress passed legislation temporarily overriding the NWFP. Assisting displaced workers and distressed communities proved difficult, and the federal workforce underwent a major contraction. By the early 2000s, timber sales collapsed as the agencies proved unable to undertake planned logging of older forests in Matrix/Adaptive Management Area (AMA). The newly elected Bush administration tried for years to reverse the decline in sales by altering the NWFP but found little success. In response, forest managers crafted new ways to harvest forests—ways that would also provide ecological benefits—and timber offerings began to rise again. This chapter tells that story.

Congress Intervenes Again: The Salvage Rider

The first and perhaps most significant challenge to the NWFP came before the ink was dry on the new plan: the political environment in Washington, DC, underwent a seismic shift as Republicans took over both houses of Congress in the 1994 midterm elections. The new Republican members seated in January 1995 were more intensely partisan than Washington had recently experienced. They were bent on reining in the power of President Clinton and his administration and countering the forest management direction enshrined in the NWFP.

The elevation of Jack Ward Thomas to the position of chief of the Forest Service on December 1, 1993, somewhat muted the effects of the changed political climate once the Republican leadership learned, as many others had before, that Thomas could not be bullied.[2] Also, from a technical standpoint, Thomas had led or participated in all the science assessments underlying the NWFP (except SAT II) and was well prepared to explain the plan's necessity, purpose, and approach. He did so in innumerable congressional hearings in his own folksy, endearing style once an atmosphere of mutual respect had been established on this political battlefield. Most importantly, he helped Congress understand that abolishing the NWFP would inevitably take federal forests in the Pacific Northwest right back into the earlier gridlock. Still, Thomas could not stop Congress's attempt, led by the Northwest congressional delegation, to assert its authority to mandate timber harvest levels on federal lands one more time.

President Clinton's bold promise to end the timber wars did not involve Congress. He depended on scientists to write a plan and agency professionals to implement it. But powerful members of the Northwest congressional delegation like Senator Mark Hatfield (Republican from Oregon) had no intention of watching from the sidelines. Hatfield, as chair of the Senate Appropriations Committee, held views on forest policy that had remarkable persuasive appeal to other members of Congress who depended on his committee for federal funds. Hatfield's views were in turn finely tuned to the needs of the timber industry, the most important employer in his state during most of the 30 years he served in the Senate.

Hatfield was well known for legislative riders, provisions added to (riding on top of) unrelated bills. In June 1995, just months after implementation of the NWFP had begun, Hatfield and Slade Gorton (Republican senator from Washington) attached a rider to a spending bill that provided relief funds for victims of the Oklahoma City bombing and ethnic cleansing in Bosnia. The legislation set off a ferocious debate within the administration over the rider's potentially destructive impacts, but ultimately it was signed by President Clinton and became law on July 27, 1995.

The Salvage Rider required the Forest Service and BLM to cut timber "to the maximum extent feasible . . . notwithstanding any other provisions of law" through the end of 1996. The rider not only ordered increased logging of dead and dying trees on the national forests but also contained two provisions directly addressing federal forests in the NWFP area. It required the relevant secretary (of agriculture or the interior) to expeditiously prepare, offer, and award timber sale contracts on federal lands described in the Record of Decision for the NWFP without reference to the standards and guidelines in that document. It also required that the relevant secretary award, release, and permit completion of the remaining timber sales from the previous congressional legislation in 1989 prohibiting environmental litigation (known as Section 318). A number of 318 sales had been held up temporarily for various reasons, and many of those remaining were now within Late-Successional Reserves (LSRs) or Riparian Reserves identified in the NWFP.[3]

By specifying that potential violation of environmental laws could not stand in the way of these sales, commonly known as *sufficiency language*, the Salvage Rider deprived activists of standing to sue the government over the matter. The suspension of environmental protections and citizen enforcement of environmental laws, as well as the inclusion of green (nonsalvage) sales in protected areas, resulted in almost instantaneous resumption of the timber wars in the NWFP area. Denied access to the courts, activists took to the streets and the forests with protests and civil disobedience (fig. 9.1).

Figure 9.1. Protest after citizen litigation under the federal environmental laws was prohibited by the Salvage Rider. Activists shut down access to timber that the Forest Service had sold as salvage after a wildfire in the Warner Creek drainage on the Willamette National Forest. (Photograph © Kurt Jensen.)

LOGGING WITHOUT LAWS, PROTESTS WITHOUT LIMITS
(James Johnston)

President Clinton signed the Salvage Rider into law on July 27, 1995. The first test of the new law in the NWFP area was the Warner Creek salvage sale east of Eugene, Oregon. The Warner Creek sale was in an inventoried roadless area that had been protected as an HCA [a reserve designated in the ISC's report on the spotted owl] when an unknown arsonist set a fire that burned 9,000 acres in 1991.* The Forest Service, claiming that the burned landscape was no longer suitable owl habitat, planned a salvage sale. Environmental groups sued, successfully arguing that the sale violated the National Environmental Policy Act (NEPA) for failing to consider the implications of rewarding arson in owl reserves with timber sales. But on September 6, 1995, Oregon District Court judge Michael Hogan ruled that, per the Salvage Rider, logging on federal lands no longer needed to comply with NEPA, the National Forest Management Act, the Endangered Species Act, or any other law. The Warner Creek sale would proceed—environmentalists had no grounds to challenge timber sales because environmental laws no longer applied.

The Salvage Rider unleashed a grassroots insurrection against federal lands logging in the Pacific Northwest. Activists occupied the only road leading into the Warner Creek salvage, erecting a log fort across the road, chaining themselves to concrete embedded in the road surface, and removing large sections of the road itself. Dozens of the remaining old-growth Section 318 timber sales became the site of road blockades and tree sits. Thousands of ordinary citizens participated in protests and hundreds were jailed.

The NWFP was a balancing act, and the plain intent of the authors of the Salvage Rider was to put Congress's hand on the scale and ensure that the NWFP met its timber production promises and then some. They badly miscalculated. After the Salvage Rider, it was impossible to frame old-growth logging as a carefully crafted compromise backed by well-meaning scientists. Instead, to activists and large swaths of the public, all old-growth logging was lawless logging, a shady deal struck between the timber industry and the politicians in DC.

James Johnston, who started a group called Cascadia Wildlands during this period, spent years of his young life fighting the Salvage Rider.

---

* Archives West, "Warner Creek Fire Collection."

The Clinton administration quickly realized it had an environmental and political nightmare on its hands.[4] Not only might this rider derail implementation of the NWFP, but it might also impact President Clinton's chances of carrying Oregon and Washington in the presidential election the following year. Katie McGinty, Jim Lyons, and others in the administration quickly moved to minimize both the environmental and political damage.[5]

Most important was the administration's interpretation of the Salvage Rider's provision to cut timber "to the maximum extent feasible . . . notwithstanding any provision of law." On August 1, 1995, President Clinton sent a memo to the secretaries of agriculture, the interior, and commerce as well as the head of the Environmental Protection Agency that stated:

> Public Law 104-19 gives us the discretion to apply current environmental standards to the timber salvage program and we will do so. With this in mind, I am directing each of you, and the heads of other appropriate agencies, to move forward expeditiously to implement these timber-related provisions in an environmentally sound manner, in accordance with my Pacific Northwest Forest Plan, other existing forest and land management policies and plans, and existing environmental laws, except those procedural actions expressly prohibited by Public Law 104-19.[6]

In addition, Secretary of Agriculture Dan Glickman provided policy directives that excluded sales from roadless and other environmentally fragile areas,[7] and Chief Thomas directed that the timber sales were not exempt from environmental constraints.[8] Also, Lyons began negotiating with firms owning Section 318 timber sales to find replacement volume in less controversial areas.[9]

An office in Portland that McGinty had set up to coordinate implementation of both the NWFP and the economic assistance initiative played a crucial role in minimizing damage from the Salvage Rider. The US Office of Forestry and Economic Development, led by Tom Tuchmann, served as the Clinton administration's primary representative on all issues related to implementation of the plan—a heady assignment for a young congressional staffer.[10] Tuchmann's office was charged with alerting McGinty to troubles brewing between federal departments involved in that effort, coordinating economic assistance, identifying budget needs, and sending weekly reports on timber sale accomplishments, among other duties. During the Salvage Rider, Tuchmann took on an additional task: personally vetting timber sales proposed under the rider, which the environmental community thought was especially egregious, to help McGinty and Lyons decide whether to let the sales proceed or find replacements.[11]

In the end, the Salvage Rider did much less damage to the Northwest Forest Plan than might have occurred. The Forest Service and BLM offered approximately their probable sale quantity in 1996 (fig. 9.2) even though the Salvage Rider could be read to allow them to offer much more. Most of that harvest came from Matrix/AMA. Warner Creek, the site of an eight-month protest over a proposed salvage sale, was never logged. Neither were most of the remaining Section 318 timber sales. Some reserved forest was cut, such as in the Umpqua watershed, but collectively, less than half of 1 percent of the forest in LSRs and Riparian Reserves was harvested as a result of the Salvage Rider, with minimal effect on the expected environmental performance of the NWFP.[12] Protest and subsequent redirection of the harvest by the Clinton administration to less controversial areas greatly reduced the environmental damage done by the Salvage Rider.

However, residual anger over the Salvage Rider made protest and litigation of LS/OG harvest almost inevitable from that time on. More generally, presidential approval of the Salvage Rider, albeit attached to a must-pass piece of legislation, ruptured the trust the Clinton administration had developed with the environmental community through adoption of the NWFP. From that point on, environmental groups scrutinized implementation of the NWFP much more critically and skeptically. When Al Gore was preparing a run for president, he was asked about the biggest mistake of the Clinton administration. "I think the biggest mistake that we have made involved an issue known in the United States as the 'Salvage Rider,'" Gore responded.[13]

Economic Assistance to Displaced Workers Proves Difficult

President Clinton had heard the stories of loggers, mill workers, and the leaders of the communities in which they lived at the Forest

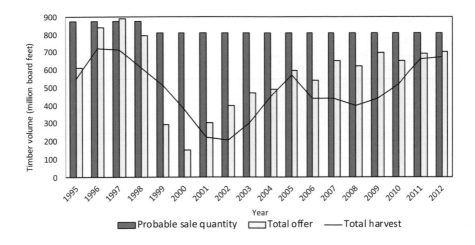

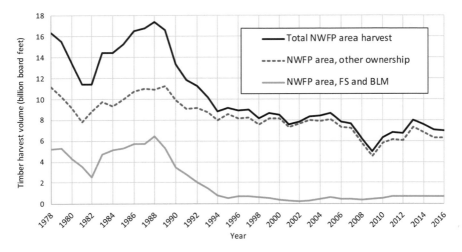

Figure 9.2. (above) Annual timber volume offered for sale and harvested from federal forests in the NWFP area from 1995 to 2012 in relation to the probable sale quantity established in the NWFP. (Adapted from Charnley et al., "Socioeconomic Well-Being and Forest Management," 641.) (below) Federal, nonfederal, and total annual timber harvest volume in the NWFP area from 1978 to 2016. (Adapted from Spies et al., "Twenty-Five Years," 517.)

Conference in April 1993. He too had grown up in a hardscrabble rural community and had said to Katie McGinty as they left the conference, "We really need to help these people." Toward that end, the Clinton administration had set up a high-level task force on labor and community assistance and committed to providing $1 billion at the same time President Clinton announced the selection of Option 9 on July 1, 1993. His staff quickly moved forward with that assistance.

In a report submitted to the president and Congress in 1996, Tuchmann (coordinator of that assistance in Portland) outlined President
Clinton's Northwest Economic Adjustment in detail. The initiative
was designed "to recognize the plight of and directly help those workers, businesses, tribes, and communities in the range of the northern
spotted owl affected by reductions in federal timber harvests" and was
"intended to ease the transitions necessary to allow dislocated workers to compete for permanent jobs, business and industry to survive
and adapt to the new federal forest policy, and affected communities
to develop the capacity to decide on and pursue a future appropriate to their opportunities and resources."[14] The initiative aimed "to
provide assistance needed to cope with 11,000 to 16,000 displaced
workers—6,000 projected to be displaced as a result of Plan adoption
and implementation and between 5,000 and 10,000 remaining from
the economic slowdown and timber sale injunctions of 1990 to 1992—
and associated effects on businesses and communities."[15] Assistance
would "favor those beneficiaries in the most affected communities"
and "incorporate a high degree of state and local participation and
leadership in providing assistance"[16] Thus, the Clinton administration took responsibility for helping timber industry workers and
communities that its decisions directly impacted and also those that
had been harmed by previous economic and legal decisions. Further,
the administration put helping workers who had been displaced at the
forefront of the economic assistance it would provide.

One controversial issue, of course, was how large a job loss the
Clinton administration would face with its economic assistance initiative. Actual employment reductions in the timber industry from 1990
to 2000 are a two-part story. A large drop in harvest in the NWFP
area occurred between 1990 and 1994 across ownerships—after the
injunctions on federal sales in owl habitat before the NWFP came into
effect (fig. 9.2 below)—and many jobs were lost.[17] Then, total harvest
in the area stayed relatively constant from 1995 to 2000, yet more jobs
were lost as the industry consolidated and retooled to process mostly
smaller, second-growth timber from private lands.

The PNW Research Station's NWFP social science monitoring
team gave its estimate of industry employment loss between 1990 and
2000 in "Rural Communities and Economies":

Thirty thousand direct timber industry jobs were lost between 1990 and 2000 in the Plan area (compared to Plan expectations of 25,000 jobs lost). Most of this job loss was in nonmetropolitan counties, with Oregon being the hardest hit of the three states. About 19,000 of these jobs were lost between 1990 and 1994, and the main cause was reduced timber supplies across ownerships. Roughly 11,400 of the lost jobs can be attributed to cutbacks in federal harvests triggered by the listing of the northern spotted owl and subsequent injunctions on timber sales. Timber supplies across all ownerships in the Pacific Northwest were relatively stable during the last half of the 1990s. Nevertheless, about 11,000 of the 30,000 timber industry jobs lost during the 1990s were lost in the last half of the decade. About 400 of the 11,000 jobs lost since 1994 can be attributed to a net reduction in federal timber harvesting. The remaining 10,600 job losses occurred during a period of increased log availability to local mills and are the result of less efficient mills closing and mills continuing to invest in labor-saving technologies. This timber industry restructuring was in response to reductions in timber log supplies from the levels at the start of the decade and the shift to harvesting smaller diameter trees.[18]

A recent, highly sophisticated econometric analysis of the lumber and wood products sector of the timber industry by Ferris and Frank in "Labor Market Impacts of Land Protection" estimated somewhat higher employment impacts, which they referred to as the effect of the owl listing under the ESA and which they measured through implementation of the NWFP. They estimated a loss of 16,000 or 32,000 timber jobs, depending on which comparator or control region is used to define basic employment behavior over the analysis period.

While people have debated and will continue to debate over how much of the job loss was from federal timber harvest reductions, it is clear that a major contraction in the timber industry occurred during the 1990s. With an estimate of one indirect or induced job for each one in the industry, the total effect on job loss (or job opportunities) would approximately double.

However, these negative effects on employment must be put in the context of the regional growth that occurred in the NWFP area in the 1990s: total employment increased by almost 1.3 million jobs during the 1990s, most of which were in the trade and services sectors in urban areas.[19] The regional economy boomed as the timber industry contracted.

The Clinton administration committed to provide $1.2 billion in assistance over five years. We analyze here the approximately $560 million provided during the 1994–1996 period covered in Tuchmann's report. Assistance was grouped into four categories (with the percentage of total assistance funding during 1994–1996):

- assistance to workers and families aimed at the immediate-term effects of retraining dislocated timber workers and supporting their families as well as similar opportunities for others dislocated by the loss of timber jobs (7 percent);
- assistance to business and industry aimed at retraining existing businesses and diversifying the economic base, such as rural economic development grants (27 percent);
- assistance in developing the community infrastructure and technical capacity to effect the transition to an economically sustainable future, such as water systems, sewer systems, county courthouses, and juvenile detention centers (48 percent); and
- ecosystem investment aimed at providing short-term jobs through a Jobs in the Woods program (17 percent).[20]

Thus the funds went overwhelmingly to communities to develop an "economically sustainable future" rather than to direct assistance for displaced workers and their families. In practice, most funds were distributed by state and local officials who emphasized building community infrastructure and technical capacity followed by assisting business and industry. Relatively little funding went to assist timber workers and their families in transitioning to other livelihoods, even though aiding them was a core aim of the Clinton administration's initiative.

Certainly, funneling money to communities to help the improve their infrastructure should help position them in the long run to welcome new businesses and entrepreneurs. However, as Ellen Donoghue, a social scientist who studied these distributions, noted: "Whether it got to communities that needed it most is debatable. . . . Support tended to fall along the Interstate corridor and wasn't necessarily distributed uniformly. It also depended on the capacity of communities to seek out those grants and opportunities . . . there wasn't a system in place to help smaller communities build this capacity."[21]

Susan Charnley, an environmental anthropologist studying socioeconomic well-being in the NWFP area, gave a useful summary of the limited impact of the Clinton administration's economic assistance initiative on displaced workers in "The Northwest Forest Plan as a Model for Ecosystem Management":

> The Northwest Economic Adjustment Initiative did little to help displaced timber workers and their families because the money that arrived was too little and too late to have much effect. . . . The Jobs-in-the-Woods program, intended to provide displaced timber workers with jobs that produced ecological benefits, was somewhat successful, but it created few long-term employment opportunities. For the most part, the initiative failed to create sustainable local jobs that were comparable to the number and quality of those lost to reductions in federal timber harvesting.[22]

One study looked at what happened to displaced timber workers. Helvoigt, Adams, and Ayre, in "Employment Transitions in Oregon's Wood Products Sector," tracked changes in timber industry employment in Oregon over the 1990s. The study included changes in employment in areas of eastern Oregon, which was outside the range of the northern spotted owl but also saw significant reductions in timber harvest during the study period. In the analysis, Helvoigt and colleagues utilized the Oregon state employment database, which provides information on the wages, hours worked, and industry for every worker covered by Oregon unemployment insurance (called *covered employment*). The vast majority of sawmill or pulp mill workers fall into the covered employment category, although some loggers and other independent contractors might be missed.

Helvoigt's study focused on the fate of approximately 35,000 workers who left the wood products industry between the base period of 1989–1991 and 1998. Importantly, the study tracked actual workers in the industry who were displaced rather than the total number of employees in the industry over time. The authors looked at all workers who left the wood products industry, but some of these were replaced; thus, the decrease in the work force of the industry was less than the number of workers who were displaced.

Approximately half the displaced workers found employment in some other covered Oregon industry, most commonly in the wholesale-retail trade, which provided relatively low-paying jobs. Nearly a third of the separated workers from the southwestern and eastern regions of Oregon who found employment had moved to the northwestern part of the state (Portland-Salem area) by 1998.

Approximately half the displaced workers disappeared from the state employment database. The authors acknowledged that workers drop out of covered employment for numerous reasons, including retirement and movement to another state. They also noted that this group might form the basis for a cadre of chronically under- or unemployed rural residents.

That virtually half the former timber workers dropped out of the employment database suggests that many of these displaced workers and their families went through hard times indeed. For some, a way of life that the federal government had said would go on forever abruptly ended. Consequently, it should not be surprising that feelings of betrayal by the federal government emerged in the timber towns of the Pacific Northwest and intensified as promised retraining showed limited results.

Charnley and her colleagues aptly described the broader effects of job loss on wood product workers in "Socioeconomic Well-Being and Forest Management," drawing on the studies in the Pacific Northwest:

> The impacts of job loss on woods product workers were not purely economic: they were also social. Existing literature finds that mill workers were concerned about economic stability and have a strong attachment to their home communities (Lee, et al., . . . ["Social Impact of Timber Harvest"]). . . . Loggers' sense of identity was closely tied to their occupation, which fostered independence, pride in their work, and the feeling of having a unique job. . . . They were also part of an "occupational community" that included other loggers, social interactions with whom strengthened their sense of

identity (Carroll, et al., . . . ["Occupational Community and Forest Work"]). . . . Thus, not only did job loss represent a loss of jobs and income; it also undermined loggers' sense of identity and personal empowerment, which were tied to working in the woods, making finding a substitute occupation difficult. Moreover, loggers and the timber industry were often vilified during the years of the so-called "owl wars," leading to occupational stigmatization, which had a negative social and psychological impact on loggers and their families (Carroll, . . . [*Community and the Northwestern Logger*], Carroll, et al., . . . ["A Response to 'Forty Years of Spotted Owls?'"]).[23]

It was also a chaotic, frustrating, and disheartening time for Forest Service professionals throughout the range of the northern spotted owl. As timber sales declined sharply, so did the need and budgets for Forest Service employees. Foresters and engineers who had spent their careers laying out old-growth timber sales and access roads were no longer in demand. In addition, they were often vilified, much like the loggers, for having spent their careers on such an effort. Many retired. Others had to transfer to other Forest Service locations or to other federal agencies or shift to other lines of work. Some found new agency employment opportunities in the complex analysis and implementation requirements under the NWFP, such as watershed analysis and Survey and Manage protocols. Still, these opportunities were small compared to the contraction in other areas, and they often required different skills: full-time-equivalent positions on the national forests of the NWFP area fell by 36 percent between 1993 and 2002, and many field offices were closed or severely downsized.[24] Those trends continued into the second decade of the plan, especially in the national forests of Oregon and Washington, which saw the number of positions decline by 60 percent in total from 1993 to 2012.[25]

Employment on BLM districts in western Oregon declined more moderately.[26] The BLM had always had a smaller workforce relative to the size of its timber program, in part because of its narrower mission. Also, Congress and the forest industry had more interest in maintaining the BLM's professional staffs, believing that future favorable court rulings would allow high harvests from these lands in years to come.

Some Communities Suffer Greatly; Others Rapidly Move On

The workday started early in communities near the national forests or BLM districts in the heyday of old-growth logging. By 4:00 or 5:00 a.m. coffee shops and diners were full of loggers eating hearty breakfasts before heading to the woods in six- or eight-passenger crummies. Later in the morning mill workers would show up for breakfast before heading to their shift at the sawmills. A mill whistle often signaled the arrival of noon. Activity picked up again near the end of the day as the loggers returned and the swing shift of mill workers replaced the day shift.

Economic and social impacts on communities from reduced federal timber harvests came from many sources. First were the displaced wood product workers, including loggers, truck drivers, and mill workers. Then came the grocers, chainsaw repair experts, and restaurant employees who might be laid off as business declined. In addition, many owners and managers of these firms were leaders in their communities, and the disruption they faced reduced community capacity to respond to the challenges brought by the abrupt decline in federal timber availability.[27]

Clayton Dumont, a social scientist at San Francisco State University, also studied the impact of the sudden sharp declines in timber harvest on the rural communities of the Pacific Northwest. His article titled "The Demise of Community and Ecology in the Pacific Northwest" emphasizes the cultural heritage and sense of community that were being lost in the small, timber-dependent communities of the region—a social crisis resulting from economic and ecological crises.

Forest Service cutbacks also hurt in many ways. The departure of Forest Service employees meant the loss of the portions of their paychecks that would have been spent in local communities. The closure of ranger districts dried up a major local source of seasonal employment. Also, Forest Service professionals had been encouraged to play important volunteer roles in their communities, such as participating in fund drives or being on school boards, as part of the federal presence; thus, their departure had not only an economic but also a social impact.

The greatly reduced payments from shared timber sale receipts were another economic blow to counties and the communities that depended on their services in the NWFP area. Counties had historically received a portion of federal timber sale receipts to compensate them for the lack of property taxes from federal lands. In the 1980s, this income sometimes exceeded $300 million annually in the area of the NWFP.

The BLM O&C payments issue was especially prominent in the news year after year. Payments to the 13 counties in western Oregon from the BLM averaged almost $175 million annually between 1970 and 1990. The payments were especially important to the six south-western Oregon counties that contained most of the BLM's western Oregon forests, both because of their amount and because they were unrestricted dollars that could be used to fund almost any public ser-vice the counties provided, including public safety (the sheriff's office), libraries, and public health. National forest payments, on the other hand, were generally restricted to funding schools and roads.[28]

Recognizing the potentially disastrous effects of the loss of these payments to counties from reduced federal timber harvests, the Clinton administration and Northwest congressional delegation developed leg-islation to temporarily make up these lost payments while the affected counties broadened their tax base and transitioned to new economies. As an example, Congress appropriated an average of a little over $100 million per year to O&C counties from 1995 to 2010 to compensate for the lost timber revenue from the BLM lands there.[29]

The overall effects of reduced federal harvest on communities var-ied greatly, depending on the relative economic importance of timber harvesting and processing, the degree to which local mills depended on federal timber, the degree to which local residents depended on federal jobs, the potential for community economic growth, and other factors.[30] Communities along the I-5 corridor from Eugene to Portland and Olympia to Seattle were part of a growing, increasingly urbanized regional economy. Oregon, as an example, remained a manufacturing state, as high tech concentrated in urban areas (the so-called silicon forest) replaced the much more widespread wood products industry. Washington also shifted strongly to urban-based high tech along with airplane manufacture.[31]

Thus, communities near or in urban areas generally recovered relatively quickly from the economic shocks of reduced federal tim-ber harvests, as did communities with major recreational attractions, such as Bend, Ashland, and Hood River. Rural communities lacking such advantages, such as Sweet Home, Alsea, and Oakridge in Oregon,

Shelton and Packwood in Washington, and Happy Camp and Hayfork in California, would continue to struggle for many years to come.

The monitoring team of the PNW Research Station assessed changing conditions in communities throughout the NWFP during 1990–2000. They found that "about a third of communities decreased in socioeconomic well-being between 1990 and 2000 while another third increased."[32] Their estimate that a third of the communities suffered a decrease in socioeconomic well-being somewhat exceeded FEMAT's estimate that 27 percent of communities were at risk. Both found that they were generally the smaller, more isolated rural communities. Economic winners and losers became clearer and clearer as descriptors like the "urban-rural divide" and the "two Oregons" emerged.

Although many workers and communities in the Pacific Northwest faced painful adjustments, Richard Haynes, a well-known forest economist at the PNW Research Station, found that US consumers saw little change in wood product availability or prices:

> The harvest decline (roughly five billion board feet*) was offset by several factors. Some were expected and included increased lumber imports; increased private nonindustrial harvests in the Pacific Northwest; and increased harvests in other regions, particularly from managed forests in the South and from mature forests in interior Canada provinces. . . . Unforeseen was the collapse of the log export market from the Pacific Northwest [because of contraction of the Japanese economy] that allowed timber managers and landowners to shift formerly exported logs (more than 2 billion board feet, log scale, per year) to the domestic market.[33]

Timber Sales in Older Forests Abruptly Collapse

Total timber sale offerings in the NWFP area approached the projected 875 million board foot PSQ in 1996 and continued at that level through 1997, with most of that harvest coming from LS/OG logging in Matrix/

---

*A harvest decline of approximately four million federal board feet and one million nonfederal.

AMA. Thereafter, timber sale offerings began to fall and dropped to approximately 150 million board feet by 2000 (fig. 9.2).

Sale volumes in the early years came primarily from harvests in mature and old-growth (LS/OG) forest in Matrix/AMA— the main planned harvest source under the NWFP for at least the first 20–30 years. After 1998, sales from these forests abruptly declined.

The NWFP's policies and requirements, particularly those for Survey and Manage, have rightly been credited as a major cause of this sudden decline.[34] While the limitations and difficulties imposed by Survey and Manage requirements certainly impacted logging of older forest, it is important to recognized two other causes of that decline. Just as critical were the loss of public, professional, and scientific support for such harvest, and successful litigation over violations of the Endangered Species Act and other laws. All three are covered here.

Survey and Manage requirements for proposed timber sales took effect between 1998 and 1999, depending on the species group to be surveyed. Much data was collected during the late 1990s and early 2000s on Survey and Manage species. "Records for some taxa doubled, and increased approximately fourfold for fungi, fivefold for bryophytes, and nearly fourfold for mollusks. The major increase in known sites was from predisturbance surveys [in Matrix]."[35]

An annual species review evaluated the findings. Teams of taxa experts compiled data on species selected for review and presented the information to panels of managers and biologists who then suggested changes in the conservation status of individual species. Through this process, the agencies justified removal of species from the survey list over time.[36]

However, surveys increased costs of timber sale preparation, reduced the area on which harvest could occur when surveyors found one of the sought-after species within the sale area, and increased the legal leverage of groups opposing old-growth harvests. A socioeconomic monitoring study on implementation of the NWFP on the Klamath National Forest (KNF) in northern California described these impacts:

> The Plan had several requirements that called for new procedural processes associated with ground-disturbing activities. . . . One of the most onerous of these on the KNF was for the survey and

manage species. . . . The cost and timing requirements of the surveys, and the amount of time required to meet the procedural requirements, were a deterrent to producing timely timber sales. Furthermore, the presence of survey-and-manage species on the forest-imposed harvest restrictions in some areas, further reduced the land base available for timber sales. . . .

Because the Plan has so many extensive and complex procedural requirements associated with timber harvest, people can find ways of opposing sales because many grounds are available on which to file appeals or lawsuits. . . .

Local environmental groups are particularly opposed to sales that included old-growth trees, that were in key watersheds, or that were on steep slopes. Appeals and lawsuits have stopped several timber sales on the KNF.[37]

Most other national forests and BLM districts in the NWFP area had similar difficulties.

Citizen groups soon realized that discovery of species like the red tree vole in a proposed old-growth timber sale could disrupt it. For example, the Northwest Ecosystem Survey Team (NEST) proved highly effective at finding red tree vole nests that government surveyors missed (fig. 9.3):

NEST is an all-volunteer group of self-organizing, tree-climbing humans. Each summer, NEST volunteers take on a role as canopy surveyors, who utilize the Northwest Forest Plan's "Survey and Manage" laws to protect ancient forests threatened by logging. . . .

Under Survey and Manage, a red tree vole nest site is required to receive a 10-acre surrounding buffer where no ground disturbing activities can occur. Following a little research, they realized just what needed to be done to do a much more thorough job, on their own.

Since this time, NEST has found hundreds of Red Tree Vole nests in areas that were slated for logging. Thousands of acres of red tree vole habitat have been saved, as a result of NEST surveys.[38]

Figure 9.3. (*left*) Northwest Ecosystem Survey Team (NEST) looking for red tree vole nests. (Courtesy of Northwest Ecosystem Survey Team.) (*right*) Red tree vole (*Arborimus longicaudus*). (Source: Stephen DeStefano.)

The impact of Survey and Manage on LS/OG harvest blindsided many people when it came fully into force in 1998–2000, especially since the Clinton administration understood that this impact was already included in harvest estimates for the NWFP. If the administration had realized that these effects had not been included, it might have authorized a major effort during the three-to-five-year grace period before Survey and Manage fully came into effect to understand better which species were adequately protected by the extensive reserve system, and thus whittle down the list of those that needed to be considered in designing timber sales. By the time the impacts were becoming clear, however, the Clinton administration's second term was coming to an end and key players in the NWFP effort were leaving the administration. Also, the Forest Service and BLM had begun trying to remove Survey and Manage from the NWFP in its entirety.

More broadly, grassroots environmental groups that fought the Salvage Rider were in no mood to let further old-growth harvest proceed after their access to administrative objection processes and the courts was restored in the late 1990s. They were organized and ready for a fight, and they were not alone. Public support for old-growth harvest had significantly declined. In *Public Values and Forest Management,*

Charnley and Donoghue highlighted studies documenting a shift in public values pertaining to old-growth harvest in the national forests in Oregon and Washington. "Whereas in 1986, 70 percent of Oregon residents surveyed supported the harvest of old growth, 75 percent of Oregon and Washington residents surveyed in 2001 believed that old growth should be protected from logging on national forests, with slightly more support for this position in urban versus rural counties" (Davis, et al., . . . [*Public Opinion about Forests*]).[39] Other studies also found that residents in the Pacific Northwest increasingly favored old-growth protection. Whether this change was the result of increased knowledge of old-growth ecological values, national media coverage, attempts by activists to stop such harvest, or other causes, the public tide had turned against old-growth logging.

Also, the Forest Service's dominance by foresters and engineers who had overseen the old-growth sales declined as many retired or transferred, while ecologists and wildlife and fish biologists gained influence as they played vital roles in helping to implement the complicated provisions of the NWFP. In addition, the Clinton administration purposely broadened agency leadership, especially at the forest supervisor level, to include biologists, hydrologists, and other professionals. With these shifts, the traditional view that the prime use of older forest was as fuel for the sustained yield engine largely disappeared.

In addition, key policy makers and scientists called for the end of old-growth logging. Mike Dombeck, who took over when Chief Thomas retired in January 1997, issued a policy barring the cutting of old-growth timber on national forests on January 9, 2001—a clear challenge to the second Bush administration on the day it took office.[40] While the Bush administration immediately reversed the policy, Dombeck's statement foreshadowed a coming major change in national forest management. Norm Johnson and Jerry Franklin testified before Congress in spring 2001 that NWFP implementation had shown that delivery of a stable supply of timber from LS/OG forests in the NWFP area was impossible and that it was time to focus on the young stands established following past clearcutting for the harvest. This statement, by two FEMAT members who had helped create the NWFP, caused an uproar in the hearing room and beyond. A few months later, Thomas and Dombeck published an op-ed in the *Seattle Post-Intelligencer* calling for an end to old-growth harvest.[41] Scientists publicly identified with leading the development of the NWFP had now been joined by a current and a former chief of the Forest Service in concluding that the days of logging old forest on federal lands were over.

Difficulties in coordinating critical habitat for the spotted owl and the LSRs also contributed to the collapse of old forest harvest. The US Fish and Wildlife Service (USFWS) considered the NWFP to be the federal contribution to survival and recovery of the northern spotted owl and other listed species throughout the spotted owl's range. The agency had concluded that the reserve system and Matrix/AMA practices were sufficient to enable survival and recovery of the owl,[42] and therefore it regularly granted incidental take permits for LS/OG harvest in Matrix/AMA even though they would degrade owl habitat. However, the agency had not realigned its 1992 critical habitat designation for the owl to coincide with the NWFP reserve system. Therefore, critical habitat still covered portions of Matrix/AMA, and law students successfully challenged the logging of LS/OG forests in such instances as a violation of the ESA.

Susan Jane M. Brown, then a second-year law student at Lewis and Clark Law School in Portland, was monitoring early NWFP implementation on the Gifford Pinchot National Forest in southwestern Washington as a member of the Gifford Pinchot Task Force. She soon realized that USFWS biological opinions for the dozens of Gifford Pinchot's timber sale NEPA documents allowed for incidental take of hundreds of spotted owls and logging of thousands of acres of critical habitat in Matrix/AMA. Using this information and that from other national forests, the Gifford Pinchot Task Force and others filed suit in the Western District of Washington accusing the USFWS of violations of the ESA for failing to protect the northern spotted owl.[43] The district court ruled against the plaintiffs in July 2002, but that was reversed at the Ninth Circuit Court of Appeals.[44]

At the Ninth Circuit, the task force argued that the USFWS regulation pertaining to adverse modification of critical habitat was unlawful, in that "the regulatory definition sets the bar too high because the adverse modification threshold is not triggered by a proposed action until there is an appreciable diminishment of the value of critical habitat for both survival and recovery" of listed species, essentially reading "recovery" out of the regulation. The court agreed with the plaintiffs, writing: "The FWS could authorize the complete elimination of critical habitat necessary only for recovery, and so long as the smaller amount of critical habitat necessary for survival is not appreciably diminished, then no 'destruction or adverse modification,' as defined by the regulation, has taken place. This cannot be right. If the FWS follows its own regulation, then it is obligated to be indifferent to, if not to ignore, the recovery goal of critical habitat."[45] Thus, the challenged biological opinions were invalid because they relied on an impermissible interpretation of the ESA.*[46] As a result, the Forest Service and BLM largely abandoned attempts to harvest LS/OG forest within critical habitat of the spotted owl, and the USFWS had to rethink the relative ease with which it distributed incidental take permits.

The Bush Administration Fights Back

President George W. Bush's election in 2000 coincided with the collapse of the federal timber sale program in the NWFP area (fig. 9.2). Soon after his election, he pledged to "make the NWFP work" during a visit to Medford, Oregon. His administration then spent many years trying to modify the plan to produce the harvests projected in it, most of which were scheduled to come from LS/OG forest over the first few decades, and to develop a new timber-oriented plan for the BLM. These efforts were largely unsuccessful, but they did establish a number of legal precedents that gave further definition to the NWFP.

To help achieve its goals, this second Bush administration settled a lawsuit brought by a timber group, the American Forest Resource Council, alleging that the NWFP violated a series of federal laws.[47] This type of agreement is sometimes known in legal circles as a *sweetheart settlement* because both sides want the same thing (in this case higher timber harvests), and the settlement agreement was a way to achieve that goal. In it, the administration agreed to consider several specific policy changes: eliminating the Survey and Manage program; amending the Aquatic Conservation Strategy; creating a forest plan for the BLM that emphasized the timber production wording in the 1937 O&C

---

* Plaintiffs also argued that the USFWS could not rely on the LSR network as a surrogate for critical habitat—that the LSR network essentially "stood in" for designated critical habitat. The court agreed with the plaintiffs on this claim, too. GPTF v. FWS at 1076.

Table 9.1. Attempted revisions of the NWFP and associated requirements during the George W. Bush administration (2001-2008) and their outcomes.

Initiative	Outcome
Eliminate Survey and Manage	*Largely unsuccessful*: Survey and Manage retained for harvests proposed in LS/OG forests, but eliminated for thinning and fuel treatments in stands up to 80 years of age
Amend the Aquatic Conservation Strategy	*Unsuccessful*
Create a forest plan for the BLM O&C lands separate from the NWFP with higher timber harvest levels	*Unclear:* Separate BLM plan adopted by the Bush administration; replaced by the Obama administration; litigation continues
Delist the northern spotted owl under the ESA	*Unsuccessful*: Status review did not result in delisting
Reduce the amount of critical habitat designated for northern spotted owl	*Temporarily successful*, but reversed by the Obama administration
Delist the marbled murrelet under the ESA	*Unsuccessful*: Status review did not result in delisting
Reduce the amount of critical habitat designated for the marbled murrelet	*Unsuccessful*

Act; delisting the northern spotted owl or (at least) reducing the extent of its critical habitat; and delisting the marbled murrelet or (at least) reducing the extent of its critical habitat (table 9.1).

The first three proposed policy changes in the table (eliminate Survey and Manage, amend the Aquatic Conservation Strategy, and create a separate forest plan for the BLM) go right to the heart of the provisions in the NWFP and will be covered in depth here. Additionally, the Bush administration worked on one other precedent-setting policy that will be discussed at the end of this section: adding economic loss as a criterion for expediting project implementation.

The Forest Service and BLM made numerous attempts to eliminate or greatly reduce the impacts of Survey and Manage on timber harvests between 2000 and 2005. A number of these proposals went through the NEPA process and the litigation that predictably followed.

The Forest Service and BLM prepared an Environmental Impact Statement (EIS) in 2000 to evaluate the effectiveness of the Survey and Manage program and then issued a Supplemental EIS and Record of

Decision (ROD) in 2001 to remove or modify the program's requirements. Environmental and timber groups both challenged this decision. The timber groups (led by the Douglas Timber Operators of
Roseburg, Oregon) subsequently entered into a settlement agreement
with the secretary of agriculture that required the Forest Service and
BLM to prepare a second Supplemental EIS to evaluate elimination of
the Survey and Manage program and consider other changes.

That Supplemental EIS, released in 2004, considered three alternatives: retain the Survey and Manage program; remove the program
entirely from the NWFP; or move some Survey and Manage species
into agencies' special-status species program. The last was a discretionary program lacking definite requirements for surveys and no-harvest
buffers, leaving it to the agencies to determine how to proceed with
protecting species moved to that list. The agencies selected the third
alternative for implementation, which again elicited legal challenges.

Environmentalists, led by the Northwest Ecosystem Alliance of
Bellingham, Washington, challenged this decision on NEPA grounds
and prevailed on many of their claims.[48] The plaintiffs successfully
argued that the Supplemental EIS failed to analyze the potential environmental consequences of the special-status species program for any
former Survey and Manage species that were not also added to the
BLM's sensitive species list and thus denied special protection.[49] The
court agreed: "Agencies have an obligation under NEPA to disclose and
explain on what basis they deemed the standard necessary before but
assume it is not now."[50] That the Forest Service and BLM could deviate
from the NWFP as written was not the issue: the issue under NEPA was
whether the agencies adequately explained their rationale for doing so.

Though the court ruled in favor of the environmental-group plaintiffs, it did not issue a range-wide injunction as Judge Dwyer had done
in the past. The judge best known for his landmark rulings on the spotted owl had died in 2002, and the torch had passed to district court
judge Marsha Pechman. Judge Pechman ordered the parties, including
both environmental groups and timber groups (who had intervened to
help defend the government's position) to enter settlement discussions
regarding a remedy. The resulting settlement agreement retained the
original Survey and Manage program but also created several exemptions, including one for projects that thinned or undertook hazardous

fuels reduction in forests up to 80 years old. This agreement became known as the Pechman exemptions.[51]

These court decisions and settlements on Survey and Manage helped direct the future of the timber harvest program on federal forests in the NWFP area.[52] They retained limitations on logging LS/OG forest while releasing them on thinning and fuel treatments in younger forest, which helped accelerate the shift in Moist Forest from variable retention harvest of LS/OG to plantation thinning and the limiting of fuel treatments in Dry Forests to younger stands.

In addition, surveys undertaken through the Survey and Manage program during the first 10 years of the NWFP enabled the agencies to whittle down the list of species of concern, as some previously poorly understood species were found to be less in need of protection. By December 2003, the list had been reduced to 298 species, including 189 fungi, 15 bryophytes, 40 lichens, 12 vascular plants, 36 snails and slugs (mollusks), 4 amphibians, 1 mammal (red tree vole), and 1 bird (great gray owl). Thus, Survey and Manage activities had greatly increased knowledge about the abundance and habitats of many species, enabling removal of over 100 species from the list.[53] However, some species affecting the ability to harvest LS/OG forest, such as the red tree vole, remained on the Survey and Manage list. After 2003, cuts in the program's budget by the Bush administration limited further survey work, except for that directly related to timber sales.

Attempts to clarify the requirements of the Aquatic Conservation Strategy (ACS) also led to litigation in the first ten years of the NWFP. In 1998, fishing and environmental groups led by the Pacific Coast Federation of Fishermen's Associations challenged a programmatic biological opinion (one that covered a large area as opposed to a particular project)[54] issued by the National Marine Fisheries Service (NMFS) on federal timber harvests in the Umpqua River basin. They argued that the NMFS improperly assumed that compliance with the ACS of the NWFP was sufficient to demonstrate compliance with the ESA for the recently listed coastal coho salmon. Judge Barbara Rothstein of the Western District of Washington rejected plaintiffs' arguments and upheld the biological opinion, stating that it was reasonable for the NMFS to presume in a programmatic biological opinion that agencies would faithfully implement the ACS as required. However, the court did invalidate several timber sales linked to the programmatic biological opinion, because they lacked a basis on which the NMFS could determine whether the agency undertaking an action, like a timber harvest, had in fact complied with the ACS. No parties appealed.[55]

This decision by Judge Rothstein that the NMFS must provide a basis for determining whether agency actions complied with the ACS became a powerful tool for further litigation. The Pacific Coast Federation of Fishermen's Associations soon led a return to Judge Rothstein's court, arguing that the ACS required compliance at four spatial scales (regional, province/river basin, watershed, and site) as well as two temporal scales (immediately upon project implementation and in the future on a timescale relevant to the life cycles of the salmon in question).

Judge Rothstein agreed that the ACS required this multiscale, multiple-time-period analysis. However, the NMFS had conducted its analysis of ACS compliance only at the watershed scale and only in the longer term—10 to 20 years after timber sale implementation. The NMFS argued that timber harvests should be allowed that might have site-specific adverse effects at the time of project implementation, but that in the longer term and at the watershed scale were not likely to jeopardize listed species. However, the court ruled that impacts at the project level, even if for a small area, must be assessed and aggregated:

> Any project that maintains or restores fish habitat presumably would not jeopardize the survival of the species. However, a project that degrades habitat at the project level must be included in any realistic study at the watershed scale . . . disregard of projects with a relatively small area of impact but that carried a high risk of degradation when multiplied by many projects and continued over a long time period is the major flaw in the NMFS study. Without aggregation, the large spatial scale appears to be calculated to ignore the effects of individual sites and projects.[56]

The court further ruled that looking only at the longer term was unreasonable because the listed species (coastal coho) was in critical condition in the short term:

The NMFS never disputes that short-term effects have the potential to jeopardize listed fish populations. On the contrary, NMFS believes that the next few generations will be critical to Umpqua River anadromous species. In the Programmatic Biological Opinion, NMFS states that "even a low level of additional impact to any life form, especially the anadromous form which is at critically low levels, may reduce the likelihood of survival and recovery of the ESU as a whole." Given the importance of the near-term period on listed species survival it is difficult to justify NMFS's choice not to assess degradation over a time frame that takes into account the actual behavior of the species in danger.[57]

In response to this decision, the Forest Service and BLM proposed to amend the ACS in 2003 to clarify that its provisions should be applied at the watershed scale and in the longer term, and that individual projects need not avoid site-specific adverse effects as long as they achieved compliance at the watershed scale and for those longer time periods. Again, the court ruled against the Forest Service, BLM, and NMFS. The court held that, as in the Survey and Manage cases, the agencies had failed to explain why they were making this clarification and what the implications would be:

Where an agency has previously made a policy choice to conform to a particular standard, and now seeks to amend that standard, "the Agencies have an obligation under NEPA to disclose and explain on what basis they deemed the standard necessary before but assume it is not now."[58] . . . Under this reasoning, the . . . assessment of the impact of the ACS amendment is inadequate and fails to conform to NEPA standards.[59]

Thus, the court's requirement for compliance at four spatial scales and two temporal scales remained in place. The court remanded the NEPA documents back to the agencies for repair, and there attempts to amend the ACS ended.

The BLM did draft the Western Oregon Plan Revision in 2007 as its own forest plan independent of the NWFP, as directed in the

2002 settlement agreement between the Bush administration and timber industry. The preferred alternative identified in the draft plan and associated NEPA documents significantly increased the timber harvest level, compared to what had been projected for BLM lands in the NWFP. After receiving and responding to public comments, the Bush administration issued the final plan and associated NEPA documents, with a slightly lower projected harvest level than the draft plan, just prior to the Obama administration taking office in January 2009. Environmental groups, led by the Pacific Rivers Council, immediately sued, challenging the plan's legality, in part because it had been adopted without consultation with the USFWS and NMFS to ensure that it met ESA obligations.[60]

New on the block, the Obama administration undertook some on-site investigation before taking a position on the BLM's new plan. They sent representatives to look at proposed sales in mature forests with BLM staff and invited Jerry Franklin to go along. Under their plan, these forests would be clearcut and planted with Douglas-fir to create the BLM's best approximation of a tree farm. Franklin argued that if such forests were to be harvested and managed for wood production, there were more ecologically beneficial ways to do so that did not involve clearcutting. The visitors also learned that old-growth stands would eventually be cut using this same method.

Well, that did it: the Obama administration was not going to defend a plan for clearcutting old-growth forests, let alone one lacking consultation with the regulatory agencies empowered to enforce the ESA. As the environmentalists' lawsuit proceeded, the administration informed the court that it would not defend the BLM's plan, and the court vacated the plan, effectively ending the plan's life before it began. Simultaneously the new administration promised to address lagging harvests from federal forests, including another attempt to create a separate plan for the BLM O&C lands. In the meantime, the NWFP would remain intact.

Litigation also challenged the Forest Service's revision in 2003 of its regulations guiding the administrative review of forest projects and expansion of the definition of an emergency situation under NEPA to include a "substantial loss of economic value to the Federal Government." This new provision would enable timber salvage following wildfires on national forests without waiting for administrative appeals of projects to be resolved.[61]

The Siskiyou Regional Education Project brought one of the first test cases over the use of this new provision, challenging its use to justify rapid harvest of dead timber from the enormous 2002 Biscuit Fire in southwestern Oregon.[62] The Forest Service EIS for the project authorized fire salvage within Matrix and LSRs (some of which were in roadless areas) to recover the commercial value of fire-killed trees. Since the commercial value of dead trees declines rapidly over time, the Bush administration argued that the situation constituted an economic emergency, and that harvest should not have to await resolution of administrative appeals. The courts agreed, although much of the timber was not salvaged for a variety of other reasons.[63]

In 2010, though, the courts came to a different conclusion that resulted in revision of the economic emergency provision. The Alliance for the Wild Rockies challenged the use of the provision for the Rat Creek fire salvage sale in Montana. While the district court's decision favored the Forest Service, the litigants appealed to the Ninth Circuit Court of Appeals, which did not find the government's argument persuasive: "With all due respect to the budgetary concerns of the Forest Service, a loss of anticipated revenues to the government of 'as much as $16,000,' or even a 'potential loss' of $70,000 in the event of no bids, is likely not a 'substantial loss to the federal government.'"[64] In response, the Obama administration rewrote the emergency salvage provision and changed the justification for action before resolution of administrative appeals to "avoiding a loss of commodity value sufficient to jeopardize the agency's ability to accomplish project objectives directly related to resource protection or restoration." This newer rationale has been used to allow immediate action after the decision is signed where the Forest Service has convinced the court that revenue from salvage sales will enable postfire recovery, such as repairing roads damaged by the wildfire.

Young Stand Thinning Saves the Federal Timber Program

Finding little success in revising the NWFP to advance harvests of LS/OG forest, the Bush administration pressed the Forest Service and BLM to find any volume they could muster to reach the promised one billion board feet per year. Both agencies quickly realized that if their projects did not provide significant commercial timber volume, their leadership would be chastised and future budgets would decline, resulting in more transfers and layoffs.

On federal forests, commercial thinning of plantations established after past clearcutting in the Moist Forests of the Coast Range and western Cascades of Oregon and Washington quickly became the primary source of timber harvest volume. Thinning these stands had been an afterthought in FEMAT's description of near-term timber harvests, with the focus on logging of LS/OG forests in Matrix/AMA. Now, though, it became the primary means for producing harvest volume in the NWFP area (fig. 9.4).

Advocates of sustainable development commonly recognize three aspects of sustainability—ecological, economic, and social[65]—and thinning younger stands might achieve all three aspects simultaneously. Arguably, thinning projects in these stands could be designed to restore compositional and structural diversity and accelerate their progressive development into forests with LS/OG characteristics. Commercial thinning could also generate merchantable volume that more than paid for the treatment once plantations reached 40 to 50 years of age, and tens of thousands of acres of these 40-to-50-year-old stands were available by the early 2000s. Finally, commercial thinning found general public acceptance because old-growth forest was not being cut, the harvested areas were not clearcuts, and scientific studies predicted that thinning would accelerate the development of more natural forest conditions.[66]

Environmental groups, who had helped shut down LS/OG harvests, warmed to plantation thinning as evidence accumulated that such activity could have positive ecological effects. These projects also allowed environmental groups to deflect charges that they opposed all timber harvests—a charge that might result in Congress acting in ways adverse to their interests. And on the other side, plantation thinning enabled the Forest Service and BLM to argue that they were simultaneously restoring forests and providing timber harvest volume. This could help them justify their budgets and rebuild their credibility and prestige with the public.

The federal agencies had found a path through the social, legal, and political minefield of the NWFP. By early 2005, Forest Service and BLM

Figure 9.4. Plantation management on the Siuslaw National Forest to increase structural and compositional complexity, with the goal of accelerating development of late-successional conditions, before (*above*) and after (*below*) thinning. (Courtesy of the Siuslaw Stewardship Watershed Restoration Program.)

districts in the Coast Ranges and western Cascades in Oregon and Washington had shifted almost entirely to thinning in younger stands in Matrix/AMA, LSRs, and (to a limited degree) Riparian Reserves. These districts then experienced a period of relative tranquility, with a nearly complete cessation of appeals, protests, and litigation.

Thus, the agencies practiced adaptive management—changing their forest management strategy based on experience gained about activities that were likely to be successfully completed and those that were not. Not adaptation based on formal hypothesis testing, as envisioned for the Adaptive Management Areas, but adaptation based on the real-life results of project experience. In places like the Siuslaw and Gifford Pinchot National Forests, broad-based collaborative groups helped the agencies move down this path. Other national forests and BLM districts did it largely on their own.

Scientists working through the series of assessments had provided a conservation framework for harvesting timber while protecting species and ecosystems that they built into the NWFP. Amazingly, a single element of the framework has enabled most federal harvest since 2000— permitting and encouraging the thinning of stands up to 80 years old in Moist Forest LSRs, a principle that was soon also applied in Matrix/ AMA. Four interrelated considerations help explain the success of the 80-year rule:

1. Recognition by FEMAT scientists that silvicultural treatments, broadly categorized as thinning, could potentially accelerate the rate at which young stands moved toward structural conditions comparable to natural LS/OG stands,[67] and scientific studies soon thereafter confirming that plantations lacked the composition and structure of natural stands.[68]

2. A simple criterion to apply in the field to identify stands eligible for thinning: "Thinning (precommercial and commercial) may occur in [LSR] stands up to 80 years old regardless of the origin of the stands."[69] The maximum age of 80 years came from the decision by Franklin and Spies (see chapter 6) to use age rather than structural characteristics to determine the beginning of older forest functionality. "Regardless of stand origin" came from the desire to eliminate debates over whether a young stand was of planted or natural origin.

3. The knowledge that forests younger than 80 years of age rarely provided high-quality habitat for the northern spotted owl. Thus, the USFWS became a willing partner in facilitating the young stand treatments. Scientists did caution, though, that

thinning large contiguous areas of young stands could be disruptive to the northern spotted owl and its prey species.
4. The fact that thinning young stands in Matrix/AMA maintained options for their future. The NWFP expected that
young stands in Matrix/AMA would eventually undergo a final
variable retention harvest. However, restoration thinning in
Matrix/AMA stands did not preclude a later decision to let
them grow into LS/OG forest instead of making a final harvest.

Forest management standards can both limit and protect discretion,
as is the case here. While the 80-year rule prevented thinning of older
stands, it also minimized arguments over thinning stands less than 80
years old. Given the contentious nature of federal forestry in the NWFP
area, this rule stands as a model for finding lasting agreements.

Five caveats, though, apply to this major advance in forest management strategies for the federal forests of the NWFP area. First,
thinning projects in younger stands, especially in Matrix/AMA where
management included timber production goals, often produced classic
tree farms of evenly spaced trees, rather than the more irregular spac-
ing characteristic of natural forest structures. Second, the NMFS has
slowed or stopped thinning within the first site-potential tree height of
streams in Riparian Reserves that contained threatened salmon populations because of concerns over short-term impacts. Third, thinning
in coastal forests could create problems for survival of the marbled
murrelet by attracting and enhancing habitat for jays and ravens, which
prey on murrelet eggs and nestlings. Fourth, a forest management
strategy based on thinning dense young stands could not last forever
unless more of these stands were periodically created through time by
harvest, wildfire, or other disturbance. Finally, the ability to turn plantations established for timber production into old-growth forests was
uncertain in terms of how long it would take and which stands were
most suitable for that transition. Still, young stand thinning provided a
lifeline to the forestry program in Moist Forests.

The situation was quite different in the Dry Forests of the eastern Cascades and Klamath provinces that contained excessive fuels,
which had accumulated through fire exclusion, past logging, and
other factors and were the location of many large wildfires (plate 9.1).

Thinning to reduce those fuels was encouraged in the NWFP[70] and by fire scientists,[71] but these treatments might call for removal of trees that were important habitat for the northern spotted owl. In addition, some scientists argued that these forests were naturally dense and that thinning to reduce fuel loads lacked sound ecological support, which made thinning in Dry Forests somewhat controversial.[72] Finally, managers and regulators often made very conservative interpretations of standards and guidelines for the NWFP that allowed fuel treatments in older forests in LSRs where unnaturally high fire hazards existed, limiting those actions largely to younger forests there.

The cumulative effect of these difficulties can be illustrated by a case study of the Okanogan-Wenatchee and Deschutes National Forests. Marty Raphael and colleagues reported an analysis of over five million acres of these forests and surrounding areas, with 80 percent of that acreage in the Wenatchee Study Area.[73] They found that the restrictions of the NWFP significantly constrained efforts to reduce the fuel loads that contributed to greater fire damage. Only approximately one-quarter of the total area studied was available for treatment. They noted that "current restrictions on the fuel treatment placement may be impeding managers' ability to protect against wild-fire and improve habitat."[74] They also found that the more aggressive fuel treatment scenarios they modeled would have negative effects on the spotted owl in the short run without many positive effects in the long run. "Particularly for the Wenatchee analysis area, we did not find larger ending . . . population sizes from aggressive fuel reduction treatments relative to the No Treatment scenario."[75] Thus, the search continued in Dry Forests for restoration strategies that benefited both forests and spotted owls.

Adaptive Management Areas Do Not Fulfill Their Promise

While managers were adapting to their changing circumstances under the NWFP, especially in Moist Forests, few of the ideas for this adaptation came from the heralded Adaptive Management Areas (AMAs). The initial goal of the AMAs, set in the FEMAT effort, was to encourage innovation in integrating timber production and forest restoration with conservation of species and ecosystems using the principles of scientific inquiry.[76] This goal was broadened in the NWFP to also encourage collaborative efforts where local people would work with each other, federal agencies, and scientists to reach agreement on innovative strategies to test, as had been done successfully in the Applegate Partnership in southwestern Oregon.

Some AMAs did succeed in undertaking experiments or demonstrations to assess new ideas for forest management. Scientists from the Pacific Southwest Research Station designed and implemented an experiment to evaluate the value of mechanical treatments and prescribed fire in accelerating late-successional conditions in the Goosenest AMA on the Shasta-Trinity National Forest.[77] That work has produced significant findings and publications that help guide Dry Forest restoration, and when a wildfire swept through the experiment in 2021, it yielded valuable information on the effectiveness of different treatments.[78] Central Cascades AMA scientists at the HJ Andrews Experimental Forest designed an alternative landscape-level approach that emulated natural disturbance processes and demonstrated it on portions of the 23,908-acre Blue River drainage near the Andrews.[79] In both cases, existing cadres of scientists in the local area had experience in working with local Forest Service managers, and neither waited for a collaboratively developed management plan to take action. Also, the Applegate Partnership, the model for the AMA social structure, quickly integrated the new management rules of the NWFP into its framework and worked creatively with the federal agencies on the Applegate AMA over the next decade.

Overall, though, the AMAs fell significantly short of the NWFP goal of collaborative development and testing of new approaches for managing federal forest lands.[80] Most AMAs lacked an existing cadre of scientists who could shift their work to the AMA. The Forest Service and BLM made few funds available to lure scientists into AMA efforts or conduct experiments, as their managers were skeptical that the work would be of much value to them. The addition of a social planning process in the NWFP standards and guidelines fundamentally misjudged the social cohesion that existed among stakeholders and with agencies and hamstrung many AMAs. Finally, and perhaps most important, NWFP standards and guidelines to protect at-risk species limited experiments and demonstrations,[81] as did the requirement that AMAs produce timber harvest levels equivalent to those of Matrix.

With all these factors working against the AMAs, they did not have a fair chance to demonstrate their power to help accelerate learning on how to best manage these forests to meet the goals set for them in the NWFP. Experimental learning is still needed for the NWFP, and the experience of the AMAs should not be forgotten in designing how to do that in the future. Successes like Goosenest provide a base to build on.

By 2010, the Forest Service and BLM placed little emphasis on the AMAs functioning as laboratories for the creation of new ideas in forest management. In general, they managed the forest within them similarly to Matrix in the surrounding area. Adaptation had occurred in the first 15 years of the NWFP as agencies figured out which projects were legally and socially acceptable, and occasionally scientists conducted small experiments in the federal forests under the plan, but relatively little of this knowledge came from the AMAs.

Managers' Experiences under the Northwest Forest Plan

EDUCATION OF A WILDLIFE BIOLOGIST ON THE WILLAMETTE NATIONAL FOREST
(Cheryl Freisen)

As I read through this book, it struck me over and over again that I could insert hundreds of hyperlinks to a parallel universe of stories unfolding in the field. The timeline laid out here is so clinical, it's easy to forget there were a whole bunch of professionals struggling to keep up in an emotionally turbulent time.

The years before the NWFP found wildlife biologists like me studying and searching for spotted owls while old forest was being aggressively harvested. There was no management strategy, no protections, no seasonal restrictions. It was noisy out there with chainsaws, donkey whistles, and big timber falling. The Willamette National Forest (WNF) in its heyday was harvesting 10,000 acres per year of old forest.

During the injunctions, we were out on ranger districts unraveling timber programs. We were tasked with identifying spotted owl habitat in existing planning projects and shifting boundaries to delete those acres from the sale. A new coworker in the fire shop said to me on a field trip, "I just met you, but already I don't like you."

The first 10 years of the NWFP were a roller coaster. I was given half a million dollars to conduct a pilot watershed analysis, which was exciting and fun for many of us young professionals. But then the Salvage Rider was dropped on us, reinvigorating the harvest of older forest. This was followed by an ugly period of ecoterrorism: the Oakridge Ranger Station was burned down, antilogging graffiti and a bomb-like item were found on the Detroit Ranger Station building, tree sitters threw urine at employees, and we staffed our buildings with armed guards and put up fencing. It was a painful time. After the rider, the reality of harvest levels under the NWFP came home to roost. We started downsizing, forced relocations, buyouts, early retirements. The WNF went from over 1,300 employees down to around 350. We closed three ranger district offices and consolidated staff. The timber cut went down 95 percent and became focused on plantation thinning. We lost our road crews, dozens of engineers, and foresters.

The second 10 years of the NWFP found some stability. The WNF was back to harvesting 10,000 acres a year but was still focused on thinning plantations that were created in the 1950s and 1960s. Instead of a timber product of 20-inch-diameter trees at the smallest, contracts were written to thin down to 5-inch-diameter trees. The mills that were left retooled, and new equipment came into play; one-log loads were replaced with 100-log loads. We made some runs at putting timber sales in the 700,000 acres of mature and old growth (LS/OG) that still existed in NWFP Matrix lands, but those were met with appeals, protests, and lawsuits. For efficiency's sake, programs were focused on projects that could be successful (i.e., low controversy), and harvest in those available old forests was deferred.

The last five years of the NWFP have seen the rise of collaborative groups and attempts at shared decision-making. Zones of agreement with these public groups have tended to focus on the low-hanging fruit, supporting the plantation thinning program and some Dry Forest restoration involving bigger trees. Biologists in the field have successfully gone after outside funding, particularly focused on stream restoration. Aquatic resources had been a large component of the NWFP, and the need for that type of restoration gained momentum with a new, young workforce of fish biologists and hydrologists eager to do work besides timber sales.

The next five years? For a wildlife biologist like me who has watched this all unfold for 30 years, it's still interesting: politically, ecologically, socially, personally. Commercial thinning of young plantations is not sustainable forever, and for some forests, the uneven flow is a difficult backdrop for maintaining a stable workforce. Big fires on the west side are shifting perspectives on community interactions with their rain forests. Citizens are choking on wildfire smoke. Climate change and the value of our forests for water and carbon sequestration are creating new questions and demanding new approaches. New timber products like cross-laminated timber (CLT) are coming into play. Gaining efficiency is the priority, while the USFS as an agency struggles to tell its story and stay relevant through initiatives like *This Is Who We Are.*

Who am I? I'm still a wildlife biologist with the USFS, doing my part to help the agency do the right thing in a constantly changing world. I have no doubt another book could be written 25 years from now to document a continuation of this drama.

Cheryl Freisen is a science liaison on the Willamette National Forest.

FINDING A NEW PATH ON THE SIUSLAW NATIONAL FOREST
(Jim Furnish)

I stood onstage at the Majestic Theatre in downtown Corvallis, Oregon, on July 1, 1993, looking at a large crowd of angry, worried, and bitter Forest Service and BLM employees. As President Clinton concluded his historic Forest Conference in April 1993, he had charged a group of scientists, led by soon-to-be-chief Jack Ward Thomas, to develop an owl solution in 60 days. And we now had a first glimpse of that solution—an audacious plan and sweeping repudiation of recent federal forest management. By inference, the new plan (Option 9) rejected the work of those gathered at the Majestic. Yet successfully implementing the strategy depended on us, the aggrieved. Rhetorically, I posed a central question: "Do you believe this new plan can work?" To my own question, I answered, "Yes."

Implementing the NWFP offered numerous challenges en route to surprising outcomes unforeseen by the architects. Not the least, the NWFP proposed continued reliance on clearcutting old-growth forest to achieve the dramatically lowered annual harvest of about one billion board feet. In a brief conversation with a triumphant Jim Lyons, undersecretary of agriculture, I noted that "even significantly reduced clearcutting of old-growth will remain highly controversial" to environmentalists. Indeed, implementation proved thorny and halting. Briefly, here is my story, told from my vantage point as Siuslaw National Forest supervisor during the tumultuous years that followed.

The Siuslaw, a relatively small national forest at 1,000 square miles, nevertheless functioned for many decades after World War II as a high-octane timber machine, clearcutting about 4,000 acres a year to reliably produce more than 300 million board feet. The NWFP delivered a knockout blow. The extensive network of reserves left very little Matrix for continued timber harvest; the new planned harvest penciled out at only seven million board feet per year—a 98 percent reduction! My blunt assessment: we were out of business.

The path forward is described in detail in a portion of my memoir *Toward a Natural Forest* (Oregon State University Press, 2015), and briefly below.

Several crises confronted us: (1) local communities and woods and mill workers faced epic disruption, (2) a large organization and budget both faced collapse, (3) a dispirited and despairing workforce contained mostly those resistant to the challenges of the NWFP (yet a number I would describe as receptive), and (4) a landscape bleeding from decades of timber-centric assaults on mature forest and salmon streams needed restoration. Maintaining an optimistic attitude became the biggest challenge of my career.

Our basic challenge from the NWFP: protect mature forest and salmon habitat and restore damaged forests and streams. I saw a dilemma. We had a significant acreage of old clearcut units that had been planted with a high density of monoculture Douglas-fir, predicated on the belief that they would be clearcut again. Our old growth and Riparian Reserves contained many old clearcuts, and I wondered whether these dense plantation forests could attain desired old-growth character if simply left alone.

Another dilemma related to our attaining even the minimal harvest proposed for our Matrix lands was that accumulated evidence of our owl and marbled murrelet surveys showed a 95 percent probability of discovery in our mature forests. Given that fact, why even bother trying to log big trees? This made little sense biologically or economically.

This circumstance led to my decision to cease clearcutting big trees and focus instead on thinning our numerous young plantations in ways that would, I hoped, accelerate their return to high-quality mature forest.

Game changer! Thus we pivoted, with much difficulty and angst, to a radically different future—one not predicted by the NWFP. But neither was it prohibited! Therein lies a tale. . . .

Agency hurdles strewn in our path proved most distressing to me. Financial solvency proved problematic in light of disappearing appropriated monies for timber harvesting. The emergence of stewardship contracting authority (which did not exist when the NWFP was issued) allowed the Forest Service to cut trees and retain revenue. It is not an exaggeration to state that Siuslaw efforts largely invented stewardship contracting in spite of agency headwinds.

Briefly, we sought to use timber money to commercially thin planta-tions mostly on (non-Matrix) reserve areas. My superiors claimed this vi-olated fiscal rules because such lands were not technically timberlands. I ultimately prevailed in arguing that any commercial sale of timber, even within a Riparian Reserve, constituted a legitimate use of timber money. In addition (after more internal fighting), we adopted dramatically sim-plified methods to reduce costs. Federal timber harvesting revenue proved critical to financing priority salmon restoration, even on private lands (thanks to Senator Ron Wyden), thus enabling holistic watershed approaches.

The Siuslaw—expected to produce only 7 million board feet per year by clearcutting mature timber—ultimately achieved a harvest of 40 mil-lion feet per year by thinning old clearcuts with virtually no controversy. At the same time, almost all other national forests encountered great difficulty meeting harvest expectations, in large measure because of re-liance on old-growth logging. But on the Siuslaw, ecological restoration became the hallmark of management.

We succeeded because we sought to apply the spirit of the NWFP, but not be bound by its strictures. We invented ways and means to make the plan work on the ground, addressing restoration needs while pro-tecting vital habitats. I dare to think the NWFP authors celebrate this outcome. In fact, they established Adaptive Management Areas through-out the owl habitat region in hopes of such outcomes.

I'll be blunt—the imposition of the NWFP by researchers sent a po-tent message to land managers: "You failed, but we can help you solve this crisis." The message was not well received. In fact, I think many of my peers resented researchers usurping their authority and responded, at best, with little enthusiasm; at worst, they secretly hoped the phenom-enon of the NWFP would simply fade away.

But the restoration and protection principles embodied in the NWFP have strengthened, not faded. Mistakes were made requiring principled correction. I am proud to have been a part of such history.

Jim Furnish was supervisor of the Siuslaw National Forest from 1992 to 1999.

FINDING COMMON GROUND ON THE
GIFFORD PINCHOT NATIONAL FOREST
(Emily Platt)

In the early 2000s, despite the solution provided by the Northwest Forest Plan, the timber wars were still ongoing in Pacific Northwest national forest management. Interest groups still advocated individually and directly to the Forest Service on whatever issues concerned them, and communication between interest groups was often through barbed comments in the local paper. Gridlock was the norm in many places, including on the Gifford Pinchot National Forest (GP). Spotted owls weren't being recovered. Local economies had tanked. Timber production had come nearly to a halt, and virtually every GP timber sale offered was litigated by the organization I worked for at the time, the Gifford Pinchot Task Force.

The GP Task Force hired me in 2001 to stop old-growth timber sales and work with rural communities around the GP. I began to meet individually with key stakeholders to try to understand the various interests. After dozens of initial meetings, it seemed that some common ground was possible, and the GP Task Force and partners organized a two-day field tour and invited a diverse array of interests, including county commissioners, loggers, mill workers, conservationists, tribes, local congressional staff, and many others. At the end of the field tour, participants agreed there was enough potential common ground to move forward, and the Pinchot Partners collaborative group was born. Collaboration was still relatively new in the Northwest and had taken root only in scattered places, supported financially and technically by organizations like Sustainable Northwest.

The Pinchot Partners' first project was the painfully small 40-acre Cat Creek plantation thin. The project tested the waters and gave group members a chance to find agreement on something tangible. Critically, it also enabled group members to get to know each other. Building trust was essential to early work and remains so today. Time in the field together and informal gatherings like a drink at the local bar after meetings helped develop working relationships between collaborative members. The Cat Creek project, while small, showed that projects on the ground could create volume and jobs for local communities while helping restore forest structure and function associated with older forests. Everyone could get behind this kind of work.

After finding agreement on plantation thinning, we worked together on other types of projects like replacing culverts to restore fish passage. After several years, we worked our way up to taking on NEPA analysis for roughly 2,000 acres of plantation thinning, over 20 miles of road removal, and associated restoration work. When the decision was signed, it represented the largest vegetation management project on the GP in over 20 years, and it included previously controversial work like reserve-land thinning and road removal. Building the trust needed to make projects like this happen took a lot of patience and time for everyone involved, and it took committed local leadership. Turnover in the Forest Service, and less often in partner organizations, was challenging because new people didn't have the context for how far everything had already come—they came in with a new reference point. Introductory meetings and briefing papers didn't capture the real change that had occurred on the ground. And yet, collaborative work gave me and many others a new and deeper appreciation of different perspectives, and I was grateful to be part of such positive change.

After years of collaborative work on the GP and in eastern Oregon, I began to see need for broader systemic change. I earned my PhD from Oregon State University's College of Forestry, where I studied landscape resilience to wildfire and Forest Service governance. After an internship at the Forest Service's headquarters in Washington, DC, and a couple of years working for the Pacific Northwest Regional Office, I became district ranger of the Mount Adams Ranger District on the GP. As district ranger, I was the primary decision-maker for land management on roughly 600,000 acres of national forest land. While I served as district ranger, we were able to apply the latest science to our projects and try new and innovative things, in large part because of a supportive forest supervisor, talented district staff, and exceptional collaborative partners working with us at every step along the way.

The Upper White project highlights some of the success we've had locally. Upper White is a large vegetation management project in the district's most at-risk forestlands in the eastern Cascades of southwestern Washington. Fires occur more frequently in this project area than in any other part of the GP, and adjacent areas had recently burned in large wildfires. Much of the project area is dominated by mixed conifer stands with extensive grand fir ingrowth because of fire suppression and past logging practices. The need for thinning, prescribed fire, and other work to reduce large-scale mortality and reshape the trajectory of these forestlands was evident. Yet the project area also incorporated the Gotchen LSR, an area set aside for species like the spotted owl that require late-successional forest habitat.

After many meetings and field tours, we created a project that protected the best owl habitat and aggressively restored the frequent-fire forests around this habitat both inside and outside the LSR. The project was supported by everyone from the logging community to the conservation community and the US Fish and Wildlife Service. Implementation is happening years before it would otherwise occur because of the financial support of Washington State and strong partnerships with the South Gifford Pinchot Collaborative and Mount Adams Resource Stewards. Strong partnerships and collaborative work have brought restoration to life on the Mount Adams Ranger District.

Emily Platt is now forest supervisor of the Helena–Lewis and Clark National Forest in Montana.

10

New Approaches but No Final Answers (2010–2020)

Achievement of the many goals President Clinton laid out at the Forest Conference in 1993 remained elusive as the second decade of the twenty-first century brought an environment awash in increasingly severe challenges, including climate change, invasive species, and wildfire. It became ever clearer that it would be difficult to resolve, once and for all, the issues associated with conservation of old-growth forests and the species within them.

This chapter begins with major trends and challenges relative to the issues that the creators of the NWFP sought to address, including conserving spotted owls, murrelets, and salmon; protecting old-growth forests; ensuring a continued supply of timber; and providing economic assistance to impacted workers and communities. Next comes a discussion of climate change, which the NWFP did not address in any detail, and its potential effect on conservation and management of federal forests under the NWFP. Then, the chapter covers plans and proposals that have been made to improve the NWFP.

Finally, the chapter briefly covers the firestorm of September 2020 in which over two million acres in the NWFP area burned, communities in the western Cascades of Oregon and southwestern Oregon were decimated, and thousands of people fled for their lives. Immediate lessons of those fires are highlighted along with a brief discussion of how the fires may reframe debates over federal forest management.

Trends and Challenges

The amount of mature and old-growth forest on federal lands stayed relatively stable in the NWFP area over the first 25 years (1994–2018) of the NWFP. Gains, mostly from Moist Forest ingrowth, offset losses, mostly from Dry Forest wildfire.[1] Even with the massive Biscuit Fire in southwestern Oregon in 2002 (plate 9.1), gains and losses generally balanced out across the area. Of course, the marked reduction in logging in mature and old-growth forest in the early 2000s certainly helped. However, even with 25 years of a fairly steady amount of older forest, the continued—and often accelerating—decline of northern spotted owl populations throughout their range[2] raised serious doubts that the spotted owl would ultimately survive. The closely related barred owl emerged as an existential threat to spotted owl survival—the northern spotted owl was in very deep trouble despite many years of massive recovery efforts.

Meanwhile, scientists found little evidence of an overall trend in the populations of the marbled murrelet across the NWFP area between 2001 and 2016.[3] Populations varied across individual states, declining in Washington and increasing in Oregon and California. In addition to the contribution of the NWFP to murrelet recovery, scientists emphasized that it would be important "to increase murrelet conservation on nonfederal lands, particularly those adjacent to NWFP lands, and in key areas (such as southwest Washington and northwest Oregon) where few federal reserves exist."[4]

Scientists and land managers increasingly recognized the Aquatic Conservation Strategy of the NWFP as a major advance in the conservation of aquatic and riparian ecosystems, and monitoring results over the first 20 years of the NWFP showed improvement in watershed and riparian conditions on federal lands there.[5] However, many populations of Pacific salmon within the range of the northern spotted owl were listed under the Endangered Species Act (ESA), a number of them soon after adoption of the NWFP. The National Marine Fisheries Service (NMFS) had listed one salmon population (Sacramento Chinook) as threatened under the ESA before 1994. By 2010, the NMFS had listed 22 more populations as threatened or endangered and identified critical habitat for many of them (table 10.1, plate 10.1).

Commercial salmon fishing off the coast of the NWFP area largely ended, impacting local communities, and recreational fishing was heavily restricted and limited mostly to hatchery fish. Also, the loss of the salmon greatly harmed restoration of tribal culture in the area. As an example, tribal communities in the lower Klamath River basin, such as the Yurok, Karuk, and Hoopa, had depended on salmon as a resource to sustain their way of life since time immemorial. Now they could take only a very few for ceremonial purposes. In addition, dams on the Klamath River totally cut off the Klamath Tribes in the upper river basin from their historical access to salmon runs (fig. 1.7).

Table 10.1. Pacific salmon populations (Evolutionarily Significant Units) within the range of the northern spotted owl listed as threatened or endangered under the Endangered Species Act during the period 1994–2010.

Species	Population
Chinook salmon (*Oncorhynchus tshawytscha*)	Puget Sound, WA
	Upper Columbia summer-fall run, WA
	Lower Columbia River, OR/WA
	Upper Willamette River, OR
	California coast
	Central Valley, CA
	Klamath River, OR/CA
Coho salmon (*Oncorhynchus kisutch*)	Lower Columbia River, OR/WA
	Oregon coast
	Southern OR / northern CA coast
	Central California coast
Chum salmon (*Oncorhynchus keta*)	Hood Canal, WA
	Lower Columbia, OR/WA
Sockeye salmon (*Oncorhynchus nerka*)	Ozette Lake, WA
Steelhead (*Oncorhynchus mykiss*)	Puget Sound, WA
	Upper Columbia, WA
	Middle Columbia, OR/WA
	Lower Columbia, OR/WA
	Upper Willamette, OR
	Northern California
	Central California coast
	Central Valley, CA

Source: Reeves et al., "Aquatic Conservation Strategy— Review of Relevant Science," table 7.1.

Thinning of young stands in Moist Forests continued as the primary silvicultural activity of the Forest Service and BLM. These successes led the agencies to implement the strategy at increasingly larger scales, as in the Lowell Country Project in the 95,000-acre Fall Creek watershed on the Willamette National Forest, which was approved in 2018 (plate 10.2 top). Thousands of acres of old-growth forest had been clearcut in that watershed between 1940 and 1990 and replaced with dense plantations of a single species (Douglas-fir) to produce high timber yields. The road system, most of which was built between 1950 and 1980, was in decline, which limited access, degraded water quality, and impeded fish passage.[6]

The Lowell Project will thin plantations, close unneeded roads, and repair much of the remaining road system over a decade (plate 10.2 bottom). It should also produce more than 200 million board feet of timber harvest that should generate $20–30 million in revenue for reforestation and restoration efforts and other uses.[7]

While the Forest Service concentrated on thinning plantations during 2010–2020, hundreds of thousands of acres of mature and old-growth Moist Forest remained in Matrix/Adaptive Management Area (AMA) land allocations as part of the allowable cut base. These forests were theoretically available for harvest under the NWFP, but few timber sales were proposed and even fewer implemented. Permanent protection of these forests was the focus of several congressional initiatives, but, despite many valiant attempts, all failed to pass into law.[8]

In the early 2020s, though, the Forest Service ramped up its attempts to log mature Moist Forest in Matrix/AMA.[9] These efforts generally have not been successfully completed and implemented for a variety of reasons, including protest against such actions. Why the agency would start offering mature forest at this time is unclear, but the level set for the Moist Forest allowable cut under the NWFP does depend, in part, on logging older forest in Matrix/AMA.

Concerns about wildfire and fuel loading increasingly dominated management discussions in the Dry Forests of the eastern Cascades, southwestern Oregon, and northern California. However, solutions—such as restoration of fire- and drought-resistant conditions—had to cope with the usefulness of these dense, fuel-loaded forests as habitat for spotted owls and their prey.[10] This conflict had deeply unsettled Jack Ward Thomas in a review he did in 2003 of northern California's national forests.[11] His frustration and dismay over the lack of action had boiled over as he drove through miles and miles of these forests on the Shasta-Trinity National Forest that were highly vulnerable to fire, drought, and insects. However, actions to address this problem were slow in coming over the next 15 years despite his review. There were a few bright spots, like the story of Emily Platt's Dry Forest thinning project on the Mount Adams District of the Gifford Pinchot (end of chapter 9), but they were exceptions.

In addition, large fires in Dry Forests on federal lands drained the energies and budgets of the Forest Service and BLM. The largest burned areas of these forests in the NWFP area lie in southwestern Oregon and northern California, and the Klamath National Forest there may be indicative of the future of federal Dry Forest management. The Klamath, overwhelmed by fire between 2010 and 2020, saw the major focus of management shift from forest restoration to wildfire preparedness by creating networks of fuel breaks along ridges and around communities, suppressing fires that did occur, and undertaking postfire actions to stabilize affected areas. Forest staff salvaged areas burned by high-severity fire to recover timber volume, generate revenue (part of which could be utilized in recovery efforts), eliminate hazards along roads, and reduce potential fuels for subsequent fires.[12] In this salvage logging, the Klamath National Forest removed significant numbers of large, old trees killed by fire.

Total timber harvests remained slightly below the adjusted NWFP timber sale goal of 800 million board feet per year (fig. 9.2 above), even including thinning volume from reserves that was not considered in the original projections. Harvest came primarily from relatively predictable and noncontroversial thinning in Moist Forests. Harvest volumes from Dry Forests, in comparison, were sporadic and often in the form of postfire salvage, especially in California.

Relatively little of the reduced stumpage revenue from federal harvest went to the national treasury or to counties in lieu of taxes, as had traditionally occurred. The Forest Service instead put that revenue into restoration activities on the national forests and, to some degree, on adjacent private lands.

A number of rural communities that had historically depended on federal timber for their mills continued to struggle, as described by Kirk

Johnson in his 2014 *New York Times* article "A Town That Thrived on Logging Is Looking for a Second Growth." Environmental protection, mechanization, and consolidation have left once-prosperous towns like Sweet Home searching for the next step. Processing timber from the nearby Willamette National Forest was once a steady source of employment but is much less so now. The Rice family illustrates the difficulties there. Dan Rice continues his family's log-trucking business as best he can, while his wife, Cindy, runs the local food bank and sees firsthand the pattern of poverty and desperation left behind when the local mills closed.

As Johnson summarizes: "Some logging families like the Rices stayed on, but many more moved away, chasing hope or fleeing calamity. An influx of retirees and commuters partly filled the gap, colonizing areas around Foster Lake a few miles from downtown, with their new white-trimmed townhouses. The patchwork that resulted—some people with money, many people without and few ways to earn a wage—now defines much of rural Oregon."

However, community leaders and county commissioners in places like Sweet Home continued to battle for economic and social change that would benefit their towns and the surrounding areas. They argued for more timber harvest from federal and state forests, they created more inviting economic conditions for firms to locate there, and they leveraged the recreational resources nearby, like the Santiam River and Foster Reservoir near Sweet Home, to create desirable conditions for second homes. They never gave up.

Hayfork, a small, isolated community in the Trinity Alps of northern California, also fit the model of a community in trouble: it lost its sawmill in 1993. As Lynn and Jim Jungwirth and other residents watched a stream of U-Hauls moving people away and the town went into a swoon, they vowed not to let the town blow away. Toward that end, they tried out many ideas and sought advice from many quarters, settling on a sawmill that used small material from thinning, and a watershed restoration institute. The mill lasted about a decade, but the Watershed Research and Training Center continues today, providing a wide variety of forest and watershed restoration services. Gorgeous scenery also helped attract retirees and others who had mastered the art of working remotely. Marijuana farming was legalized, which proved an economic boon to some, although it also brought its own headaches. Over time, Hayfork reversed its population decline. It did not blow away.[13]

Many communities near urban centers or recreation areas continued to diversify economically, and their dependence on federal timber became a distant memory. Philomath, Oregon, went from ten sawmills to one but benefited from proximity to Oregon State University's booming growth, which sent people scurrying to adjacent communities for affordable housing. Hood River and Bend, Oregon, continued to grow and became world renowned for wind sailing on the Columbia and skiing on Mount Bachelor, respectively. The Mount Baker–Snoqualmie and Olympic National Forests in western Washington became major recreational resources for the growing Puget Sound region.

Management of private industrial forests, which often border federal lands, was increasingly dominated by investment firms that utilized intensive timber management to optimize financial return to the extent permitted by law (fig. 10.1). The shift in forestland management from integrated forest product firms to investment firms began in the mid-1990s because of changes in tax laws and, more broadly, in response to global capital markets. The change in ownership shifted management objectives away from providing long-term timber supplies to mills to

Figure 10.1. A checkerboard of corporate lands and BLM lands west of Eugene in Oregon's Coast Range. Most intact forest is on BLM land. (Source: National Agriculture Imagery Program, NAIP).

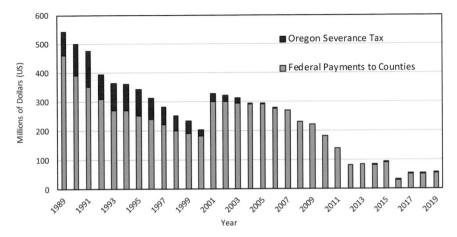

Figure 10.2. Annual payments to counties in western Oregon for the period 1989–2019 from the federal government (mostly from federal stumpage before 1995 and appropriations after 1995) and from local property taxes on timber (state severance taxes). (Source: Schick et al., "Big Money.")

obtaining income from timber sales, accomplished mainly by intensive tree farming with short (30-to-40-year) rotations.[14] Integrated firms that survived also felt financial pressures to shorten rotations. This shift created even more pressure on federal lands to maintain habitat for owls, murrelets, salmon, and hundreds of other species that thrive in older, natural forests.

Rural counties increasingly faced shortfalls in funding for county services, from libraries to public health to sheriff's offices, as federal payments to counties declined (fig. 10.2). These shortfalls added to the pressure to increase federal timber harvest levels, especially on BLM lands in southwestern Oregon, where harvest had historically returned high levels of county payments. Additionally, changes in Oregon's property tax laws greatly reduced taxes on private forest landowners (fig. 10.2). These losses caused hard financial times for many county governments.

A Warming Pacific Northwest

Scientific concern about the effects of climate change intensified by the early 1990s, as reflected in publications like "Effects of Global Climatic Change on Forests in North America" by Franklin and his colleagues. While FEMAT did not directly address climate change, it did view ensuring forest ecosystem resilience to disturbances as a major objective. Thus, FEMAT incorporated several features to help provide this resilience, including creation of an LSR network sufficiently redundant that it could withstand anticipated disturbances, and recognition of the need to reduce tree densities in Dry Forests to minimize the impact of wildfire, drought, and bark beetles.

By the second decade of the twenty-first century, global warming had become an overarching concern in natural resource policy and management. Halofsky and colleagues, in "Changing Wildfire, Changing Forests," summarized expected temperature increases: "Compared to the historical period from 1976 to 2005, 32 global climate models project increases in mean annual temperature for the middle and end of the twenty-first century in the Northwest. These projected increases range from 2.0 to 2.6 °C for mid-century (2036 to 2065) and 2.8 to 4.7 °C for the end of the century (2071 to 2100), depending on future green-house gas emissions."[15]

They noted that warming is expected to occur during all seasons, with most models projecting the largest temperature increases in summer, and all models suggesting a future increase in heat extremes—that is, days with maximum temperature above 90°F. Predictions of changes in precipitation are less certain, with some models predicting slight decreases and some predicting large increases. However, most models project a decrease in summer precipitation and an increase in the number of days with extreme amounts of precipitation.[16]

Global warming affects ecosystems directly by altering the metabolism and population dynamics of organisms, and indirectly by altering disturbance regimes.[17] Extended, drier summers impact some plant and animal species more than others. Species that have persisted during times of rapid change have life-history traits that allow for survival in stressed and frequently disturbed environments. For the Pacific Northwest, these species include red alder, Douglas-fir, Oregon white oak, ponderosa pine, and lodgepole pine, suggesting that these species will be more successful than others in a rapidly warming climate.[18]

Western hemlock, on the other hand, is relatively sensitive to hot, dry conditions even on moist soils. Since hemlock provides much of the canopy in many old-growth Moist Forests, its widespread decline would significantly alter the structure and function of these forests. One benefit of such change might be an increase in understory shrubs and herbs and associated fauna, but it could also result in reductions in the gross primary productivity (the carbon fixed during photosynthesis) of that ecosystem.[19]

Carbon stocks, particularly in older Moist Forests noted for large quantities of woody debris, are also a concern. Logs on the forest floor in these forests are saturated with water for much of the year, which slows decay processes. Decay rates accelerate as logs dry during the summer months and increase carbon dioxide emissions. Longer and drier summers could result in gradual decreases in dead wood carbon stocks in these forests and might shift older Moist Forests to being net sources of carbon (as CO_2 emissions) to the atmosphere until a new equilibrium is reached.[20]

At the landscape level, warming trends will likely cause the arid fringe of Dry Forests to contract, as sites shift to nonforest vegetation over time, and enable higher Moist Forests to expand above the existing tree line.[21] In the western Cascades of Oregon, recent simulations suggest that Dry Forests will creep north and upslope over time, shifting (Moist) Douglas-fir/western hemlock forests to (Dry) Douglas-fir forests.[22] Wildfire could accelerate this shift, as seedlings of some species may not be able to reestablish themselves in the severe environments that follow such disturbances.

Changes to disturbance regimes will likely include more frequent and severe wildfire, more severe storms of both oceanic and continental origin, and more outbreaks of insects and diseases.[23] Increases in wildfire would be a direct result of increased temperatures and moisture deficits in a region already characterized by relatively dry summers; the resulting lengthening of the fire season has already started.[24] More frequent and extended droughts, more hot, dry east winds, and increased potential for lightning could make extraordinary burning conditions more common.[25]

Halofsky and colleagues, in "Changing Wildfire, Changing Forests," and Reilly and colleagues, in "Climate, Disturbance, and Vulnerability to Vegetation Change," project significant future increases in wildfire severity and frequency in Dry Forests. Some may be converted to shrub fields as a result of repeated fires, especially in southern Oregon and northern California.[26]

In Moist Forests, on the other hand, Halofsky and colleagues project only a moderate increase in wildfire frequency, extent, and severity with climate change.[27] However, that conclusion depends on little or no increase in the frequency of hot, dry east winds that drive major wildfire events. Otherwise, large, high-severity fires in Moist Forests may indeed become more frequent.

Outbreaks of pests and pathogens are more likely as warming alters the biology of pest organisms or the vulnerability of the host tree species or both simultaneously. For example, drought effects have been observed through outbreaks of bark beetles in the Southwest and California, which are reflections of reduced tree vigor from water stress.[28]

Climate change impacts on streams in the NWFP area are already apparent. Winter and spring stream temperatures have warmed over the past 50 to 70 years. Spring snowpacks are smaller and melt-out occurs earlier, so low summer flows occur earlier in the season. These changes, along with rising summer air temperatures, have increased water temperatures in many streams, although effects vary with local conditions, such as stream orientation and amount of topographic shading. These effects are expected to accelerate as temperatures continue to rise along with an increase in the frequency and magnitude of flood events and other disturbances.[29]

Warming conditions in streams will alter the distribution and abundance of many native fish in the NWFP area. Most salmon populations will suffer although the degree of risk varies among species (fig. 10.3), depending on physiological requirements, life-history and migratory patterns, habitat preferences, and other considerations.[30]

Overall, climate change has become widely recognized as a context and driver for decisions about conservation and management of federal forests in the NWFP area. Managers and scientists see definite trends but also face considerable uncertainty about where it will all lead.

Climate Mitigation and Adaptation Strategies Needed

Many suggestions have emerged about how federal forest management in the NWFP area might help mitigate climate change and adapt to it.

Mature and old-growth Moist Forests contain immense carbon stocks in both living and dead biomass, which harvesting disrupts and releases to the atmosphere. Carbon storage capacity (sequestration) has often justifiably been cited as a reason for reserving remaining older Moist Forests because of their massiveness, stability, and potential for additional carbon storage.[31]

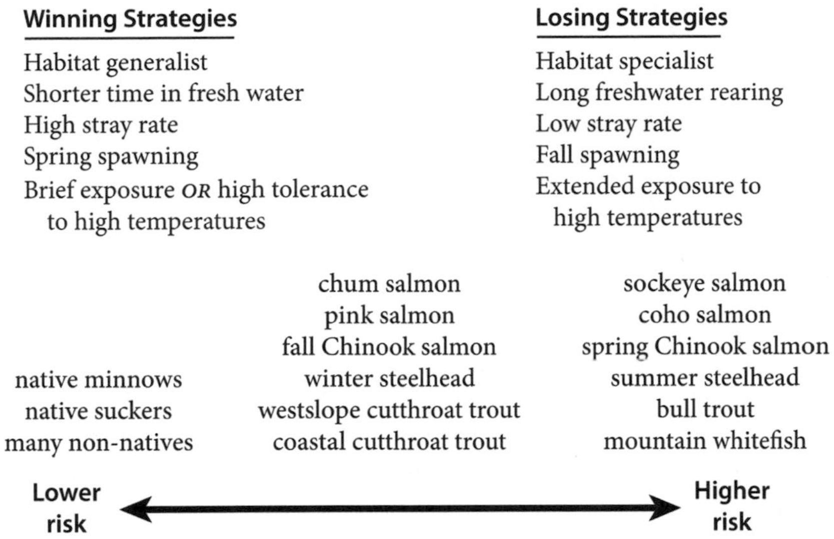

Winning Strategies		Losing Strategies
Habitat generalist		Habitat specialist
Shorter time in fresh water		Long freshwater rearing
High stray rate		Low stray rate
Spring spawning		Fall spawning
Brief exposure OR high tolerance to high temperatures		Extended exposure to high temperatures

	chum salmon	sockeye salmon
	pink salmon	coho salmon
	fall Chinook salmon	spring Chinook salmon
native minnows	winter steelhead	summer steelhead
native suckers	westslope cutthroat trout	bull trout
many non-natives	coastal cutthroat trout	mountain whitefish

Lower risk ⟵————————————————⟶ **Higher risk**

Figure 10.3. Life cycle and habitat preference strategies of freshwater fish that may determine whether they are winners or losers with a warming climate in the NWFP area. (Source: Reeves et al., "Aquatic Conservation Strategy: Relevance to Management.")

In addition, continuing to log older forests would not be a carbon-smart way to meet wood product needs. The shift in harvest location caused by the contraction in federal timber harvest in the Pacific Northwest resulted in a net saving in carbon sequestration in forests. Setting aside older forests under the NWFP diverted harvests to other regions, especially the southern United States, where the forests cut were typically younger plantations, with the result that carbon loss per unit of volume harvested was much less.[32]

While a strong case can be made for reserving older Moist Forests as a climate mitigation strategy, making a case for logging older Moist Forests as a climate adaptation strategy is more difficult. Fuel loadings are naturally high because of their productivity, and efforts to reduce them by thinning would not only reduce stored carbon but could easily produce unnatural ecosystems with unpredictable consequences for native fauna, including listed species.[33] If the hotter and drier conditions predicted by climate models materialize, though, fuel management in the southern part of these forests will need consideration.

Adaptive climate strategies are promising, though, for young Moist Forest stands, especially plantations. These strategies include diversifying forest tree composition, through incorporation of deciduous hardwoods and other species as important stand components, and developing resilient forest structures such as mixed-age stands.

Restoring Dry Forests to conditions more comparable to those of their historical state would be a major first step in adapting them to a warmer, drier, and more fire-prone future.[34] Not surprisingly, scientists increasingly recommend fuel reduction and forest restoration strategies for these forests.[35]

Dry Forest restoration also increases resistance to drought and insect outbreaks, which is significant because unnaturally dense Dry Forests are often dominated by short-needled species, such as grand or white fir. Such stands can suffer severe mortality in their second century of life from infestations of insects and disease, even if they do not burn; hence they are unstable (boom-and-bust) ecosystems insofar as carbon storage is concerned. Finally, and most profoundly, restoration in Dry Forests produces ecosystems much more characteristic of the historical forests in terms of function and associated plants and animals.

Adaptation strategies have also been suggested to help fish populations survive in the warming world, although most needed changes (such as providing shade along streams) are on nonfederal land, as the application of the Aquatic Conservation Strategy (ACS) ushered in an era of protective management on federal lands in the NWFP area. However, the developers of the ACS and their colleagues did emphasize that more active management on federal land to increase stand diversity in plantations in Riparian Reserves would be beneficial. Also, shifting federal management and research from increasing the size of salmon populations to increasing their life-history diversity would help salmon survive in a changing world, as would keying management to the variable effect of climate change across the landscape.[36]

Taking an Ecological Forestry Approach

The decade of 2010–2020 saw many proposals and plans to better meet the goals set forth for the NWFP. One called for taking an ecological approach to forest management on federal forests, as suggested by Franklin and Johnson in a 2012 *Journal of Forestry* article, "A Restoration Framework for the Federal Forests of the Pacific Northwest." Members of the Oregon congressional delegation, especially Congressman

DeFazio and Senator Wyden, requested new approaches to managing the federal forests within the NWFP. Franklin and Johnson responded with a strategy reflecting two trends that had developed through years of wrangling over implementation of the NWFP. First, federal forestry projects that provided ecological benefits passed most easily through consultation with the regulatory agencies and the gauntlet of protest and appeal from environmental groups. Second, ecological projects that produced commercial timber harvest, such as plantation thinning, were widely supported by the timber industry and the communities historically dependent on federal timber. Thus, Franklin and Johnson searched for federal forest management strategies that would provide both ecological and economic benefits, which they labeled *ecological forestry*.

They began by recognizing that Moist and Dry Forests required fundamentally different approaches, building on distinctions recognized in FEMAT. Their Moist Forest strategy called for reserving mature and old-growth forests. Moist Forest timber harvests on federal lands would focus on the expanses of young forest, with two objectives. Thinning younger stands would continue, with silvicultural prescriptions designed to accelerate development of structural and compositional complexity. Variable retention harvest in younger stands in Matrix/AMA would be used to create openings that retained significant biological legacies of forest patches, trees, snags, and logs and nurtured development of early successional (preforest) ecosystems, which would include plants such as forbs and shrubs that required sunlit environments.[37] Over time, trees would regain dominance of the site, and a forest more characteristic of historical ecological conditions would emerge.[38]

The Franklin/Johnson ecological forestry strategy in Dry Forests called for forest restoration across the landscape, with the goal of reducing stand densities to protect old trees within those forests and restoring more fire- and drought-resistant forest conditions. They argued that the highest priority should be placed on stands with concentrations of old trees, as the unnatural buildup of younger trees around the older ones would likely doom the old trees through the effects of wildfire, drought, and insects unless action was taken.

Franklin and Johnson collaborated with the BLM in demonstrating ecological forestry approaches in southwestern Oregon from 2012 to 2014 in a series of pilot projects at the request of members of Oregon's congressional delegation and Secretary of the Interior Ken Salazar, who oversaw the BLM. Projects were conducted in both Moist Forests and Dry Forests.

Several variable retention harvests were undertaken as pilot projects in young (50-to-80-year-old) Moist Forest stands on previously clearcut sites in the Roseburg and Coos Bay BLM Districts. They drew considerable discussion and comment but no litigation. However, a proposed Moist Forest pilot harvest in an unmanaged (unharvested) older stand, called White Castle, generated an outpouring of protest and litigation from local environmental groups, whose members occupied the stand for many months.[39] Eventually, the courts rejected authorization of the sale, utilizing, somewhat ironically, the work of Franklin and Johnson about scientific controversies surrounding the ecological benefits associated with harvests in natural stands like White Castle that contained spotted owl habitat. Most importantly, the White Castle case reminded everyone how controversial the inclusion of older natural stands in plans for timber production from federal lands would be.

The Medford BLM District demonstrated the Dry Forest strategy in Pilot Project Joe by reducing forest densities and accumulated fuels that had resulted from past fire exclusion while retaining and nurturing existing old-growth trees. Thinned areas were part of a landscape-level design in which some forest areas used by northern spotted owls were left unharvested. In this way, the owl reserves should have their potential survival (or hang time) enhanced by thinning around them to reduce fuels and make fires easier to fight.

Saving the Northern Spotted Owl One More Time

After President Obama's election in 2008, the US Fish and Wildlife Service (USFWS) revised the recovery plan and critical habitat designation for the northern spotted owl in 2011 and 2012, respectively.[40] In many ways, these two actions, and the analysis behind them, constituted the sixth science assessment of federal forest management that would be compatible with conservation of species and ecosystems in the owl's range. The USFWS's Portland office under the leadership of Paul Henson led the effort and utilized staff scientists and professionals, a panel of outside owl experts, and sophisticated modeling algorithms.

With the owl's precarious status, it was not surprising that the USFWS moved beyond the NWFP in terms of recommendations and restrictions on federal forest management activities as well as the amount of forest considered important to owl recovery.

The *Revised Recovery Plan* recommended, among other things, that the Forest Service and BLM conserve spotted owl activity centers and associated high-value spotted owl habitat, regardless of whether the sites were currently occupied by spotted owls (Recovery Action 10); conserve structurally complex, multilayered conifer forests across the landscape (Recovery Action 32); and conserve and restore large trees and snags after wildfire in LSRs (Recovery Action 12).[41]

The *Revised Recovery Plan*, as summarized by Henson and colleagues in "Improving Implementation of the Endangered Species Act," also attempted to address a fundamental issue in restoration of Dry Forests—the conflicts that can occur when ecosystem restoration goals do not fully align with the more narrow or short-term conservation needs of individual species:

> For example, even in situations in which there is a general scientific consensus that some degree of active intervention is necessary to restore fire-prone, ecologically departed forests in the West to a more natural and resilient state . . . there is still reluctance to take action if such actions might harm listed species in the short term. . . . Ironically, this deferral may put those same listed species at greater risk in the long term.
>
> To reconcile this conflict for northern spotted owls, the USFWS developed an overarching vision in the owl recovery plan (USFWS 2011) that can best be summarized as *If it's good for the ecosystem, it's good for the owl.* The plan explicitly encourages active management for maintenance or restoration of forest ecological processes (e.g., uncharacteristic wildfire, managing invasive species) even if there are short-term impacts to spotted owls. Of course, this vision depends on a robust scientific understanding of the ecological processes and the relative risks and rewards of taking action. . . . But it was appropriate to make these recommendations explicit in the recovery plan to overcome a tendency for risk-averse land managers to forego

taking actions that may be controversial, even though they are scientifically and ecologically justified.[42]

An important and potentially game-changing concept: if it's good for the ecosystem, it's good for the spotted owl. However, it must be acknowledged that taking risks with management of a threatened or endangered species for the greater good always presents challenges and will face many obstacles. Still, a focus on ecosystem function potentially set the USFWS on a new course in conservation of the northern spotted owl.

A year later, in 2012, the USFWS expanded the amount of critical habitat for the spotted owl and included significant amounts of Matrix/AMA (plate 10.3). Most of the Late-Successional Reserves (LSRs) were included in the critical habitat designation, but the agency concluded that NWFP reserves and standards would not be sufficient to recover the owl—conservation of more habitat would be needed.

The USFWS's modeling underlying its identification of critical habitat for the spotted owl also showed that more was needed than expanded habitat protection—barred owl control would be essential.[43] While this conclusion was tentative given the limited amount of scientific work on interactions of the two owls by 2012 when the modeling was done, the analysis solidified USFWS understanding that competition between spotted owls and barred owls could sink any chance of recovery of the northern spotted owl.

Barred owls are native to eastern forests in the United States and were historically separated from western forests by the treeless Great Plains. European occupation of the Great Plains led to fire-exclusion policies, fundamentally altering the ecological character of those systems. Forests filled in as the burning stopped and trees were planted in towns and farmsteads, which may have facilitated the bird's movement westward. The barred owl arrived in British Columbia and Washington around the beginning of the twentieth century and remained largely a novelty in the Northwest until the 1980s and 1990s, when its numbers exploded and it spread into California.[44]

In the Pacific Northwest, barred owls and northern spotted owls select similar habitat for nesting and roosting.[45] However, their diets differ. The northern spotted owl is relatively specialized, particularly in

the northern part of its range where it feeds primarily on flying squirrels and typically utilizes large home ranges of 2,000–5,000 acres.[46] In
contrast, the barred owl utilizes a very broad range of prey, including
shrews, moles, salamanders, fish, crayfish, and snails. Consequently,
barred owls have much smaller home ranges than spotted owls and
occur at much higher densities. Scientists have reported three to eight
barred owl pairs per spotted owl territory,[47] and recent USFWS control
efforts have found up to 25 barred owls per spotted owl territory.[48] For
a spotted owl trying to compete with this new invader, the odds are
overwhelming.

     While both owls are territorial, barred owls are very aggressive in
creating and defending territories and are 15 to 20 percent larger than
spotted owls. When they encounter each other, spotted owls invariably
lose the battle and either leave the area or learn to stay silent and avoid
further encounters. Since spotted owls rely mainly on vocalizations to
establish and maintain territories, find and communicate with mates,
and feed their fledged young, silence means low rates of survival and
reproduction.[49]

     Creating additional habitat suitable for northern spotted owls
would not provide a solution to this competitive relationship. As
Lowell Diller, a practicing wildlife biologist in northern California and
barred owl expert, has succinctly noted, "There is no known habitat
exclusive to spotted owls . . . [which] means there is no known habitat
solution for conserving spotted owls, and the most likely outcome from
setting aside more habitat will be to have even more barred owls."[50] All
these considerations caused Diller to conclude: "With their size and
numerical advantage, the inescapable conclusion is that barred owls
are capable of taking over and excluding spotted owls from all available
nesting habitat."[51]

     The dire state of the northern spotted owl was summarized in 2021
by Allen Franklin and other leading spotted owl experts:

> We conducted a prospective meta-analysis to assess population
> trends and factors affecting those trends in northern spotted
> owls using 26 years of survey and capture-recapture data from
> 11 study areas across the owls' geographic range. . . . We found
> that northern spotted owl populations experienced significant

declines of 6–9% annually on 6 study areas and 2–5% annually on 5 other study areas. Annual declines translated to ≤35% of the populations remaining on 7 study areas since 1995. Barred owl presence on spotted owl territories was the primary factor negatively affecting apparent survival, recruitment, and ultimately, rates of population change.... Our analyses indicated that northern spotted owl populations potentially face extirpation if the negative effects of barred owls are not ameliorated while maintaining northern spotted owl habitat across their range.[52]

Barred owl impacts on other species in northwestern old-growth forests, however, go far beyond its ability to drive the northern spotted owl to extinction. It is an omnivorous bird with a high reproductive potential, and a new top predator in these forest ecosystems that feeds on many species that have never been subjected to such intensive predation. Diet analyses of barred owls indicate that they feed heavily on a variety of native fauna in these forests, including small birds (such as screech owls), small mammals, crustaceans, amphibians, fish, insects, and bats. Since barred owls feed on a such a wide variety of prey, which allows them to have small home ranges, they can and do densely populate northwestern forest landscapes, with potentially severe long-term negative consequences for prey species.[53]

In sum, the barred owl has taken over much prime northern spotted owl habitat in the NWFP area and has the northern spotted owl on the run, leading to a precipitous decline in spotted owl numbers and the likelihood that it will disappear from much if not all of its range. The impact of the barred owl on numerous prey species provides another reason for grave concern.

To explore and document the effectiveness of removing barred owls, the USFWS began a 10-year experiment in 2013 on the effect of barred owl removal in four study areas.[54] Knowing that removing one raptor (albeit an invasive species) to save another was certain to be controversial, the USFWS secured support for the experiment from a wide variety of groups, including the Humane Society, before moving ahead.

Barred owls can be removed by either shooting or trapping. Aggressive and highly territorial, they have a biological imperative to fly up and challenge intruders to their territories, which quickly puts them within shotgun range. Capturing barred owls is tougher and raises the question of what to do with hundreds or thousands of captive owls. Thus, the USFWS focused on shooting barred owls in its study of experimental removal of the bird.[55]

A major publication in 2021 by barred owl and spotted owl experts, summarizing the results of the removal study, provided strong evidence that such removals make an enormous difference to spotted owl survival:

> After removals, the estimated mean annual rate of population change for spotted owls stabilized in areas with removals (0.2% decline per year), but continued to decline sharply in areas without removals (12.1% decline per year). The results demonstrated that the most substantial changes in population dynamics of northern spotted owls over the past two decades were associated with the invasion, population expansion, and subsequent removal of barred owls.[56]

Barred owl removal helps reverse, or at least substantially slow, the precipitous decline that spotted owl populations have experienced in recent years—no doubt about it. The scientists have presented their evidence—a convincing story it tells, indeed. Next come the economics, politics, and ethics of killing one owl to save another.

Creating a Separate Forest Plan for BLM Lands

The BLM's western Oregon forests were held to the same environmental standards as the national forests during development of the NWFP. All federal forests were required to provide well-distributed habitat for viable populations of vertebrates and diversity of plant and animal communities, even though the law on which these requirements were based applied directly only to national forests. Also, the Clinton administration gave no special credence to the 1937 Oregon and California Revested Lands Sustained Yield Management Act (O&C Act), which defines management goals for most of the BLM's western Oregon lands, and did not mention that law in instructing FEMAT. This approach has continued to be a bone of contention both inside and outside the BLM ever since the NWFP was created.

The forest industry and southwestern Oregon counties have been especially persistent in seeking a separate plan for these BLM lands in western Oregon that better reflected (in their eyes) the BLM's authorized mission in the 1937 O&C Act, which applies to most of these lands. The act specified that O&C timberlands "shall be managed for permanent forest production, and the timber thereon shall be sold, cut and removed in conformity with the principle of sustained yield for the purpose of providing a permanent source of timber supply, protecting watersheds, regulating stream flow, and contributing to the economic stability of local communities and industries, and providing recreational facilities."[57] Consequently, numerous attempts have been made to extract the BLM lands from the NWFP by administrative rule, plan revision, legislation, and litigation. These extraction efforts finally hit pay dirt in 2016.

While Judge Dwyer ruled that adoption of the NWFP was within the discretion of the secretaries of agriculture and the interior, he did not comment on whether all NWFP provisions were necessary or required equal application across different federal ownerships. The judge did endorse cross-boundary ecosystem management, however, and warned that unilateral changes in land management by one federal agency could require compensating measures by others.

Both the second Bush administration and the Obama administration committed to the BLM developing its own forest plan. While the Bush administration's effort to develop and implement a plan for the BLM lands failed, partially because of lack of consultation with the USFWS and the NMFS about protection of species listed under the ESA, the Obama administration successfully developed a plan that passed muster with the two regulatory agencies. In 2016, the BLM approved new forest plans for its western Oregon lands, separating management of these lands from the NWFP, and began implementing them as litigation challenging the plans proceeded on multiple fronts.[58]

The BLM developed separate Resource Management Plans (RMPs) for two planning areas: northwestern and coastal Oregon (predominantly Moist Forests) and southwestern Oregon (predominantly Dry Forests). These RMPs reflected the BLM's focus on conservation of species listed as threatened and endangered under the ESA and other species of concern. They also clarified that management of these forests would not be bound by the Forest Service's species viability and diversity requirements.[59]

The landscape design of the 2016 RMPs still resembled that of the NWFP in many ways, with the forests divided between reserves (LSRs and Riparian Reserves) and timber production areas (the Harvest Base). In total, almost 80 percent of the forest was placed in reserves, slightly more than under the NWFP, including all stands in Moist Forests over 160 years of age and most stands in Dry Forests over that age (fig. 10.4).

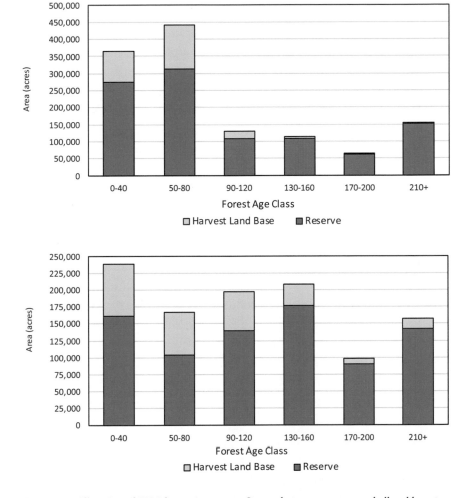

Figure 10.4. Allocation of BLM forests in western Oregon between reserves and allowable-cut land base (Harvest Base) in the BLM's 2016 Resource Management Plan for (above) northern and coastal forests (mostly Moist Forest) and (below) southern forests (mostly Dry Forest). (Source: BLM, "Resource Management Plans" and related files.)

Within this overall similarity in landscape design between the NWFP and the 2016 BLM RMPs, some differences did occur. The BLM worked with the USFWS to identify complex stands in the Harvest Base that were turned into small LSRs, reflecting recommendations in the northern spotted owl *Revised Recovery Plan* of 2011. The BLM also contracted the Riparian Reserves on fish-bearing streams from two tree heights to one tree height, reflecting its focus on federally listed fish species as compared to broader species concerns, and substituted its own aquatic protection standards for those of the ACS. Those standards were less restrictive in a number of ways, including the requirement of cumulative effects analysis at only one scale (watershed) and one time period (a few decades into the future) instead of the four scales and two time periods that were needed under court interpretations of the ACS.[60]

The BLM introduced two new harvest restrictions in its 2016 RMPs that represented potentially major improvements in ecosystem conservation. First, it required the retention of large, old trees in stands in the Harvest Base. Trees over 40 inches in diameter that were established prior to 1850 in Moist Forests and trees over 36 inches (ponderosa pine and Douglas-fir) or 24 inches (maple and oak) that were established prior to 1850 in Dry Forests must generally be retained during harvest. This decision represents the first time large, old trees in the Harvest Base or Matrix/AMA had been specifically reserved on federal forests in the NWFP area. Second, the 2016 RMPs prohibited salvage logging in Moist Forest reserves following wildfires, except where necessary to protect public safety or to keep roads and other infrastructure clear of debris, which greatly stiffened the BLM's salvage standards relative to those in the NWFP.[61] In Moist Forests, natural recovery is the surest way to create high-quality mature and old-growth forests in the future.

Silvicultural prescriptions for Moist Forests in the 2016 RMPs generally followed those of the NWFP, with young stand thinning prescribed in reserves and thinning and variable retention (final) harvest prescribed for the Harvest Base. Thus, the 2016 RMPs proposed reinitiating final harvests in Moist Forests on BLM lands—a practice that had largely ceased on federal forests in the NWFP area in the early 2000s. Overall, the RMPs more strongly embraced ecological forestry principles in Moist Forests than did the NWFP, including provisions for a biologically rich preforest stage. Still, variable retention harvest, especially of the small amount of mature forest in the Harvest Base, could be problematic.

Silvicultural prescriptions for Dry Forests in the 2016 RMPs relied mostly on partial cutting across all age classes to maintain an uneven-aged, fire-resilient condition throughout the landscape, while retaining large, old trees, expressing a greater concern about wildfire effects than did the NWFP. Timber production, within this overall goal, remained an objective in the Harvest Base. However, a high density of historical northern spotted owl sites occurs in much of the BLM's Dry Forests. Given the declining owl populations, it may be a difficult harvest strategy to implement unless it is linked to a comprehensive and successful barred owl removal program by the USFWS.[62]

The BLM's allowable cut (projected harvest volume per year from its Harvest Base) in the 2016 RMPs remained at about the same level as in the NWFP,[63] to the disappointment of many who had urged the creation of a separate plan for the BLM. Moist Forests would provide over 75 percent of the allowable cut from a Harvest Base composed primarily of young forest. However, variable retention harvest of the 10 percent of Moist Forest Harvest Base in mature forest would be a priority in the short run.

The plans also projected harvest volume from thinning in reserves, which raised the annual harvest target for the first decade by one-third.[64] That harvest level will decline over time in Moist Forests as the need for ecological treatments there declines, but continue in Dry Forests. The NWFP did not estimate harvest volume from reserves.

The BLM's 2016 RMPs shifted the geographic location of proposed harvests to the north compared to expected locations under the NWFP. Substantial increases were made in allowable cuts on the Salem and Eugene Sustained Yield Units, offsetting substantial declines in the Medford and Coos Bay Units. The northern units contain most of the BLM's young Moist Forest, the source of its future wood basket, while the southern units have experienced difficulties in integrating its timber harvest strategy with protection of the northern spotted owl (in the Medford Unit) and marbled murrelet (in the Coos Bay Unit).[65]

The increased reserves, salvage prohibitions, and extensive use of ecological forestry principles enabled Paul Henson, Oregon state supervisor of the USFWS, to call these plans "an improvement" for

BLM lands over their management under the NWFP.[66] The NMFS issued a biological opinion concluding that the plans were not likely to jeopardize endangered or threatened fish populations or adversely modify their critical habitat.

The Pacific Rivers Council sued all three federal agencies involved (BLM, USFWS, and NMFS), arguing that these agencies had failed to assess the consistency of their approach with the ACS. In its opinion upholding the district court ruling in favor of the three agencies, the appeals court affirmed that the ACS was not the only method for them to satisfy their obligation to determine a project's effect on listed fish species, and that the court deferred to agency expertise on these matters.[67]

The BLM's approach to plan development was an impressive aspect of the revision. As an example, the BLM developed its forest management strategies collaboratively with the USFWS, including the use of variable retention harvest in Moist Forest, both inside and outside critical habitat for the spotted owl. Also, after sharp criticism of the riparian buffer strategy in its draft plan by the NMFS, the BLM adopted a buffer width of one tree height on most streams in its final (2016) plan, with controls in thinning, which enabled the NMFS to issue a biological opinion that was supportive. The interagency approach championed by Katie McGinty and Jim Pipkin in development of the NWFP shone through in the BLM's development of its 2016 RMPs.*

All things considered, many of the BLM's 2016 RMPs for Moist Forests might be more feasible to implement than the NWFP (as it was written) because harvest in those plans is less reliant on mature and old-growth forests, and Survey and Manage requirements have been dropped. Also, the planning effort included more of the professional staff from the BLM, USFWS, and NMFS who would implement the plan than did the NWFP.[68] As of 2020, the BLM has been able to move forward in meeting its allowable cut from Moist Forests under its new plan.

Implementing the BLM's Dry Forest strategy should be challenging, despite the involvement of staff from many agencies. High numbers of historical spotted owl nests, the need to harvest in stands with significant numbers of large, old trees, and an organized citizenry in southwestern

---

* Jim Lyons, counselor to the secretary of the interior by 2015, also urged such an approach. The BLM and USFWS are both in the Department of the Interior.

Oregon opposed to many of the proposed harvests under the 2016 RMPs make for many potential roadblocks. As of 2020, relatively few timber sales have moved forward in Dry Forests under the 2016 RMPs.

Just as BLM staff began plan implementation, a court decision threatened to blow the 2016 RMPs out of the water with this statement: "Of this there can be no doubt: the 2016 RMP's violate the O&C Act." With these words, on November 22, 2019, Judge Leon of the District Court for the District of Columbia agreed with claims by the American Forest Resource Council, a trade association representing firms in the Pacific Northwest dependent on federal timber, that the BLM's O&C timberlands must be managed for timber production. He argued that

> the Act imposes two relevant, mandatory directives on BLM's management of all O&C land that has "heretofore or may hereafter be classified as timberlands. . . ." [That land] "shall be managed for permanent forest production, and the timber thereon shall be sold, cut and removed in conformity with the principle of sustained yield." . . . Use of the word "shall" in a statutory directive to an agency "signals mandatory action." . . . The 2016 RMPs violate these mandatory directives by excluding portions of the O&C timber land from sustained yield timber harvest.[69]

Judge Leon cited the precedent-setting case *National Association of Home Builders v. Defenders of Wildlife* (2007) to bolster his decision, in which the Supreme Court ruled that the ESA's Section 7 requirement that federal actions do not jeopardize the survival of a species simply "does not attach" to nondiscretionary mandates,[70] such as the requirement in the O&C Act that once "O&C land is classified as timberland, BLM is required to harvest the timberland pursuant to sustained yield principles."[71] Thus, Judge Leon concluded that "according to *Home Builders*, BLM cannot justify a refusal to abide by those statutory commands by pointing to Section 7 of the ESA."[72]

In a few short paragraphs, Judge Leon potentially created an existential threat to the philosophical foundation of the NWFP. The Clinton administration had crafted an interagency ecosystem plan for federal lands across the range of the northern spotted owl that would

meet federal laws for protecting species and ecosystems, especially the
National Forest Management Act (NFMA) and the ESA. The BLM's
2016 RMPs had pulled away from the NFMA requirements for species
viability and diversity, although the agency's need to meet the requirements of the ESA maintained many elements of that ecosystem plan.
Judge Leon's decision, on the other hand, called for a forest plan for the
BLM that provided sustainable supplies of timber volume from its timberlands, notwithstanding the requirements in the ESA for threatened
species.

Judge Leon's decision about the BLM's sustained yield mandate
opposed previous legal decisions, including Judge Dwyer's affirmation
of the NWFP, that the BLM must fulfill conservation duties imposed by
other statutes in managing the O&C lands. However, that court ruling,
and others that affirmed this responsibility, came before the Supreme
Court ruling in the *Home Builders* case that exempted mandatory
requirements in law from the ESA protections for threatened and
endangered species.

As of the end of 2022, BLM's forest management on its lands in
western Oregon remained unsettled. A new and radically different forest plan might be needed for those lands, one that would threaten the
scientists' vision of a conservation strategy for the northern spotted
owl—a vision that began with the ISC's report, was embraced by the
four science assessments that came after it, and formed the foundation
of the NWFP. Conservation of other species would also be impacted.

Judge Leon ruled that once "O&C land is classified as timberland,
BLM is required to harvest the timberland pursuant to sustained yield
principles."[73] If this view of the BLM's mandate under the O&C Act
is upheld upon appeal, how might any resulting ecological damage to
the BLM's forest resources be minimized, especially to mature and old-
growth forests? An ecological forestry approach, expanding on measures in the BLM's 2016 RMPs, would reduce the ecological impacts
of shifting forest from reserves to the sustained yield timber base.[74]
Mature and old-growth Moist Forests would potentially suffer the most
severe damage since they provide their highest ecological value by
being left intact. However, it would certainly be better to manage them
on a multiaged basis, harvesting individual trees and small groups, than
to clearcut them on short rotations. BLM"s uneven-aged management

stategy from its 2016 RMPs could be utilized in Dry Forests to reduce threats to old trees, reduce overall stand density, favor early seral species, and increase average stand diameter. Higher stand densities could be left after harvest in some patches of Moist and Dry Forests to improve their habitat usefulness for the northern spotted owl and other species that need dense, multilayered forests.

These suggested approaches would have different ecological effects in mature and old growth in Moist Forests and Dry Forests. In Moist Forests, the actions suggested would generally degrade older forests. How much degradation occurred would depend on harvest intensities and whether harvests were focused on younger trees and left mature and old-growth trees as legacies. These actions could actually help restore and maintain Dry Forests and be ecologically beneficial if old trees were retained; in fact, some partial cutting and/or prescribed fire would be needed on a continuing basis to maintain Dry Forest resistance to wildfire and drought-related insect outbreaks.

National Forest Planning Once Again

National forest planning under the National Forest Management Act began in 1979 in the NWFP area, and plans for most national forests in the range of the northern spotted owl were released in the late 1980s. Forest management strategies in those plans were then superseded by the NWFP in 1994, which was technically an amendment to individual national forest plans.

In 2015, the Forest Service announced plans to revise the national forest plans of the owl's range, as they had not been updated for many years. However, the agency would need to do that in a different planning world than that of the 1980s. The 1980s plans had been developed under the 1982 Planning Rule, which emphasized calculation of maximum sustainable timber harvest levels over time, constrained by the need to provide habitat for viable populations of native vertebrate species, general biodiversity, and watershed protection considerations. Further, the 1982 Planning Rule was framed largely in terms of the language of economics, with concepts like net public benefit, trade-off analysis, and minimum cost solutions guiding the instruction of what to do.[75]

By 2015, though, national forest planning had undergone a seismic shift—a once-in-a-generation change that caught forest planning up with the special role of the national forests in American life. In a policy sense, this new direction first surfaced in the late 1990s, toward the end of the Clinton administration, as Undersecretary Jim Lyons attempted to modernize the Planning Rule guiding Forest Service implementation of the NFMA. Lyons want to leave office with an ecosystem approach to forest planning, such as was developed in the NWFP, firmly embedded in national forest policy. Toward that end, he asked Norm Johnson to chair and help organize a second Committee of Scientists to advise the secretary of agriculture on changes to the Planning Rule. The committee issued its report in March 1999.[76]

At the time, the purpose of natural resource planning was often defined as finding a balance among the three aspects of sustainability (ecological, economic, and social).[77] The second Committee of Scientists, though, took the concept one step further—it identified ecological sustainability as the foundation of stewardship, with the conservation of native species and the productivity of ecological systems as the surest path to maintaining this sustainability. To help achieve that goal, the committee further called for maintaining conditions necessary for ecological integrity—that is, maintaining the characteristic composition, structure, and processes of ecosystems. Contributing to economic and social sustainability, while recognized as vitally important, would occur within the context of sustaining ecological systems.

The Clinton administration released a new planning rule based on the committee's concepts in the last few months of 1999 just before it left office. Not surprisingly, the second Bush administration, upon taking office in 2000, immediately withdrew the rule, and there things languished until the Obama administration came on the scene eight years later.

By 2012, the Forest Service had developed, and the Department of Agriculture had approved, a new Planning Rule. It called for the national forests to provide for ecological sustainability, measured through the maintenance and restoration of ecological integrity, while contributing to economic and social sustainability to the degree feasible—much as the Committee of Scientists had recommended.[78] Planning directives to implement this approach importantly described how to achieve ecological integrity in an era of climate change: "Plan components [should be designed] to provide ecological conditions to sustain functional ecosystems based on a future viewpoint. Functional ecosystems are those that sustain critical ecological functions over time to provide ecosystem services."[79] Legal challenges to the new Planning Rule did not succeed.

The Planning Rule also dealt with the two provisions of the 1982 rule that had given the Forest Service so much trouble in the Pacific Northwest. Regional guides, whose appeal set off the spotted owl litigation, disappeared. The blanket requirement to protect habitat for viable populations of all native vertebrates was no more. Rather, the 2012 Planning Rule and 2015 Planning Directives relied on ecosystem planning to protect biodiversity through the maintenance and restoration of ecosystem integrity, with viability analysis limited to those species (vertebrates and nonvertebrates) not adequately protected by such an approach. Planning was framed largely in the language of ecology. Detailed timber harvest scheduling to find maximum sustainable harvest levels over time, so important to implementing the 1982 Planning Rule, was replaced with short-term estimates of likely timber harvest levels. A whole new planning world awaited forest plan revision in the NWFP area.[80]

In addition, national forest plan revision efforts in the NWFP area almost immediately got high-centered over what to do with the NWFP itself. The Forest Service considered revising or even retiring it. However, such a change could require a major science assessment and planning effort covering the entire NWFP area such as occurred with FEMAT—a truly arduous task. Unsure about whether it wanted to head down that road, the Forest Service undertook two major efforts to prepare for plan revision however it might occur. A scientific synthesis, updating what had been learned in the last 20 years, would inform forest plan revision. Then a bioregional assessment would integrate that science and associated monitoring results with managerial experience and public feedback.

The Forest Service PNW Research Station compiled and published a review of science developed after the NWFP in a 2018 report titled *Synthesis of Science to Inform Land Management within the Northwest Forest Plan Area*. Forest Service managers selected the questions the report would address, focusing on those they viewed as most relevant to plan revision, and many of the federal scientists who had been involved in creating the NWFP were senior authors of its chapters. A number of the report's findings and graphics appear in this book.[81]

In general, the synthesis endorsed the approaches, standards, and guidelines of the NWFP while acknowledging that the plan had not achieved all the goals set for it. Where the NWFP fell short, the synthesis made modest but important recommendations for clarifying some NWFP standards. A particularly important recommendation was to make sure that Forest Service silviculturists understood that they should apply fuel treatments in LSRs in Dry Forests to more of the forest than just young stands, as the synthesis viewed heightened fire risk as a general condition of these forests.[82]

Unlike the five science assessments used to create the NWFP, however, this one stopped short of turning its findings into conservation frameworks that could guide plan revision. As an example, the authors of the synthesis emphasized that the reserve approach of the NWFP was inconsistent with the dynamics of Dry Forests and called for revisiting the reserve design and developing new landscape-level conservation strategies for these forests.[83] However, the authors did not outline what these new designs and strategies might look like. A special effort was made by PNW Research Station leadership to make sure that this assessment did not prescribe conservation frameworks that would limit what managers thought best.

The forest planning departments in the Pacific Northwest Region and the California Region then completed a joint bioregional assessment of the national forests of the NWFP area in 2020 to provide a big-picture context for the planning that would follow. This assessment was intended to help "explore innovative planning strategies . . . while considering community and stakeholder interests."[84] The Forest Service emphasized that the bioregional assessment would not make land management planning decisions, would not have details on specific solutions, and would be intentionally nonprescriptive. Rather, decisions, solutions, and prescriptions were to be left to individual forest plans and to the projects that would be derived from them.

The assessment summarized the state of the forests, watersheds, economies, and societies in the assessment area, drawing largely on the scientific synthesis, 25 years of monitoring the ecological and social outcomes of the NWFP, and managers' experiences. It pointed out the many limitations, inconsistencies, and other difficulties associated with the NWFP and its implementation that drove managers mad, especially relative to where, and under what terms, timber could be harvested on the national forests.

Certainly, managers may have chafed at the 80-year age limit on thinning in LSRs and at the NMFS's dim view of thinning plantations near streams that contained ESA-listed fish. However, the most significant difficulties surrounded limitations on management in Matrix/AMA. These allocations in the NWFP attempted to outline where timber production on a continuing basis would be a primary goal. To say that achieving that goal has been less successful than intended would be an understatement. Critical habitat designations, recovery plan recommendations, and Survey and Manage substantially limit the portion of Matrix/AMA across the NWFP area where timber production can be the primary goal.[85]

As an example, almost 40 percent of Matrix/AMA falls in the critical habitat for the northern spotted owl designated by the USFWS in 2012,[86] substantially defeating the goal of the developers of the NWFP to provide critical habitat for the owl in the NWFP reserves. While the USFWS has been immensely cooperative in working through its regulatory responsibilities, the added layer of critical habitat does impact forest management.

The 2020 bioregional assessment made a number of overarching recommendations to guide plan revision, including:

- Recognize more completely the substantially different ecology of Dry Forests and Moist Forests and the dynamic nature of those forests in setting and shifting reserve boundaries.
- Shift from a species approach to one that focuses more on ecosystems with additional species considerations as needed.
- Recognize more fully that fire is a natural process and is important in restoring ecosystem integrity.
- Use ecological forestry to expand the role of active management and timber harvest in ecosystem restoration.

The assessment, though, generally left the details of how these recommendations might be implemented to planning teams on the national forests.[87]

328   CHAPTER 10

The assessment also set priorities for revising forest plans in the NWFP area and adjacent forests. With the perilous condition of Dry Forests, the bioregional assessment rated the national forests in southwestern Oregon and northwestern California of highest urgency for plan revision.

The Firestorms of September 2020

Wildfires swept through the forests of the NWFP area in just a few days in mid-September 2020, burning almost 2 million acres in total, approximately 1.5 million acres of that on federal land (plate 9.1). Both Moist and Dry Forests burned severely in intense wildfires, with the federal acreage burned almost equally split between NWFP lands in Oregon and California. While wildfire in Dry Forests had been relatively commonplace in the 25 years since adoption of the NWFP, massive fires in Moist Forests near population centers were a shock to most residents.

In Oregon, these firestorms swept through forests and down river valleys, burning up small communities as people ran for their lives, and advanced on the cities of Portland, Eugene, and Salem. Pihulak and colleagues described how this happened:

In the late summer of 2020, unusually low humidity and high temperatures in western Oregon dried out the thick layer of duff, branches, and sticks that lay on the forest floor. This set the stage for a rare and powerful wind event that erupted on September 7th, 2020, bringing dry, hot winds up to 75 miles per hour that quickly spread any fire that started or was already burning. Humidity was measured in the single digits, previously only seen in particularly dry areas in the height of summer.

In the first twelve hours following Labor Day, the flow of easterly winds originating from the southwest pushed existing fires with hot temperatures and extremely low humidity at great speed, immediately creating a fifty mile run of flame down canyons and to the valleys below. This hot, dry airmass flowed like a river of flame below the cooler air above, hugging the ground level.[88]

Communities in the Santiam River canyon east of Salem and in the McKenzie River corridor east of Eugene were especially hard hit.

In addition, the Almeda Drive Fire, which started in a field in a suburban area, destroyed large portions of the towns of Talent and Phoenix and put the entire city of Medford on evacuation alert. In total, approximately 40,000 rural residents in western Oregon fled their homes, often with very little notice or time to gather belongings. More than 500,000 people, including in neighborhoods in southeastern Portland, were alerted for evacuation. After a few days, the east winds died down and the fires were contained, largely sparing metropolitan areas. However, dense smoke lingered in western Oregon for weeks, creating some of the worst air pollution in the world during that time.[89]

The western Oregon fires attracted national media attention largely because they had decimated many rural communities and threatened metropolitan areas—highly unusual events for the state. The apparent novelty of the fires in the western Cascades further heightened interest in their causes. One immediate suggestion in media inquiries to Franklin and Johnson was that these severe megafires were the result of a combination of climate change, with its higher temperatures, drier fuels, and longer fire seasons, and forest management practices, particularly the suppression of wildfires, that presumably allowed a substantial buildup of fuels.

As terrifying as the western Cascades wildfires were for those in their path, though, these fires were actually characteristic of fires in the Moist Forests of the western Cascades and coastal regions in Oregon and Washington, where infrequent fires containing large areas of high severity are a dominant feature of the natural fire regime.[90] The Riverside Fire southeast of Portland, the Beachie Creek and Lionshead Fires in North Santiam Canyon east of Salem, and the Holiday Farm Fire east of Eugene all demonstrated this behavior, as did the Archie Creek Fire in a mixture of Moist and Dry Forest east of Roseburg (plate 9.1).

Large, severe fires in Moist Forests historically occurred every one to three or four centuries when strong, hot, dry east winds caused intense, uncontrollable firestorms. In fact, a series of such fires long ago created most of the old-growth forests in western Oregon and Washington. Also, millions of Coast Range acres in Oregon burned in this fashion early in the 1800s and then again later in that century. In more recent times, massive fires in western Washington and Oregon burned in 1902 (including the well-known Yacolt Burn) and 1929 with east winds, as did the Tillamook Fire in the northern Oregon Coast Range in 1933.[91] A study by Abatzoglou and colleagues, described in "Compound Extremes Drive the Western Oregon Wildfires of September 2020," found that 10 of the 13 major wildfire events in western Oregon dating back to 1900 were associated with hot, dry summers, and all coincided with considerable easterly winds.

Fire suppression has not fundamentally altered the character of Moist Forests in these landscapes, nor the severity with which they burn.[92] However, some researchers such as Abatzoglou expect climate change to play an increasingly important role in accentuating fire risk in these forests in the future. They argue that climate change will decrease humidity in late summer and early fall in Oregon, a major factor in large wildfires, which will increase the severity of droughts and the dryness of potential fuels. Thus, easterly winds would then blow over drier, more flammable fuels, increasing the risk of large, severe fires.[93]

The western Oregon fires severely burned private industrial forests, especially the Holiday Farm Fire. Proponents of industrial forest management may promote plantation management as a way to reduce fire severity or extent, but the Holiday Farm Fire certainly roared through those forests in the McKenzie watershed. In general, industrial forest management has created a more wildfire-vulnerable environment in western Oregon and Washington, with large landscapes covered with uniformly dense, closed-canopy plantations of Douglas-fir and other resinous evergreen conifers that tend to burn more intensely and uniformly than the natural forests.[94]

High-severity fire is a part of western Cascades and coastal forest ecosystems and, in many ways, resembles infrequent (but huge) windstorms, rain-on-snow flood events, volcanic eruptions, and earthquakes: we can prepare for them and reduce their impacts, but we cannot totally stop them. The frequency and magnitude of these wildfires can be affected by controlling ignitions and quickly and forcefully applying suppression resources when ignitions do occur. How much they impact people and communities can be limited by controlling where and how towns and homes are built and by creating buffers around them. However, it is extremely difficult to alter the basic nature of these infrequent wind-driven Moist Forest fires.

The fires in the Dry Forests of southwestern Oregon and northern California in 2020 tell a different forest story. Those areas saw a replay of large fires experienced in recent years, although their severity has substantially increased in many cases (plate 9.1). One of them, the fire complex centered on the Mendocino National Forest in northern California, grew to more than one million acres and became the largest fire in the state's history. Past fire exclusion and other management practices had converted a fairly open forest there, dominated by ponderosa pine and black oak, into a dense forest dominated by Douglas-fir. These changes greatly increased the potential for high-severity fire and the loss of the old-growth pines and oaks (plate 10.4) and enabled a megafire of unprecedented size and severity.

In this case, the media rightly identified past forest management practices and climate change as major forces in altering fire behavior in Dry Forests, significantly increasing fire severity and difficulty of control. These practices and conditions also contribute to threats to these forests from drought and insect attack. Reducing stand densities and fuel levels remains a vital forest management strategy to restore the resistance (ability to withstand disturbance) and resilience (ability to recover from disturbance) of these forests.

The 2020 fires started much soul searching and debate about how to control them, how to prepare for them, and the role of forest management in taming them. Fortunately, the discussion generally recognized the need for different strategies for managing Moist and Dry Forests in preparing for the next fires.

The fires also brought the issue of postfire salvage and the role of managed wildfire on federal forests front and center once again. Whether old-growth trees will be protected during salvage is sure to be a source of debate, as will which part of the burned forest should be salvaged and which left alone. Whether to let seemingly benign wildfires burn in the wilderness areas of the western Cascades, where they might be picked up by the infrequent but possible east wind and then threaten communities, will also be debated.

The firestorms of 2020 also ended the pattern of a stable level of mature and old-growth forest on federal lands under the NWFP.

Firestorms that year caused the first significant (6-8%) decline in the total amount of mature and old-growth forests and associated spotted owl habitat since adoption of the NWFP, making the long-term survival of the northern spotted owl even more perilous.[95]*

The US Fish and Wildlife Service (USFWS) formally announced in December 2020 that listing the northern spotted owl as an endangered species was warranted throughout all its range, with high-severity fire as one of the reasons: "We find that the stressors acting on the subspecies and its habitat, particularly rangewide competition from the non-native barred owl and high-severity wildfire, are of such imminence, intensity, and magnitude to indicate that the northern spotted owl is now in danger of extinction throughout all of its range.[96]

The fires of 2020 illustrated how life confounds the best-laid plans, as have many other events that have occurred over the past 25 years since adoption of the NWFP. Once again forest managers will need to pick themselves up and adapt the NWFP to the conditions they find before them. And once again, the many people and groups who care passionately about these forests will be watching, criticizing, and advising as best they can.

* Ray Davis of the PNW Station uses an "old-growth structure index" (OGSI) with two age thresholds: ≥80 and ≥200 years to make estimates of the amount of mature and old-growth forest as part the NWFP Interagency Monitoring Program. The ages represent when forests commonly attain stand structure associated with late-successional forests (OGSI 80) and old-growth forests (OGSI 200).

PART 5

Significance and Future

11
Why the Northwest Forest Plan Matters

We have tried in this book to write a history of the development and implementation of the NWFP along with the scientific, legal, and social forces that shaped it. However, we are not just interested bystanders—we have been living professionally and personally within that history for the last 30 years as scientists, critics, and crafters of conservation frameworks. At various times we have also served as advisers to a president, cabinet officials, congressional committees, the Northwest congressional delegation, federal agencies, and groups interested in the issues and problems that the NWFP attempted to address. We have discussed and debated potential solutions to these issues and problems in countless forums and field trips and expect to do so in the future. For better or worse, we are identified with the NWFP and will be for the rest of our lives.

In these last two chapters, we reflect on both the past and future of federal forests within the range of the northern spotted owl. In doing that, we leave our role as recorders of history to provide personal reflections on what the NWFP accomplished, where it fell short, some of its key lessons, and how we might deal with the challenges ahead.

President Clinton announced the President's Plan to end the decade-long gridlock over management of old-growth forests in the range of the northern spotted owl on July 1, 1993. He selected that plan from a set of alternatives outlined in a 1,000-page scientific assessment, the result of three months of work by scientists building on three previous assessments that had tackled this issue—the ISC, Gang of Four, and SAT.

The President's Plan, as described in the report authored by President Clinton and Vice President Gore and issued when the president announced his selection of that plan, would protect watersheds and the most ecologically valuable old-growth forests, ensure sufficient federal habitat for the spotted owl and threatened salmon populations along with scores of other species, deliver a timber harvest level of over one billion board feet per year, and help displaced woods workers shift to other livelihoods.[1] In addition, experiments in Adaptive Management Areas and interagency planning would chart the path forward.

After 25 years, has the NWFP reached those goals? To address that question, we summarize some of the key findings from earlier chapters.

The NWFP did reserve much of the ecologically valuable old-growth forests. Most of the rest was protected by restrictions on Matrix in the NWFP, the recovery plan and critical habitat for the northern spotted owl, and evaporation of social license to log those forests. As a result, the amount of old-growth forest (and also mature forest) remained relatively stable for the first 25 years of the plan rather than decline as projected in the NWFP. Old growth no longer provided the fuel for sustained yield—a historical shift in federal forest management. Large fires, in 2020, though, caused the first significant decline in the amount of old growth forest and it was increasingly recognized that the plan unduly limited fuel treatments needed to save older Dry Forests.

Maintaining old-growth forest also helped conserve the habitat of the northern spotted owl. Saving that habitat is essential for the owl's survival, but the definitive issue may be competition from the barred owl—something that the NWFP did not address.

Implementation of the Aquatic Conservation Strategy improved federal watersheds and the aquatic and riparian habitats within them— a major accomplishment. However, despite this success, the state of habitat on private lands lower in the watershed, and other issues, held back recovery of salmon and other fish.

Greatly limiting the logging of older forest in Matrix prevented the attainment of the projected timber harvest level and associated employment for most of the plan's first 25 years. Also, funding for economic assistance to aid the economic transition of displaced workers and the communities in which they lived had limited success.

Federal managers adapted over time to the changing circumstances, but Adaptive Management Areas contributed few new ideas. Interagency planning functioned mainly through the interaction of

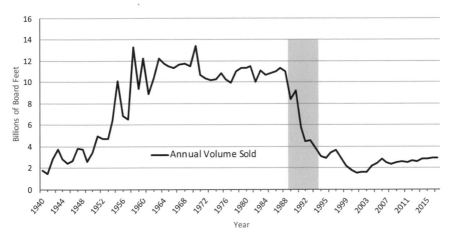

Figure 11.1. Timber sales on the national forests from 1940 to 2016 overlaid with the period from imposition of court-ordered injunctions on sales in northern spotted owl habitat through development of the NWFP (1989–1994). (Source: USFS; Robbins, *Place for Inquiry*.)

USFWS with the Forest Service and BLM—a significant change but short of the Clinton Administration's aspirations.

In summary, the NWFP achieved some, but not all, of the goals that the president had set for it and controversies over the soundness and utility of the NWFP have continued for its entire life.

Nevertheless, the NWFP has fundamentally changed the context of federal forest management. Forests are no longer to be viewed simply as collections of trees, but rather as complex ecosystems carrying out multiple functions, including (but not confined to) production of wood.

The protests, litigation, and science assessments that led to the NWFP brought forth new ways of thinking about conservation of biodiversity, old-growth forests, and the linkages between forests and embedded aquatic ecosystems. The overlap in time between these actions and the rapid decline in national forest timber harvests across the United States is striking (fig. 11.1). Wild doings in the Pacific Northwest sent shock waves through national forests across the country.

More broadly, the NWFP helped fundamentally transform how we think about conservation and management of our forests regionally, nationally, and globally. We cover a number of its contributions and lessons in this chapter and go on to discuss the many remaining challenges in chapter 12.

Scientifically Sound, Ecologically Credible, and Legally Responsible

Some may argue that the greatest significance of the NWFP lies in its longevity, especially for the national forests, where it has been the law of the land for 25 years. Neither economic upheaval nor changes in presidential administration or Congress have put much of a dent in it. That longevity might seem especially surprising, as no one seemed to like it much when it was adopted. Why has it had such staying power?

When President Clinton announced the formation of FEMAT at the conclusion of the Forest Conference in 1993, he identified five principles to guide the effort.[2] One principle stood out: *our efforts must be, insofar as we are wise enough to know it, scientifically sound, ecologically credible, and legally responsible.* We argue that the three goals in that principle, more than any other, guided development of the NWFP and help explain its influence and longevity. It truly provided the scaffolding on which the NWFP was built.[3]

Scientifically sound: The NWFP development process was scientifically sound in both appearance and reality. The Clinton administration fulfilled its commitment to avoid political interference in the development of the management alternatives and their implications. The scientists and resource professionals developing the plan adhered closely to what they understood about the structure and function of old-growth forest ecosystems as well as what would be needed to maintain and restore those ecosystems and the species within them. They also developed hypotheses about relationships where the science was weak, such as the importance of non-fish-bearing streams to salmon populations and other species, that proved prophetic. The robustness of the plan in the face of repeated litigation challenging its premises further demonstrates the soundness of the science used.

Ecologically credible: Federal forest planning had traditionally been built on an economic foundation, as represented by sustained yield planning. These plans included modest adjustments to accommodate ecological considerations as long as they did not significantly reduce timber harvest levels. The ISC's process for developing a conservation strategy for the northern spotted owl, the first of five science assessments underlying the NWFP, reversed that approach. The ISC crafted a scientifically credible conservation strategy, without compromising it because of potential impact on timber harvest. The ISC did try to apply the strategy in ways that would reduce those impacts, such as using wilderness areas as reserve anchors, but only as long as it did not threaten the strategy's integrity. FEMAT utilized and generalized that approach: scientists used concepts from conservation biology, landscape ecology, and wildlife and fish biology as well as knowledge of species habitats to develop ecologically credible conservation strategies for old-growth ecosystems and the species within them. Within that context, they applied the strategies in ways that would reduce the impact on timber harvest, such as by shifting Late-Successional Reserves (LSRs) into Key Watersheds.

Legally responsible: President Clinton insisted on a legally acceptable forest plan to address the owl/timber stalemate in the Pacific Northwest. That meant, more than anything else, that his administration, led by Katie McGinty, had to develop a forest plan under existing environmental law that could pass muster in Judge Dwyer's court; they did not want their plan to be cast aside by the judge, as had happened repeatedly during the first Bush administration. They were highly successful in this effort, as the judge ruled that the NWFP was within the authority of the administration to adopt. Also, Judge Dwyer was obviously impressed with both the scientific soundness and ecological credibility of the plan and often utilized the language from the plan and supporting documents to help explain why he was upholding it. Thus, the Clinton administration met its goal of developing a plan for a large, complex federal landscape with threatened species and a multitude of other issues that met federal environmental law—no small accomplishment.

The watchwords for national forest planning soon thereafter became, in many ways and in different forms, that planning efforts must be scientifically sound, ecologically credible, and legally responsible. While it can be difficult to trace cause and effect in planning and policy changes, there can be no doubt that the NWFP forged a new path in conservation and management of these and other forests.

This new path also reflected a different worldview from that of previous federal forest plans. A worldview can be defined as "the overall perspective from which one sees and interprets the world, a collection of beliefs about life and the universe held by an individual or a group."[4]

It has a strongly motivating and inspiring function. A socially shared view of the whole gives a culture a sense of direction, confidence, and self-esteem. Six key components of a worldview are the following:

1) theories and models for describing the phenomena we encounter;
2) an understanding of how the world functions and how it is structured;
3) explanations of why the world is the way it is;
4) descriptions of more or less probable future developments;
5) values addressing what is good and what is evil; and
6) descriptions of how we should act to solve practical problems.[5]

As an ecological perspective took hold in the early 1990s, foresters lost their dominance in federal forest planning. New science and new kinds of professionals were needed, and ecologists and conservation biologists answered the call. This changing of the guard, from foresters to ecologists and conservation biologists, is apparent in the appearance and domination of new policy players, new kinds of policy reports, and the staffing of the federal agencies.[6]

The changing of the guard, though, brought with it more than a shift in objective from maintaining commercial timber harvests to protecting ecosystems and species. In addition, we saw a shift in worldview from those who tend to emphasize managed systems (like many foresters) to those who tend to emphasize natural systems (like many ecologists and wildlife biologists) (table 11.1).[7]

These differences in worldviews affect forest management decisions. As an example, resource professionals who believe that nature is somewhat fragile in the face of their actions and that uncertainty about the effects of those actions is relatively high will tend to be more cautious than those who believe that nature is robust in the face of their actions and that uncertainty about the effects of those actions is relatively low.

An interagency approach to planning also helped forge this new path. In studying the spotted owl controversy before the Forest Conference in April 1993, President Clinton was mortified to learn how often "the departments were at odds with each other, so that there was no

Table 11.1. A generalized comparison of worldviews of those whose frame of reference is managed forest systems and those whose frame of reference is natural forest systems.

Category	Emphasis on managed systems	Emphasis on natural systems
Greatest threat	Disruption of production	Loss of species and ecosystems
Attitude toward disturbances	Suppress	Emulate
Effect of human actions on natural processes	Nature generally robust	Nature somewhat fragile
Confidence in technical progress to solve problems	Relatively high	Relatively low
Uncertainty about effect of human actions on outcomes	Relatively low	Relatively high
Burden of proof	On those who want to stop actions that might harm species and ecosystems	On those who wish to undertake actions that might harm species and ecosystems

Source: Adapted from J. Franklin et al., Ecological Forest Management, chap. 17.

voice of the United States."[8] Therefore, he and Katie McGinty made development of an interagency approach a key component of resolving that controversy and delegated Jim Pipkin to lead the effort.

Pipkin summarized his thoughts on the contribution of the effort to establishing an interagency approach to NWFP implementation:

Obviously, science was the key to the NWFP, but I believe that agency coordination was a very valuable part of the plan's success— and that it paved the way to a different and broader approach to agency relationships on other issues. As Katie McGinty pointed out, agencies had been operating "in their own silos" and the NWFP should show "how government should work."

I would not suggest that these lofty goals were achieved 100 percent. The agencies chose not to give decision-making authority to the regional interagency groups, and over time they moved much of the interagency work back to themselves. But the establishment of the interagency structure, and the successful work within it, created a model for how the agencies could function on a collaborative basis and encouraged agency personnel to broaden their vision of how to work together in the public interest.[9]

Yes, it is true that the dream of the Clinton administration of joint Forest Service and BLM planning by province and watershed did not come true, as the two agencies largely went their separate ways. However, the interagency approach to planning initiated by the NWFP continued through the consultation process required under the Endangered Species Act and provided a new planning model. This new model is seen most clearly in the development of the 2016 Resource Management Plans in which the US Fish and Wildlife Service (USFWS) and BLM worked together to craft a forest plan, with the USFWS ensuring that the BLM conservation strategy for the owl and murrelet fit into broader range-wide recovery plans. The National Marine Fisheries Service (NMFS) also contributed guidance on protecting threatened salmon populations.

Making Policy Formation More Than the Art of the Deal

The Forest Service and BLM embarked on major forest planning efforts in the 1980s as they grappled with the impact of new environmental laws and changing demands on forest use. In those initial efforts, managers working with their resource specialists chose species protection strategies. Scientists were not directly involved, although their reports were consulted. That approach collapsed because the Forest Service and BLM were unable to convince the courts that the plans they produced would conserve the species under their care.

Scientists then plunged into detangling the gridlock over management of old-growth forests in the range of the spotted owl. It took five sequential science assessments and the development of a regional forest plan before the courts were satisfied. This series of events brought scientists front and center in recommending credible conservation strategies and acceptable risks in the range of the spotted owl and beyond.

William Dietrich, in *The Final Forest*, captured Jack Ward Thomas's view of how the ISC's report of 1990 (the first science assessment) helped change the political process for making natural resource decisions:

> The biologists were not used to issuing prescriptions that could cost thousands of wood industry jobs, Thomas said. "Scientists would rather just do their thing and print the results in the Journal

of Esoteric Results. Now they were being thrust center stage." As human population pressure builds on the planet's ecosystem the trend will increase, he predicted—and perhaps that is not a bad thing. To Thomas, it makes more sense to base political land use decisions on scientific facts, the evolving truth, than simply to split the difference between competing interest groups. "I am pleased [he said] science is being built more and more into the decision-making process. This will not be the last time."[10]

Prophetic words, indeed. Federal forest decision-making had become more than the art of the deal.

In the two decades before the battle over conserving the northern spotted owl, most national forest politics focused on two issues: how much timber would be offered for sale and how much of the roadless areas in the national forests would be designated by Congress as wilderness. Those decisions were, by and large, made without science and scientists having a central role.

The interjection of science and scientists into the decision process for national forest management in development of the NWFP altered and constrained this classical deal making. No longer was it sufficient for key congressional committees to work out a deal among different interest groups and land management agencies. Rather, scientific findings set limits on what could be claimed and what would pass legal muster, with scientists often interpreting their findings in publications, the media, and the courts. In addition, the very nature of scientific inquiry did not allow scientists to guarantee that anything they proposed gave the final answer.

These changes significantly reduced the power of Congress and the executive branch to mandate timber harvest levels or lands for timber production on the national forests or to settle things once and for all. Unless they were willing to limit the ability to challenge decisions in the courts, as Congress had done with the Salvage Rider in 1997–1998 and had not tried since, they had lost the unilateral power to control timber harvests on the federal forests. Scientists' advice on which management plans met requirements in federal law became, in some ways, more important than politicians' desires to placate different interests.

Giving Wild Science and Wild Scientists the Lead

Scientific inquiry into the function, structure, and composition of natural forest ecosystems provided critical knowledge necessary for the NWFP to meet conservation requirements. While most of the science was *wild*, in the sense of not being focused on advancing particular management goals, policy makers and managers ultimately found it highly useful. Ecosystem science, along with studies of the life history of high-profile species, provided critical knowledge for managing forests and streams as ecosystems. The knowledge of how ecosystems are constructed and carry out their functions (how they work) was relevant to many different challenges and needs, rather than just one, such as how to maximize wood production.

Wild science is especially useful when it evaluates the soundness of management and policy assumptions. Can spotted owls prosper in relatively small home ranges? Are old-growth forests biological deserts? A more recent example is the justly famous work of Suzanne Simard, a forest ecologist from British Columbia, in which she showed that different trees—and even different tree species—continually exchange resources and information via underground fungal networks, known technically as mycorrhizae, upending an axiom of forest production that trees only compete with each other, rather than also cooperate.[11] Closer to home, Julia Jones, a scientist working at HJ Andrews, showed that plantations can create summer streamflow deficits as compared to mature and old forests, raising questions about the desirability of plantation management in water-limited landscapes.[12]

Wild in the case of the NWFP also meant that scientists were untethered from the institutional constraints under which they often labor. The scientists who led the different assessments underlying the NWFP leapfrogged traditional institutional paths for knowledge integration. Symposia, journal publications, and organizational reviews were bypassed as the scientists directly interjected their findings and hypotheses into the policy arena when the opportunity presented itself. This happened first via the ISC's development of a strategy for conservation of the spotted owl that was commissioned by the chief of the Forest Service, then at the request of Congress, and finally through the invitation of a presidential administration that needed scientists to help fulfill a campaign pledge.

Why and how did this happen? First, the policy makers and scientists needed each other. Chief Robertson, Congress, and then the Clinton administration had a political and policy crisis that demanded rapid knowledge synthesis that only scientists had the credibility to provide. The scientists needed policy makers who would work directly with them to craft new policies. Second, a number of these scientists thrived on being in the middle of political turmoil—they liked to shake things up and go against the tide with their latest ideas and present them to an appreciative audience of power brokers.

Under normal circumstances, the Forest Service and BLM would have been the agents addressing the unfolding crises, but they had lost their credibility in addressing the legal mandates of environmental laws. Thus, the bureaucratic seas parted, political leaders took over, and scientists marched in to help plan the future. Jack Ward Thomas artfully described the role of scientists in the five science assessments: "This was not 'science' per se. Science is a testing of a hypothesis. These efforts were assessments, management planning and development of alternatives by a group of scientists. But it wasn't science in and of itself. We hired the best-qualified people in the world to collaborate and go through an extremely important planning exercise. That's not science. That's scientists doing planning."[13]

Direct connections between scientists and policy makers have been needed in the past to address major natural resource problems[14] and will be needed again someday, perhaps in the form of independent science assessments and construction of new conservation frameworks as done here. Addressing the still-unresolved future of mature and old-growth forests in the range of the northern spotted owl or coming to grips with the barred owl invasion of spotted owl habitat are two possibilities. Certainly, the challenges of dealing with climate change will likely require new linkages between science and policy.

Implementing Ecosystem Planning

The five science assessments, capped off by the NWFP, redefined and greatly deepened our understanding of how to conserve forested ecosystems in the range of the northern spotted owl. In the span of only four years, from development of the ISC's owl strategy to issuance of the NWFP, a torrent of new ideas for conservation planning emerged.

In designing a regional forest management plan, the NWFP's developers had a primary goal of protecting species associated with old-growth forests and the streams that ran through them by maintaining and restoring the ecosystems on which these species depended. Achieving that objective meant that planning units would use ecosystem boundaries rather than administrative boundaries. For some species such as the northern spotted owl, that planning boundary covered the entire NWFP area. For other species such as salmon, the planning unit might be a particular set of watersheds.

Ecosystem planning also meant seeing the forest as more than a collection of trees, seeing it rather as an infinitely complex and interconnected suite of plants, animals, and fungi and the environment in which they lived. In many ways, this changed perspective can be viewed as the inevitable consequence of congressional direction for conservation of biodiversity in the Endangered Species Act and the National Forest Management Act and their implementing regulations. As Judge Dwyer said: "Given the current condition of the forests, there is no way the agencies could comply with the environmental laws *without* planning on an ecosystem basis."[15] Nevertheless, its achievement took a major effort by a large group of scientists, and political direction from no less than the president of the United States.

Ecosystem-based planning helped bring about a nearly 180-degree reversal in management emphasis on federal forestlands in the NWFP area. This change had consequences for essentially all other federal forests in the United States and, by example, throughout the temperate regions of the world. Prior to the NWFP, federal forest plans overwhelmingly emphasized the sustained production of wood products by individual administrative units, with other values viewed as secondary. After development of the NWFP, the conservation and restoration of forest, watershed, and stream ecosystems and the species within them became the dominant land-use goal for federal forests.

Forest planning changed deeply and abruptly on the national forests in light of these new goals. An approach to planning focused on ecosystem integrity spread throughout the agency and became the basis for the 2012 Planning Rule for the national forests that guides forest planning under the National Forest Management Act. As described in the Forest Service's 2020 bioregional assessment for the NWFP area, that rule guides national forest management "so that the lands are ecologically sustainable and contribute to economic and social sustainability."[16] The 2012 Planning Rule makes clear, with its many pages on how to restore and maintain ecosystem integrity, that ecological sustainability is the foundational goal for management of the national forests.[17] As such, this rule has rightly been described as the "the most important change in federal forest biodiversity policy nationwide over the past 30 years."[18] The five science assessments underlying the NWFP solidified such an approach and certainly deserve credit for its codification in the 2012 Planning Rule, as does the 1999 Committee of Scientists report that recommended such an approach to the secretary of agriculture as the basis for national forest planning.

The BLM developed its own plan for its western Oregon lands in 2016, breaking off from the NWFP but not from its responsibilities under the Endangered Species Act (ESA). Thus, an emphasis on range-wide ecosystem planning continued on those lands through the cooperation and oversight of the US Fish and Wildlife Service and National Marine Fisheries Service, with a focus on recovering threatened species.

Conserving Old-Growth, Riparian, and Aquatic Ecosystems

The science, events, and extensive publicity associated with development of the NWFP brought about major changes in how old-growth forests were perceived by resource professionals and society at large in the United States and ultimately throughout the temperate world. Historically, forest managers and educators typically viewed old-growth forests as decadent, unproductive, and overmature. However, scientific studies in the 1970s and 1980s found that old forests were highly productive (albeit not necessarily in terms of providing significant additional merchantable wood volumes), structurally complex, and biologically diverse and helped protect watersheds.

As this knowledge emerged and policy debates progressed regarding the future of the federal forests in the PNW, it became increasingly evident that the greater issue was not the fate of an owl subspecies but that of old-growth forests. This emerged most clearly in the Gang of Four exercise, when congressional committees clarified that the future of old-growth forest ecosystems, including their inhabitants, was their fundamental concern: they wanted to know the locations of the best remaining old-growth forest so they could see to its conservation.

The new focus on conserving old-growth forests resulted in a long-overdue debate within the forestry profession and the federal and state forest management agencies. Some foresters and managers challenged the science and the scientists about the ecological importance of these forests, but such efforts were not successful. The importance of old trees and old-growth forests and their conservation became and remains a major consideration in forest policies for public lands.

In addition, the new scientific knowledge of the intimate relationships between forests and the streams and rivers running through them that emerged during development of the NWFP revolutionized the views of professional resource managers, scholars, and society at large. Scientific studies during the decades preceding the NWFP had increasingly quantified the numerous and profound relationships between forests and streams and riparian areas and the multitude of fish and other species that live in these environs. For example, forests play critical roles in controlling light and temperature and providing energy and nutrient bases for stream ecosystems, and they provide critical structural elements (logs and other woody debris) throughout entire river networks.

Resource managers, stakeholders, and policy makers began recognizing that forest practices had to fully account for the inseparability of streams and rivers from the forests through which they flowed. Furthermore, this concern for forest-stream-fish linkages was applicable to all forestlands—federal, state, and private. Adequate protection of streams (including those without fish), other aquatic ecosystems, and the riparian areas around them is now considered a critically important conservation measure across all ownerships.

A key element in this emerging understanding was that all parts of the stream network and associated riparian areas were ecologically important in any strategy that sought to protect aquatic and riparian ecosystems. Thus, starting with the Gang of Four report, this strategy included headwater streams, which can make up two-thirds or more of the stream network and had been previously ignored in federal forest policy and management. Then, the NWFP made their importance loud and clear. Additionally, it called for riparian areas to be of sufficient size to address the suite of ecological processes and habitats that occur within them rather than being limited to an arbitrarily defined distance, as they had often been in the past.

Protection of streams and watersheds on federal forests was no longer seen as a relatively unimportant issue easily handled in standard forest planning processes. Thus, the regional aquatic conservation strategies developed in the Gang of Four analysis and refined in the SAT report and the NWFP soon spread under various labels far beyond the area of the NWFP: the Aquatic Conservation Strategy in the NWFP is now seen as the gold standard for aquatic and riparian conservation throughout the western United States and beyond.

Presenting an Alternative to Tree Farms

National forest managers had seen the tree farm as their vision of progressive forest management since the early days of the national forests. Both Gifford Pinchot and William Greeley, the first and third Forest Service chiefs, wrote about managing forests like fields of corn in their attempt to convince landowners to reforest and provide timber supplies into the future.[19] After loggers cut the old growth, foresters would replant rows of young trees of commercial species. Then, they would carefully tend the crop up to rotation age, which would be set at the age of maximum average growth, and provide abundant supplies of wood for the nation forever.

Weyerhaeuser led adoption of this approach in the Pacific Northwest, establishing the first tree farm on its lands in southwestern Washington in 1937.[20] Years of experimentation on how to maximize yield ensued from the 1950s to 1970s as the industry learned techniques of fire control, reforestation, site preparation, and suppression of competing vegetation. These approaches soon spread to the national forests as well.

With development and implementation of the NWFP, though, the primary goal of national forest planning shifted from maximizing timber production to protecting biodiversity. As this new goal became ascendant, so did the professionals (such as ecologists and wildlife and fish biologists) who had the knowledge and trust to pursue it; foresters were no longer solely in charge. Also, these new leaders had a worldview grounded in the conservation of natural systems as opposed to one grounded in the economic productivity of managed systems that foresters had long pursued through the creation of tree farms.

Viewing forests as ecosystems rather than simply as collections of trees lay at the heart of the shift from tree farms to natural forests as models for management, and leading that shift was one of the major and ongoing outcomes of the NWFP. Thinning in Moist Forest reserves has an overall goal of accelerating the development of young stands, mostly plantations, toward the composition and structure of natural forests. Final harvest in Matrix/AMA must leave significant biological legacies, and tree seedlings planted after harvest grow up among native plants that create a diverse early successional ecosystem. As might be expected, tensions still exist over how to integrate wood production with natural forest development, and much experimentation with variations of ecological forestry has occurred and will continue.

Federal managers of Dry Forests have largely abandoned the tree farm model in both reserves and Matrix/AMA. Instead, they work on increasing the resistance of these forests to disturbances by shifting the composition, structure, and density back toward those of historical forests while readying them for a changing climate.

These approaches fit well with the 2012 Planning Rule for the national forests, which emphasizes conservation and restoration of natural ecosystems on these lands as a primary responsibility. They also fit much better than the tree farm model with recovery of at-risk species.

Illustrating the Difficulty of a Species Approach

FEMAT conducted the most extensive evaluation of species viability ever undertaken in a forest planning effort in an attempt to conserve all living things in old-growth forests, ultimately considering more than 1,000 species in total. That analysis covered vertebrate animals, like the northern spotted owl, which had been the traditional focus of such analysis, as well as many other organisms, such as shrubs, grasses, forbs, ferns, fungi, lichens, mollusks, and arthropods.

Few in this larger array of organisms were known to be at risk, but little was understood about their distribution and habitat needs. Attempts to make sure that the President's Plan would pass legal muster led the Clinton administration to request that scientists develop protections and protocols to ensure federal habitat for their viability after public criticism of that plan. This request led to development of Survey and Manage provisions in the NWFP.

Once Survey and Manage requirements came into effect in the late 1990s, they created turmoil in the Forest Service and BLM, affecting both timber harvest feasibility and cost. The agencies then embarked on attempts to remove Survey and Manage provisions, first on their own and then under the guidance of the second Bush administration in the early 2000s. While none of those efforts fully succeeded, they created a continuing interest in the agencies to remove the requirements.

However, it must be acknowledged that much was learned in the late 1990s and early 2000s about many Survey and Manage species. This knowledge enabled removal of many species from the Survey and Manage list as concerns diminished about their protection under the NWFP. Budget cuts starting in 2003 largely eliminated broad surveys for the remaining species on the list, and learning about their habitats and needs greatly slowed.

Ultimately both the Forest Service and the BLM shifted to planning approaches that reduced the need for species-based viability planning. The Forest Service in its 2012 Planning Rule[21] and the BLM in its 2016 forest plan[22] contracted species-viability requirements to those species shown to be at risk—species that needed additional protection beyond what would be provided by a proposed ecosystem plan. Thus, the land management agencies reverted back to placing the burden of proof of the need for a viability assessment on those who sought the additional protection. Both vertebrate animals and other organisms would be included in these assessments. The BLM has implemented such an approach under its 2016 plan, but implementation by the Forest Service awaits forest plan revision. The FEMAT/NWFP experience surely was a major impetus for these changes.

Illustrating the Futility of Sustained Yield Commitments

The Forest Service's desire to provide a sustained yield of timber harvest over time sprang from the noble goals of reducing the economic and social disruptions caused by the migratory nature of the timber industry and preventing timber famine. However, the promise of sustained yield also led to timber-dependent communities. The well-meaning people who lived in those places believed in the Forest Service's promise of a steady flow of timber forever. Changes in harvest levels—if they were to come at all—would be gradual so that communities could adjust and evolve, as Pinchot preached in his instructions accompanying birth of the national forests.[23]

The sudden collapse of federal timber harvests in the NWFP area in the 1990s had profound immediate and long-term consequences. It abruptly eliminated employment opportunities in rural communities across the area of the NWFP. Perhaps more importantly, it shattered the faith of those communities in the promises made by the federal agencies and their representatives. The inability to deliver even the much-reduced harvest levels projected under the NWFP, or much alternative employment, further reinforced this feeling.

The Forest Service learned from this experience as well as similar ones in other parts of the country. Under the 2012 Planning Rule and associated 2015 Planning Directives for the National Forest Management Act,[24] the agency stopped making long-term sustained yield commitments. In their place, it called for 20-year estimates of likely timber harvests as part of an ecologically grounded management plan. Even those short-term levels are not guaranteed.[25] Sustained yield management as the driving force in national forest management is over, as are the associated long-term estimates of particular timber harvest levels. Certainly, the experience under the NWFP was an important catalyst to making this fundamental change.

Illustrating the Limits and Promise of Democratic Participation

In *Federal Ecosystem Management*, James Skillen studied different approaches to ecosystem management in federal planning efforts of the 1980s and 1990s and concluded that the NWFP was among the most durable and had achieved significant ecological goals. However, he also noted that "what drove this model of ecosystem management was not democratic participation, deliberation, and collaboration; what drove the NWFP were the fairly rigid regulatory requirements of the Endangered Species Act and the National Forest Management Act and by extension the strict procedural requirements of the Aquatic Conservation Strategy and Survey and Manage."[26] Further, he noted that this process matched the findings of one of the more thorough empirical studies of ecosystem management in which political scientist Judith Layzer argued that "only in cases where public officials circumscribed the planning process by articulating a strong, pro-environmental goal and employing regulatory leverage are the resulting policies and practices likely to conserve biodiversity or restore damaged ecosystems."

Layzer's findings generally describe the approach taken in the five science assessments. While there was some initial interaction with the public in the early days of the ISC's analysis, the strategies and choices developed by the ISC and the four assessments that followed were completed largely without the back-and-forth bargaining, discussion, and review that occurs in the democratic participation idealized in the National Environmental Policy Act (NEPA) planning process. Rather, political leaders relied on scientific authority to interpret the meaning of key environmental laws, such as the level of protection that would meet viability requirements, which left little room for traditional public planning processes.

One partial exception to the sole reliance on scientific authority came when the Clinton administration utilized the NEPA process to legitimize the president's selection of Option 9 as his plan. Intense public criticism to proposing a plan that received low viability scores for many species led to grave concern by the legal advisers of the administration as to whether such a plan would pass muster with Judge Dwyer. This concern led to the last of the five science assessments (SAT II), changes in Option 9, and emergence of the NWFP. However, SAT II was also completed in isolation, following the model of the other assessments that came before it.

Perhaps a political decision to take planning away from the Forest Service and BLM and give it to scientists should not be surprising, as the agency-based, decade-long forest planning effort resulted in half measures to protect species that did not survive court tests. The agencies attempted to satisfy the disparate demands by members of the public through slight and marginal adjustments. Much more significant change was needed, and the agency's forest planning process could not deliver it.

In addition, the attempt by the first Bush administration to shape the outcomes only made the courts madder. Thus, the Clinton administration turned to scientists to purify the process of crafting a forest plan that would break the federal forest gridlock and satisfy the courts.

The leading role of science and scientists in crafting the NWFP gave renewed life to a long tradition of relying on technical experts to chart the course that goes all the way back to creation of the national forests. As Samuel Hays discusses in his landmark 1959 book *Conservation and the Gospel of Efficiency: The Progressive Conservation Movement, 1890–1920*: "Who should decide the course of resource development? . . . Since resource matters were basically technical in nature, conservationists argued, technicians rather than legislators, should deal with them. . . . Conservationists envisaged . . . a political system guided by the ideal of efficiency [in the sense of avoiding waste] and dominated by the technicians who could best determine how to achieve it."[27] Almost 100 years later, a group of technicians (now research scientists and associated resource professionals rather than foresters) were asked once again to plan the future of federal forests. And again, efficiency (in terms of least-cost solutions) was central to the analysis. The Gospel of Efficiency still lives!

Sure enough, the NWFP did satisfy the courts sufficiently to begin its implementation. However, it did not generate broad support or acceptance from the warring parties. Rather, the scientists' work mainly reshaped the battlefield on which federal forest policy would be fought.

Put another way, the conservation frameworks that emerged from the NWFP science assessments were more successful in establishing rules limiting activities than in authorizing them. The boundaries on LSRs, widths of riparian buffers, and the 80-year upper age limit for thinning in LSRs provided direction and structure for the forest management that followed. However, the scientists' authorization of actions, especially those involving timber harvests, was less successful, as those actions had to go through legal and political validation that often overrode what the scientists had proposed.

Allocation of a portion of the remaining mature and old-growth forests to Matrix/AMA implied that they were not needed to protect species and ecosystems and thus could be logged. Regardless of that judgment, few of those forests were felled. It might be said that such actions did not pass the test of social acceptability. In an important article titled "Social Acceptability in Forest and Range Management," social

scientists Bruce Shindler and Mark Brunson argued that grand proposals such as the NWFP flounder if their designers ignore the power of
citizens to influence, direct, and limit actions by administrative appeal,
litigation, lobbying federal legislators, and attracting media attention
for their cause.[28] All that happened during early implementation of the
NWFP.

The Forest Service and BLM began testing the types of timber
harvest that had sufficient social acceptability to be implemented and
settled on thinning younger forests. They then proceeded with that
approach for many years. However, Shindler and Brunson also noted
that "public judgments are always provisional, never absolute or final,
with each situation, each context, producing a unique set of circumstances affecting the formation of public acceptance."[29] With that
understanding, federal agencies continue to assess the social acceptability of logging mature forests in different forms and different places.

A Never-Ending Cycle

The NWFP pioneered the incorporation of ecological goals and principles into management of a large, forested region. While the NWFP
only partially succeeded in achieving its goals, the plan ended a legal
impasse, clarified the legal consequences of key environmental laws,
and provided a new starting point for the evolution of regional and
national forest policy. The NWFP also produced major and necessary
readjustments in basic assumptions and conditions underlying planning and management of federal forestlands. These changes influenced
the forest policy dialogue not only in other parts of the United States
but in many other places as well.

As the NWFP also grandly demonstrates, federal forest policy
development is a never-ending story, rather than one that arrives at
a permanent and stable endpoint. In other words, federal forest planning does not provide long-term certainty; rather, it redesigns the game
board in ways that alter the further evolutionary course of forest policy
and management.

The federal forest policy process that resulted in the NWFP fits into
the famous conceptual, heuristic model of the *adaptive cycle* of Lance
Gunderson and Buzz Holling, which they first applied to biological
systems. Their biological model recognized four stages: (1) exploitation

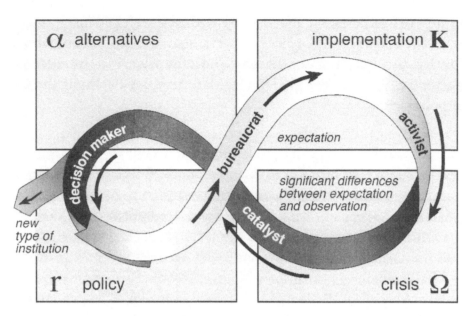

Figure 11.2. Four-phase heuristic of the adaptive cycle applied to policy development and change, indicating the dominance of different groups at different stages of the policy cycle. (Source: Janssen, "Future Full of Surprises"; copyright © 2002 Island Press; reproduced by permission of Island Press, Washington, DC.)

(rapid colonization of recently disturbed areas); (2) conservation (slowing accumulation and storage of energy); (3) release (creative destruction); and (4) reorganization (processes reorder the system for exploitation once again).[30] In "A Future Full of Surprises," Marco Janssen pointed out that the dynamics of natural resource policies in the United States can exhibit similar phases and transitions (fig. 11.2): (1) a decision-maker chooses a policy and begins robust policy implementation; (2) policy rigidity increases over time as bureaucracies and groups become committed to it; (3) challenges to the policy by activists based on differences between expectations and observations create crises that lead to policy collapse; (4) catalysts for change take action and help create a bridge to a new policy, and the cycle then begins again. Also, in more extreme cases, policy collapse can lead to entirely new institutions.

Federal forest policy development over time in the range of the northern spotted owl exhibited the characteristics of Gunderson and Holling's adaptive cycle through policy selection, implementation, collapse, and rebirth. The Forest Service and BLM embraced sustained yield of timber as their guiding forest management policy. Then, initially robust allowable cut policies became rigid over time as the agencies became so invested in them, and the social forces that benefited from them were so powerful, that they could not adjust as problems and circumstances changed. Environmental groups and their lawyers took actions that resulted in policy disintegration. Scientists provided new conservation frameworks that served as catalysts for change. A decision-maker (the president of the United States) utilized these ideas to select a new set of policies (the NWFP) and the cycle began again.[31]

12

A Path Forward

In this chapter, we look again at many of the issues that drove development of the Northwest Forest Plan (NWFP)—old-growth forests, the northern spotted owl, threatened salmon populations, timber production, and conservation planning processes. We and many others made our best effort to address them while helping to create that plan. Over the last 25 years, scientific knowledge, environmental conditions, and social values have all evolved. Here, we take a fresh look at where we are and changes we would recommend to achieve the goals of the NWFP in the twenty-first century.

As part of this effort, we suggest how consideration of the goals and rights of tribal nations who once lived in these lands should be improved, especially those related to hunting, fishing, and gathering on national forest and BLM lands. Those goals and rights were not explicitly considered in development of the NWFP. Project planning and plan revision at the regional or forest level should recognize them and develop mechanisms to ensure their full incorporation into future management.

This is the second time we have participated in a review of the NWFP. Fifteen years ago, the Gang of Four analyzed the status of the NWFP and made suggestions for change.[1] Key conclusions from that review included recognizing that

- the NWFP has evolved into an integrative strategy across the landscape, which requires conserving keystone ecological features, such as old-growth forests and trees, wherever they are found;
- conserving Moist and Dry Forests requires very different approaches; and

- NWFP forests are dynamic, and management needs to recognize disturbances and other changes as fundamental aspects of these ecosystems.

The conclusions from that review still ring true today. We build on them, while also adapting them to changing environmental and social conditions.

We also recognize that whether the Late-Successional Reserve (LSR) network retains its current form or is redrawn is itself a significant policy issue. Consequently, we initially make recommendations without LSR boundary revision or major redesign, and then we address LSR modification directly. Delineation of the LSRs initially required a major scientific effort commissioned by the president of the United States. Major changes in LSRs would be controversial and undoubtedly litigated and could put any plan revision on hold for many years. We should avoid repeating past errors where an inability to agree on a conservation strategy for the northern spotted owl resulted in at least a decade's continuance of past, out-of-date plans with much cost to forest ecosystems. We can make significant progress in forest, wildlife, and aquatic conservation without modifying current LSR boundaries, as desirable as that might be.

We generally utilize the national forests to illustrate our recommendations in this chapter. However, they also apply to the BLM as part of the ecosystem plan created in the NWFP.

Our recommendations seem more provisional than those that flowed from previous efforts. Certainly climate change plays on our minds along with related scientific and policy uncertainties associated with how it will transform the natural world. Also, we realize disruptive phenomena such as the barred owl will undoubtedly appear along with new social and political movements. Wild science will question management assumptions. We and others will continue to rethink and revise over time how best to conserve the magical resources of these precious lands. Based on our current understanding, we offer recommendations for improving achievement of NWFP goals (table 12.1).

We make one important qualification to these recommendations: they are for federal forests away from the urban fringe associated with communities. Near communities, human safety and protection

Table 12.1. Recommended changes to help achieve the goals of the NWFP.

**CONSERVING OLDER FORESTS AND TREES ON FEDERAL LANDS—
A PRIMARY GOAL AND FOCUS**

Moist Forests
- Cease logging of older, unmanaged forests and of remnant older trees in younger forests.
- Prohibit salvage logging in LSRs.
- Adopt a policy of aggressive detection and suppression of wildfires in LSRs.

Dry Forests
- Utilize treatments, including thinning and burning, to restore older Dry Forests to approximations of their historical states.
- Place the highest priority on retaining, protecting, and nurturing older trees across the landscape and revise LSR standards to recognize that priority.
- Consider salvage logging in LSRs where it will assist fuel management.

CONSERVING THE NORTHERN SPOTTED OWL
- Kill barred owls to protect northern spotted owls on at least a portion of the spotted owl's range.

CONSERVING SALMON POPULATIONS
- Restore salmon habitat in the ranches, farms, cities, and towns of the lower watersheds
- Work toward landscapes that allow for expression of a variety of life histories.
- Increase the amount of riparian thinning in federal forest plantations.

HARVESTING TIMBER ON FEDERAL FORESTS

Moist Forests
- Continue thinning younger forests across the landscape, conduct variable retention harvests in younger Matrix, and salvage younger trees in burned Matrix.

Dry Forests
- Reduce stand densities across the landscape, salvage younger trees in burned Matrix, consider salvage in LSRs to assist future fuel management.

REVISING LATE-SUCCESSIONAL RESERVES
- Revise the entire network at the same time, using critical habitat as a guide.
- Consider an alternative LSR conservation strategy for Dry Forests.

ASSISTING TRIBAL NATIONS IN ACHIEVING GOALS AND RIGHTS
- Greatly increase emphasis in forest plans and projects on their goals and rights.
- Listen to them on how to achieve these goals and rights.

---

of property should take precedence. There, we need fuel treatments in the federal forests that help buffer the effects of firestorms such as occurred with the 2020 Beachie Creek Fire east of Salem, Oregon. While most wildfires that burn communities originate on nonfederal lands,[2] some do come from national forests; mitigating that danger should be part of making rural communities near national forests safe. In addition, reducing fuels on private lands and around dwellings near national forests helps prevent wildfire from rolling onto the national forests themselves.

Conserving Older Forests—Core Issue of the NWFP

The NWFP focused on conserving older forests because of their high ecological value and because they occur almost exclusively on federal lands. The older forests, along with wilderness, represent the unique contribution of these lands to forest and wildland conservation in this region.

For the last 40 years, federal forest policy debates in the NWFP area have centered on whether to conserve or log older forests. On April 22, 2022 (Earth Day), President Biden reframed that debate through an Executive Order titled "Strengthening the Nation's Forests, Communities, and Local Economies." President Biden declared, "It is the policy of my Administration . . . to conserve America's mature and old-growth forests on Federal lands," while describing the primary threats to these forests as including "climate impacts, catastrophic wildfires, insect infestation, and disease."[3]

President Biden's Executive Order went on to state: "To further conserve mature and old-growth forests . . . the Secretary of the Interior and the Secretary of Agriculture (Secretaries) . . . shall continue to jointly pursue wildfire mitigation strategies, which are already driving important actions to confront a pressing threat to mature and old-growth forests on Federal lands, . . . within 1 year of the date of this order, define, identify, and complete an inventory of old-growth and mature forests on Federal lands . . . and develop policies, with robust opportunity for public comment, to institutionalize climate-smart management and conservation strategies that address threats to mature and old-growth forests on Federal lands."[4]

At last, a president has formally called for conservation of mature and old-growth forests on federal lands. It is true that policies expressed through an Executive Order can be reversed by a future president. Still, this statement recasts the debate over the future of mature and old-growth forests from whether to log them for economic gain to how to best conserve them.

Interestingly, the logging of mature and old-growth forests did not get mentioned either as a threat to them or as a way to help save them.

However, we expect that the president's conservation goals will rule out timber sales that terminate mature and old-growth federal forests as long as it is in effect.

We give our recommendations in this section on how to conserve mature and old-growth forests within the NWFP area to achieve the goals of the president. As we have often done in this book, we discuss Moist Forests and Dry Forests separately, since conservation of older forests and the trees within them requires different approaches for these two kinds of forest. Additionally, existing older Moist Forests have generally experienced little harvest during their life, while extensive partial cutting has occurred in many older Dry Forests.

Reserving Older Moist Forests

The NWFP protected mature and old-growth Moist Forest through LSR and Riparian Reserve land allocations, and this protection was later augmented by recovery plans and critical habitat designations for threatened species under the Endangered Species Act (ESA). However, the NWFP left significant older Moist Forest in Matrix/Adaptive Management Area (AMA) and, therefore, potentially available for harvest.[5]

After years of shying away from harvest in those forests, the Forest Service in the early 2020s once again surfaced plans to log significant amounts of mature Moist Forest in Matrix, with proposals to log mature stands that are 100–150 years old.[6] So far, these efforts have generally been unsuccessful. *Unmanaged (generally unharvested) mature Moist Forests that have largely developed through natural processes should be reserved.*

We understand the importance of mature forests much better today than when the NWFP was developed in 1994, and that knowledge clarifies the desirability of retaining rather than logging them. Douglas-fir forests undergo dramatic structural and functional changes during their second century, largely because of natural thinning and gap-forming processes, which include mortality of some dominant trees. The canopy gaps stimulate development of intermediate canopy layers and understory communities, ultimately producing the vertically continuous, bottom-loaded canopies of old-growth forest.

Accumulation of wood and stored carbon (a.k.a. stand growth) continues at very high levels in mature Douglas-fir forests, despite death of some overstory trees. Mortality of larger trees structurally enriches these stands by opening the forest canopy and also rebuilds stocks of dead wood, including snags and logs, which decay very slowly. Hence, mature forests continue accumulating more carbon in both live and dead forms while simultaneously increasing their value as wildlife habitat. The rapid development of complex structure in mature forests enables them to fulfill many of the same ecological roles as old-growth forests, including habitat for old-growth dependent species. Timber harvesting in these stands terminates development of older forest structural conditions and their associated functional capabilities.

Also, most mature forests within the NWFP are in the transient snow zone. This zone is the primary contributor to rain-on-snow events that cause major floods on the west side. Mature forest canopies help reduce the magnitude of such floods; clearcuts and young forests exacerbate these events.

Logging of mature and old-growth Moist Forests lacks ecological justification. Fire suppression has not resulted in a significant buildup of fuels, as has occurred in Dry Forests. What Forest Service managers characterize as overstocking and justification for harvest actually contributes to natural developmental processes. These overstocking characterizations reflect traditional timber production perspectives and use density measures (such as stocking tables) developed for artificial young stands— i.e., those for plantations.

In addition, foresters sometimes use the goal of creating complex early successional habitat to justify cutting mature forests, but creating such habitat should occur through harvesting plantations rather than older natural forests. As stated by Phalan and colleagues in their important article on the NWFP's impact on forest composition and bird populations:

> Overall, the early-seral ecosystem area was stable, but declined in two ecoregions—the Coast Range and Cascades—along with early-seral bird populations. Although the NWFP halted clearcutting on federal land, this has so far been insufficient to reverse declines in older-forest–associated bird populations. These findings underscore the importance of continuing to prioritize older forests under the NWFP [for conservation]and ensuring that the recently

proposed creation of early-seral ecosystems does not impede the conservation and development of older-forest structure.[7]

Further, the dynamic nature of the Pacific Northwest forest ecosystems means there is the potential for major losses of existing old-growth and mature forests—another argument for keeping the mature forests that now exist. The Oregon wildfires of 2020 were a strong reminder of the need to conserve existing mature forests as replacements for the hundreds of thousands of acres of older forest habitat that have been, and will be, lost to disturbances.

It is time to *cease* the logging of older, unmanaged Moist Forests on national forests—both mature and old growth. The current extent of these forests is far below the historical levels found in the region and they are nearly absent except on federal lands. Amazingly, they are the only fully functional Moist Forest ecosystems that remain on federal forests in the NWFP area, since the plantations that have replaced them are highly artificial, as we discuss below. These unharvested older forests are too valuable ecologically, socially, and spiritually to allow their elimination though logging.

There are two caveats to these recommendations: (1) They are meant to be compatible with the growing of managed forests on longer rotations; thus it is important to avoid creating incentives for cutting stands before they reach the mature stage. It might be best to define the stands to be reserved in terms of birth year, such as all stands in Moist Forests born before 1920 or some other date reflective of the desired threshold age, and then use a structural index for further determination. Much like the 80-year rule for thinning in Moist Forest LSRs, this approach would make the threshold age for older forests easier to determine. (2) It is important to minimize damage to older forests if a decision is made to log them despite these recommendations. In that case, logging should concentrate on younger trees in the stand and on small gap creation, to emulate natural processes to at least some degree.

Timber harvests in Moist Forests should focus instead on previously harvested forests, which are primarily plantations established over the last 100 years. These plantations were created with the singular objective of producing wood. They are highly artificial systems that lack the structural complexity, species richness, and most functional capabilities of the young forests that historically developed after natural disturbances. In addition to thinning the plantations, the Forest Service should greatly expand variable retention harvest in them with the goal of establishing new young and mixed-age forests that more closely resemble the forests that historically originated following natural disturbances. Such an approach would provide substantial timber volume while simultaneously creating species-rich, mixed-age forests that would have greater resistance and resilience to wildfire and other disturbances and greater adaptability to altered environmental conditions.

In addition to reserving remaining older Moist Forests, logging older tree remnants in younger stands undergoing harvest should cease. Trees were often left in past harvest units for various reasons and tower over the planted younger trees. Amazingly, there are no restrictions on the age or size of trees cut in Matrix/AMA harvests under the NWFP. Yet, when retained, these old trees provide extraordinarily valuable wildlife habitat in harvested areas and help create multiaged stands that are more resistant and resilient to wildfire. Age and size limits on logging large, old trees in younger stands should be established, as was done in the BLM 2016 RMPs for Moist Forests, which require retention of large trees that were established prior to a specifed year.

Saving Older Dry Forests

Conserving mature and old-growth Dry Forests requires a strategy different from that for Moist Forests. The Dry Forest strategy should use a combination of thinning and burning to restore these forests to approximations of their historical state in which they are dominated by populations of large, old trees and have high resistance to wildfire, drought, and insects. Thus, treatments on a continuing basis are required to achieve and maintain disturbance-resistant older Dry Forests, in contrast with our recommendation for hands-off reservation of older Moist Forests.

A history of excluding fire from Dry Forests, along with other activities such as grazing, logging, and tree planting, has produced dense forests full of younger trees that surround and compete with any remaining older trees.[8] These actions have transformed these forests from a very stable fire-resistant state, created and maintained by frequent, low-to-moderate-intensity fires, into boom-and-bust ecosystems extremely vulnerable to high-severity wildfire, drought, and insects.

An *ecological catastrophe* is unfolding in Dry Forests of the NWFP area, and remaining green forests require swift and comprehensive action. Recent devastating megafires, such as the Mendocino Complex Fire in northern California in 2020 (plate 10.4), demonstrate the bleak future for Dry Forest ecosystems, and constituent old trees that are their ecological backbone, without intervention (plate 10.5). However, these wildfires also demonstrate that we can save these forests through careful action. Restoration thinning followed by prescribed fire saved a portion of the old forests on the Fremont-Winema National Forest as the Bootleg firestorm swept through just east of the spotted owl's range (plates 1.4 bottom and 10.6). With investment and perseverance, similar results can be achieved elsewhere.

Uneven-aged management is urgently needed in Dry Forests containing mature and old trees to reduce threats to the old trees, reduce overall stand density, increase average tree diameter, and favor early seral species, such as ponderosa pine, western larch, Douglas-fir, and white and black oak, where each was a historically dominant tree.[9] The starting point and highest priority should be to retain older trees within these forests and enhance their survival by reducing surrounding competition and ladder fuels.[10] These old trees and their derivative snags and logs are core structural elements and social icons of natural Dry Forest ecosystems.

Trees in Dry Forests take longer to develop old-growth characteristics than they do in Moist Forests. Thus, we generally use 150 years as the threshold age for defining when a tree reaches the mature stage. Again, it is important to avoid creating incentives for cutting trees before they reach the mature stage. Therefore, it might be best to define maturity in terms of birth year, such as all trees in Dry Forests born before 1870 or some other date reflective of the desired threshold age.

Unfortunately, as Spies and colleagues make clear in the science synthesis reports to support revision of the NWFP, "The goals, standards, and guides for LSRs in dry forests that prioritize conservation of existing dense older forests appear to be inconsistent with management for ecological integrity and resilience to climate change and fire."[11] The standards and guidelines for the NWFP (appendix 4) do in fact recognize special fire risk and fuel considerations for Dry Forests: "Given the increased risk of fire in these areas . . . additional management activities are allowed in Late-Successional Reserves."[12]

However, these standards and guidelines get the priorities for action out of order by endorsing treatments in younger forests while setting a high bar, with triple justification needed, to undertake vital treatments in older forests:

> Silvicultural activities aimed at reducing risk shall focus on younger stands . . . activities in older stands may be appropriate if: (1) the proposed management activities will clearly result in greater assurance of long-term maintenance of habitat, (2) activities are clearly needed to reduce risks, and (3) activities will not prevent the Late-Successional Reserves from playing an effective role in the objectives for which they were established.[13]

In addition, the guidelines do not recognize risks from drought and insects, which cannot be effectively addressed without actions in the old stands themselves. NWFP standards and guidelines should be altered to focus on risks to the old trees. Whether the action will be taken still depends on spotted owl conservation needs, but the threats to the old trees in Dry Forests are clear, and addressing them should be given the highest priority.

Climate and other environmental changes have already increased threats to Dry Forests, and their impacts will continue and probably accelerate. *Restoring these forests to conditions comparable to their historical resistant state is the first step in adapting them to this warmer, drier, higher-fire-risk future.*

Figuring out how to restore Dry Forests while still providing habitat for the northern spotted owl is a truly thorny problem, even without considering the invasive barred owl. Northern spotted owls thrive in dense, multilayered forests (fig. 1.6). In the Dry Forest landscapes, such dense forests are largely artifacts of past forest policies, particularly fire exclusion. Spies and colleagues effectively capture the dilemma we face:

> The reserve approach of the NWFP—which focuses only on fire-risk reduction and conservation of dense, multilayered forests—is

inconsistent with management for ecological integrity in the dry, historically fire-frequent forests. . . . In these forests, northern spotted owl nesting and roosting habitat (dense, multilayered forests) were likely limited in the past to fire refugia determined largely by topography. The abundance of northern spotted owl nesting and roosting habitat in Dry Forests increased during the 20th century as a result of fire suppression and exclusion . . . , which has led not only to larger patches of high-severity fire but also to a greater proportion of high-severity fire than had occurred historically.[14]

Restoration of the Dry Forests in the NWFP area needs swift and major action, especially where older trees are concentrated. While we realize the complications associated with conservation of spotted owl habitat, without swift action little of these older forests may be left to argue about in a few decades. This dilemma will surely test the principle in the recovery plan for the owl: *if it's good for the ecosystem, it's good for the owl.* That is, management to maintain or restore ecological processes (e.g., uncharacteristic wildfire, invasive species management) is encouraged even if there are short-term impacts to spotted owls. Much depends on the outcome.[15] We are not alone in this quest. Hagmann and colleagues, in "Evidence of Widespread Changes"; Hessburg and colleagues, in "Wildfire and Climate Change"; and numerous other scientists and managers (Spies and colleagues, *Synthesis of Science to Inform Land Management*; USFS, *Bioregional Assessment of Northwest Forests*; and USFWS, *Revised Recovery Plan for the Northern Spotted Owl*) have reached similar conclusions.

How might a balance between restoration and owl habitat protection be achieved? The generic answer to this conundrum is to restore the majority of the forest landscape to a fire-resistant condition, while retaining patches of denser forest embedded within that landscape. The denser forest patches would provide owl nesting, roosting, and foraging habitat and could be located on north-facing slopes and valley bottoms where denser forest refugia would most likely have occurred historically. Embedding these denser forest patches in a restored, more fire-resistant landscape should increase their probability of persisting as well as reflect the historical Dry Forest landscape pattern more closely than the continuous high-density stand conditions in these forests today.

Some of these dense patches of suitable owl habitat will burn or otherwise be lost, but development of replacement patches would have a head start since the restored matrix is already populated with larger and older trees. When needed, areas could be identified and managed to allow development of a patch of denser, multistoried forest.

The 2017 *Rogue Basin Cohesive Restoration Strategy* by Metlen and colleagues provides a potential model for addressing the apparent contradictions in providing northern spotted owl habitat in a Dry Forest landscape needing broad-scale restoration.[16] It addresses a 4.6-million-acre area encompassing the Rogue River basin of southwestern Oregon and overlaps several NWFP Dry Forest provinces, including parts of the southern Cascade Range and Klamath-Siskiyou Mountains. The Southern Oregon Forest Restoration Collaborative worked with The Nature Conservancy and many local partners in developing the strategy and initiating its implementation.

The collaborative simulated consequences of different treatment strategies that varied the scale and placement of forest thinning. Its most effective strategy requires a fivefold increase in federal agency activities fueled by a major infusion of funds. Also, substantial work is needed by private landowners, as most wildfires that cause significant home loss originate on private lands, and excess fuels on private lands and around dwellings can increase the chances of wildfires escaping into national forests.[17] The collaborative's simulations suggest that such a strategy would reduce wildfire risk to homes by 50 percent and to high-quality spotted owl habitat by 47 percent. That result was achieved with a modest reduction of owl nesting, roosting, and foraging habitat on ridges and warm midslopes where resistance and resilience objectives called for restoring more open stands. Full implementation of the Rogue Basin Strategy would require efforts by the USFWS, Forest Service, BLM, state forestry agencies, private landowners, and communities.

Malcolm North and colleagues, in "Pyrosilviculture Needed for Landscape Resilience of Dry Western United States Forests," also proposed a landscape strategy for spotted owl forests—this time for 13 million acres of national forest in the Sierra Nevada where the California spotted owl is a species of concern. They call their fire-based strategy *pyrosilviculture*; it utilizes prescribed fire and managed wildfire, augmented by a modest amount of thinning, to restore Dry Forests. They break the forest into zones—community wildfire protection (near communities), wildfire maintenance (in wilderness areas), and wildfire restoration (remainder of the forest, which generally falls between the other two zones)—and tailor their proposed treatment mix to the legal and social considerations in each zone. They call for a major increase in the acres treated through pyrosilviculture, and the beneficial fire at the heart of the strategy, arguing that it would expand the footprint of low-severity fire preferred by spotted owls, among many other benefits. They also acknowledge the challenges of such an approach near densely populated areas and under smoke restrictions. In summary, North and his coauthors emphasize that fire is inevitable; the key question is whether we wish to have a say in when, where, and how severely it occurs.

It is clear from the Rogue basin and the pyrosilviculture strategies—and those of Scott Stephens and colleagues in "Is Fire 'For the Birds?'"—that fire and mechanical thinning will both play critical roles in restoring Dry Forests within spotted owl country. It is also clear that major investments will be needed, and coordination among many parties will be key to the success of these proposals. In addition, involvement of the USFWS at every step is essential, as the proposed treatments would reduce the total amount of owl habitat. The challenges are great, but time is of the essence.

After the Wildfire

The NWFP gave vague and potentially conflicting guidance on protecting old trees and mature and old-growth forests during salvage in LSRs after wildfire. The plan's salvage guidelines in LSRs were intended to prevent immediate and long-term negative effects in late-successional habitat, while permitting removal of some commercial wood volume. Commercial harvest there would need to be compatible with the NWFP's direction for LSRs to protect and enhance late-successional ecosystems, including redevelopment of late-successional forest attributes.[18]

Salvage guidance was not given for Matrix/AMA, except to note that management after stand-replacing events would differ from that in LSRs in allowing more freedom in commercial salvage of dead trees and making replanting of disturbed areas a higher priority. Retention of biological legacies at harvest was not required.

Many wildfires burned in LSRs during the first 20 years of the NFWP.[19] Public protest and litigation generally stopped attempts at salvage there except along roads for public safety. The court decision over the BLM salvage sale called Timber Rock was especially important here. The court made clear that if postfire logging within LSRs occurred, it had to retain snags likely to persist until the stand was again producing snags (about 80 years in the future) to meet the NWFP's objectives of developing old-growth forest characteristics in LSRs postfire and providing late-successional forest habitat.[20] Since permissible logging would be limited to removing smaller snags, the financial value of the sale declined, and it did not proceed. More recently, though, national forests, especially in California, have argued that removing dead trees, including large ones, will reduce the potential severity of future fires and provide funds for restoration. Consequently, substantial volumes of large, old trees have been removed in salvage sales, which runs counter to ecological advice on postfire restoration.[21]

The firestorms of 2020 that burned almost 1.5 million acres of federal forest in the NWFP area may bring salvage issues in LSRs back to the fore. Much of the acreage burned is in California, where there have been aggressive federal efforts at salvage on some national forests. In addition, much acreage that burned in Oregon was Matrix/AMA, where salvage harvest is easier to justify than in LSRs.

In total, recent wildfires have exposed weaknesses in protection of mature and old forest and older trees after wildfire under the NWFP. With the expectation of more wildfire under climate change, this issue will only increase. Thus, we offer the following recommendations for protecting both Moist and Dry Forests during salvage.

In Moist Forests there is little ecological justification for salvage following wildfire or other disturbances, as fire suppression has not substantially altered fuels and fire behavior there as it has done in Dry Forests. Hence, burned forests in Moist Forest LSRs should be allowed to recover without salvage, to provide the best chance for development of comparable mature and old-growth forests in the future. The BLM's 2016 RMPs can serve as a guide: salvage is prohibited in LSRs except for road access and safety.[22]

In addition, a policy of aggressive detection and suppression of wildfires is desirable for Moist Forest LSRs. Much older forest habitat has been lost in these LSRs to wildfire; hence, decisions to allow wildfires to burn there without active suppression should be made only after serious consideration of potential consequences. Allowing wildfires in these forests poses a risk to forests both within and adjacent to the LSRs because of the potential for ignitions to become sources of firestorms when an episode of strong, hot, dry east winds occurs.

Moist Forest in Matrix/AMA includes the objective of providing timber supplies over time, so salvage there may be justifiable for that purpose even though it will have negative impacts on natural recovery processes. However, this salvage should retain large, old, live and dead trees.

In Dry Forests, an ecological rationale exists for some salvage after wildfire in LSRs, in contrast to Moist Forests. Without salvage, excess tree densities that often existed before the fires remain in the form of large numbers of burned and dead stems. While those no longer create competitive pressures, they represent fuels that could make subsequent prescribed burns difficult to carry out and could potentially make future wildfires more severe. Consequently, salvage to reduce fuels may be justifiable. If salvage does occur, large, dead trees should be retained on-site during these operations.[23] Also, salvage proposals need to consider and limit potential damage to soil, water, and wildlife.

Efforts to speed reforestation of burned sites through seeding and planting should be limited to areas where sufficient seed sources for natural regeneration no longer exist. Encouraging a variety of species, including hardwoods, can help develop a more natural forest than the traditional single focus on commercial species and can also help create a more resilient forest in the face of climate change. Emphasizing the clumpy structure of natural forests should help, too.[24]

Owls, Shotguns, and the Choices before Us

The future of the northern spotted owl appears predictably grim without sustained reductions in barred owl populations. Barred owls have taken over most prime northern spotted owl habitat in the NWFP area, causing spotted owl populations to decline precipitously. The spotted owl will likely disappear from much of its range without intervention to reduce its encounters with barred owls. Also, the barred owl's impact on many other native species gives another reason for grave concern—an aggressive new top predator has arrived to which most prey species have never been subjected. Major impacts on them and associated food webs can be expected.

Most probably, the barred owl is too well established in the Pacific Northwest to be extirpated, so the policy question generally revolves around the practicality and acceptability of establishing a permanent program for removing barred owls from portions of the northern spotted owl's range. The results to date of the USFWS study of barred owl removal, begun in 2013, suggest the magnitude of even a partial program of barred owl removal. Through 2020, the USFWS had killed over 3,000 owls on four study areas, which cover a mere one-twentieth of 1 percent of the spotted owl's range.[25] Scaling that effort up to even a small portion of the range—given the density of barred owls removed in the trials—could require shooting many tens of thousands of barred owls. In addition, such a program would have to be repeated periodically, since barred owls would almost certainly repopulate the treated areas from both untreated adjacent areas and residual populations within the treated areas.

Some aspects of the future seem clear. First, removal of barred owls will most probably be attempted only in certain portions of the northern spotted owl's range, given how well the birds are established and the cost of attempting total removal. Elsewhere they will dominate and for all practical purposes exclude spotted owls. If this assumption is correct, the USFWS will need to decide which portions of the northern spotted owl's range are most important to its survival and are most amenable to management via removal of barred owls. Designated critical habitat would be a place to start, but even that area is so large that, in all practicality, the barred owls could be removed from only a portion of it.

In areas that lack barred owl removals, northern spotted owls may truly become the rare birds that people believed they were before Eric Forsman started studying them 50 years ago. A key question will be whether the USFWS will evaluate forestry projects that remove habitat in uncontrolled areas the same way it does in areas that have barred owl removals.

Second, we face the ethical issue of whether removing the barred owl is the right thing to do. Are we interfering with a natural evolutionary process, or are we trying to fix a problem that we humans caused when we started excluding fire from the prairies? Since the Forest Service called the barred owl invasive in its recent bioregional assessment[26] and the USFWS called it nonnative in its determination that the northern spotted owl deserved elevation to endangered status under the ESA,[27] it is highly likely that both agencies support barred owl control. Certainly, the USFWS has sometimes killed one abundant species to help recover a threatened or endangered species. Examples include killing raccoons to protect the eggs of threatened snowy plovers on the Oregon coast and killing sea lions to protect threatened salmon in the Columbia River.

Some people will undoubtedly argue that we should let nature take its course as has happened for millennia—that we should allow survival of the fittest and stand back as the northern spotted owl gradually disappears. Or they might argue that removing the barred owl will be too expensive at this stage or logistically impossible, like trying to hold back the tide.

Others, such as Diller in "To Shoot or Not to Shoot," see the choices here as difficult at best:

For me, the issue of lethal removal boils down to a sort of "Sophie's Choice." Shooting a beautiful raptor that is remarkably adaptable and fit for its new environment seems unpalatable and ethically wrong. But the choice to do nothing is also unpalatable, and I believe also ethically wrong. If human actions—including major alterations of spotted owl habitat and paving the way for the invasion of its eastern cousins—have put spotted owls at risk of extinction, don't we have a societal responsibility to at least give them a fighting chance to survive? Despite some protests that this is an unfair choice of one owl over another, the real choice is to conserve both species of owls or only one.[28]

We side with Diller: we should conserve both species. We need some landscapes in which this aggressive predator is constrained, particularly since the barred owl threatens many animals beyond the northern spotted owl.

The barred owl is here to stay, and the northern spotted owl is fighting for its existence. We expect that the USFWS will understandably attempt to throw spotty a lifeline on at least part of its range through barred owl removal. In addition, owl habitat outside the portion of the range subjected to barred owl removal will need protection as the federal agencies study how resilient the endearing spotted owl really is. Finally, we should not forget that many other species need old forest as well.

In many ways, this latest effort by the USFWS is the seventh science assessment underlying the conservation architecture for federal lands in the range of the northern spotted owl. It might well be the most challenging of these assessments, now that the issues have moved beyond habitat to include another species at war with one the agency is trying to save. However, it is also true that the shape and future of federal forest management in the range of the NWFP still depends on a scientifically credible conservation strategy for the northern spotted owl just as it has for the last 30 years.

Changing Our Ways: The Key to Saving Salmon

The Aquatic Conservation Strategy (ACS) of the NWFP was designed to reverse the trend of aquatic ecosystem degradation on federal lands, and that it did: monitoring showed improvement in habitat quality and watershed conditions over the first 20 years of the NWFP.[29] More can still be done on these lands, such as increasing the amount of riparian thinning to accelerate development of large trees that can fall into streams and diversify plant species, which would bring riparian areas closer to their historical condition.[30] Also, more should be spent on fixing problem roads, as is occurring in the Lowell Country Project described in chapter 10.[31] Still, much has been accomplished through the ACS.

A major goal of the ACS was to enable federal lands in the NWFP area to assume primary responsibility for the recovery and preservation of habitat of at-risk fish, especially Pacific salmon, while recognizing that habitat degradation on nonfederal land also contributed to the state of fish populations. However, three developments since the ACS was adopted suggest that habitat protection on nonfederal land is more important for recovery of these fish than was presumed at that time.

First, many Pacific salmon populations, which we will focus on here, have been listed as threatened under the ESA since the implementation of the NWFP, with most of those listings occurring in the late 1990s and early 2000s. Only one salmon population was listed as threatened under the ESA at the time of adoption of the NWFP in 1995, but 22 more populations were listed by 2010 (table 10.1), and they generally have not recovered. While many factors contribute to this dismal record, habitat quality is a core issue.[32]

Second, scientists have increasingly realized that much potentially high-quality salmon habitat is on nonfederal land.[33] Federal lands are generally in the middle to upper portions of watersheds, which tend to have steeper gradients and more confined valleys and floodplains, making them inherently less productive for many salmon populations. More gently flowing streams in the wider valleys on nonfederal lands lower in the watersheds were historical centers of salmon production. However, these areas have been extensively degraded and altered by urbanization, agriculture, and other human activities and today do little to support these fish. Federal lands, of course, are important sources of wood, gravel, food, and high-quality water for these downstream areas, but such inputs will not be sufficient to improve conditions on nonfederal lands downstream without focused habitat restoration there too.

Third, research since the ACS was developed has shown that the freshwater life histories of Pacific salmon, such as coho and Chinook, are diverse and not of a single type. Some of the life histories involve movement to and residence in different portions of rivers and streams, especially lower in the watershed. This life-history diversity is critical to the production and persistence of the species in dynamic and variable environments, including the ocean. Providing a landscape across the watershed where this diversity of life histories can be expressed will be essential to the conservation and recovery of salmon populations in the future.

Private and state forests, of course, are an important part of the salmon landscape. Historically, logging and roadbuilding on these forests degraded salmon habitat through increasing sedimentation and stream temperatures. In addition, removing large trees along streams eliminated their potential to fall into waterways and contribute to spawning and rearing habitats. With passage of the first Forest Practices Act in 1971, Oregon became a leader in protecting waterways from the negative impacts of forest management on private and state lands but has since fallen far behind Washington, California, and British Columbia in terms of conserving riparian and aquatic ecosystem functions. State, private, and tribal landowners in Washington, as an example, negotiated a major improvement in stream protection in the early 2000s that the National Marine Fisheries Service (NMFS) accepted as sufficient for state and private contributions to listed fish species.

By 2020, conservation groups in Oregon turned to threats of citizen initiatives to reform water protection rules in the state, which drove the forest industry to enter negotiation with these groups under the leadership of Governor Kate Brown. First, in 2020 the groups negotiated an increase in herbicide restrictions. Then, in November 2021, they reached the Private Forest Accord, which substantially increases stream protections.

Kelly Burnett, a retired PNW fish biologist who participated in the negotiations, summarized major features of the Accord: it widens existing riparian buffers; requires buffers on more of the stream network, including on many small, non-fish-bearing streams; identifies and protects many of the unstable slopes that can deliver sediment to fish-bearing streams; and requires road culvert design changes to improve fish passage, among many other provisions. Applauded and endorsed by the Oregon legislature in early 2022, it will underpin the state's application for a Habitat Conservation Plan and Incidental Take Permit under the ESA to cover populations of salmon, steelhead, bull trout, and cutthroat trout and numerous amphibian species in western Oregon.[34]

Protection of streams and riparian areas on private forests has come a long way over the last 20 years; these rules can and should continue to be improved, such as in control of logging and roadbuilding in burned forests. However, recovery of salmon habitat will depend most importantly on major improvements in habitat protection along streams in the lower portions of watersheds, which have been converted into farms and ranches and places where people live and work. These areas historically contained the most productive habitats, such as off-channel areas on floodplains and complex channels with large amounts of wood. They have been severely degraded or lost through the ditching and draining of floodplains, straightening of streams, and removal of large wood from within them and tree buffers along them.

Drive the coastal valleys of western Oregon, as an example, and you will often see a pastoral scene with cattle grazing in the distance. Many of these pastures were once wetlands where coastal coho spawned, juveniles were reared, and smolts overwintered before moving to the estuary. Look at the streams in the suburbs of Portland, if you can find them, and imagine how they might have looked before their banks were covered with houses and roads. Restoring a critical mass of these areas in a coordinated fashion and making them available to salmon is essential for expression of life-history diversity and overall population productivity.

Certainly, providing incentives for nonfederal landowners to participate in restoration will be important to the successful recovery of listed fish. As part of this effort, it will be imperative to recognize that not all stream segments have the same capacity to provide productive habitat, and to use this knowledge to guide policy and incentives.[35]

Research will also play a key role in the recovery effort. Salmon populations, to a large extent, are resilient to the broad range of disturbances and environments that they historically experienced because of their life-history diversity. Research that focuses on understanding potential effects of these disturbances and environments over the life history of these populations is critical moving forward. Utilizing this knowledge to move beyond providing habitat for more salmon to providing an environment that allows for the expression of a variety of life histories in terms of the particular streams and conditions that the salmon populations seek will also be important.

Federal land management certainly has a continuing role in recovery of threatened salmon populations by protecting and restoring streams, riparian areas, and watersheds in federal forests and providing resources for downstream habitats. Still, other ownerships and governments must make major contributions for that recovery to be successful. The political will to improve habitat protection on farms and ranches and in suburban and urban areas holds the key to restoring the habitats of Pacific salmon—an iconic species of the Pacific Northwest.

Sources of Timber Harvest Volume over the Next 20 Years

The 2012 Planning Rule and associated 2015 Planning Directives for the national forests,[36] which will guide forest plan revision, shifted timber harvest estimates for the national forests away from long-term sustained yield projections or commitments. In their place, the Forest Service called for 20-year estimates of likely timber harvests as part of an ecologically grounded management plan. We will maintain the spirit of that new approach here—even 20 years is a long time in the tumultuous physical, economic, and political world in which the national forests exist. We again consider Moist and Dry Forests separately.

Thinning younger forests (mostly plantations) across the landscape will remain a major source of timber harvest volume for the next few decades on national forests containing Moist Forests. However, even that source may run short in the near term. The recent emphasis on thinning younger forests may have created a bottleneck until those still in the pole stage grow a little more, making it difficult to maintain harvest levels solely from those types of stands.

Increased variable retention harvest of older plantations in Matrix/ AMA of Moist Forests can and should be another important timber volume source. Such harvests would help reset plantations that are poor candidates for developing late-successional forest attributes while providing harvest volume.

Fire salvage from the September 2020 wildfires in Matrix/AMA could be an important part of Moist Forest harvests for years to come. Care should be taken during salvage to avoid significant damage to soil and water resources and to retain large, old trees.

On the other hand, reserving mature and old-growth Moist Forest, as recommended here, will reduce the acres of Matrix/AMA by at least one-quarter and thus reduce Matrix/AMA harvest volume over time compared to that assumed in the NWFP.[37] That change will lower the projected harvest level over time that can be achieved from Moist Forests.

Dry Forest restoration tells a different story—needed actions should create a bump in timber harvest volume. Time is running out to save Dry Forests in the NWFP area. Most are choked with trees established during the last 100 years, which create conditions that threaten their future and that of the old trees within them.

Many of these problems are in northern California, where most of the Dry Forests in the NWFP are located (plate 1.3). Major restoration efforts are long overdue for the federal forests there, which will both help improve resistance and produce significant timber volume for local mills.

Even with such restoration efforts, fire-related salvage will receive much attention in NWFP Dry Forests, especially in California in the aftermath of the 2020 fires. Salvage can contribute to the recovery of burned Dry Forest, if approached along the lines of what we discussed previously where large, old trees are left.[38] It also provides needed logs for local mills.

Recognizing the Goals and Rights of Tribal Nations

Treaties between tribal nations and the United States of America were approved in the mid-1800s by Congress for major portions of Oregon and Washington, which included some of the NWFP area (fig. 1.3).[39] Many treaties included the right for tribal nations to continue traditional uses on their historical lands that they ceded to the federal government and that became national forests and BLM lands. Those traditional uses commonly included hunting, fishing, and gathering for subsistence goods like deer, salmon, huckleberries, roots, medicines, and basket materials.

In addition, tribal nations in other parts of the NWFP area, such as western Oregon and northern California, negotiated or tried to negotiate similar treaties with the federal government. However, they were ignored, reached some sort of settlement but the results were not turned into a treaty, or saw the treaty later reversed as the land became too valuable. Thus, they often got neither lands nor rights: no reservation and no rights to continue traditional uses on ceded lands. Their goals and wishes should also be respected.

The district court decision in *Klamath Tribes v. United States of America* in 1996 (which was not appealed) forcefully describes the federal government's responsibilities to protect a tribe's treaty rights.[40] The district court started with a precedent-setting Supreme Court decision from 1942:

This Court has recognized the distinctive obligation of trust incumbent upon the Government in its dealings with these dependent and sometimes exploited people. In carrying out its treaty obligations with the Indian Tribes, the Government is something more than a mere contracting party. Under a humane and self-imposed policy which has found expression in many acts of Congress and numerous decisions of this Court, it has charged itself with moral obligations of the highest responsibility and trust.[41]

And then the district court described the practical implications of these responsibilities:

In practical terms, a procedural duty has arisen from the trust relationship such that the federal government must consult with an Indian Tribe in the decision-making process to avoid adverse effects on treaty resources. . . . A determination of what constitutes compliance with treaty obligations should not be made unilaterally; rather, the Tribe's view of the hunting, fishing, gathering, and trapping activities protected by the treaty must be solicited, discussed, and considered.

Moreover, the federal government has a substantive duty to protect "to the fullest extent possible" the Tribes' treaty rights, and the resources on which those rights depend. . . . In a written policy regarding management responsibilities related to the Tribes' treaty rights, the Forest Service acknowledged its duty to manage "habitat to support populations necessary to sustain Tribal use and non-Indian harvest," including "consideration of habitat needs for any species hunted or trapped by tribal members."[42]

Some key points in these and other court rulings are the following:

• Treaty obligations are moral obligations of the highest responsibility and trust.
• The federal government must consult with tribal nations in decision-making to avoid adverse effects on treaty resources, such as in providing the resources needed to enable hunting, fishing, and gathering, and must listen to the tribes' view of how to achieve that goal.
• The federal government must protect treaty resources to the fullest extent possible, including managing habitat to sustain harvest.

• Tribal nations are not "stakeholders" or part of a collaboration. Rather, they are sovereign nations, and consultation with them is required within a government-to-government relationship.

Judged by these standards, the Forest Service and BLM could do much more to fulfill their trust responsibilities.

The highly useful chapter in the PNW Research Station's science synthesis titled "Tribal Ecocultural Resources and Engagement" by Long and his colleagues provides a template for how these two agencies can move forward to integrate tribal goals, values, knowledge, and perspectives into project and forest planning and implementation. It is truly a guidebook to a better relationship with tribes and to fulfilling federal responsibilities to tribal nations.

The Forest Service's recent bioregional assessment to support forest plan revision includes a progressive statement about the scope of its relationship with tribal nations: "More than 70 federally recognized American Indian Tribes have lands or ancestral territory within the BioA area [NWFP area + adjacent forests] and maintain a government-to-government relationship with the Forest Service."[43]

Further, the assessment acknowledges that the "Tribes hold deep connections to ancestral lands managed by the Forest Service and rely on effective forest management of Tribal resources to maintain those connections."[44] It also acknowledges the importance of traditional foods (called *first foods*) to tribes and the role of the Aquatic Conservation Strategy in enhancing one of those foods (salmon and other fish).[45] However, it provides little information on the extent or condition of traditional foods that grow in federal forests and meadows, such as huckleberries and camas, or what the federal government can do to maintain and restore them. The assessment does mention big game as a traditional and continuing source of sustenance for tribes. However, providing cover for big game is described mainly as a restraint on achieving restoration goals. While it may be true that cover requirements need adjustment, the need to provide sufficient cover and forage for big game to meet federal responsibilities to tribal nations should also be acknowledged.[46]

Also, the assessment missed an opportunity to emphasize the procedural duty to listen to a tribal nation's view of how to protect its treaty resources and the substantive duty to protect, to the fullest extent possible, a tribe's treaty rights and the resources on which those rights depend. Surely, it is crucial to forest plan revision across the NWFP area that planners recognize these duties, along with a responsibility to acknowledge and advance the goals of other tribes with ancestral lands there, in completing both required forest-level bioregional assessments and plan revision.[47]

In fact, this recognition does not need to await forest plan revision. It can and should be embedded more fully in project plans. Such an approach could further the aspirations of the many tribes in the NWFP area and help fulfill the spirit of the 1942 Supreme Court decision that the federal government "has charged itself with moral obligations of the highest responsibility and trust." Recent commitments to co-stewardship are a good beginning. More is needed.

In addition, as has become increasingly clear, we have much to learn from the historical practices of tribes in sustaining and restoring wildland resources on federal lands, as described in depth in "Tribal Ecocultural Resources and Engagement." Certainly tribal use of fire to manage forests and wildlands has gained much attention; other tribal management and harvest methods should also be of interest. Perhaps most important would be gaining a deeper understanding of the Indigenous land ethic and traditional ecological knowledge. Enabling cooperative conservation and management of federal lands between tribal nations and the federal government could and should be of high priority in the range of the northern spotted owl.

Revising the LSR Network—A Challenging Task

The Interagency Scientific Committee (ISC) in 1990 proposed the first scientifically credible conservation strategy for the northern spotted owl. It divided the federal forest into two parts—Habitat Conservation Areas and Matrix. The HCAs were to provide nesting, roosting, and foraging habitat sufficient to support self-sustaining populations of spotted owls and be sufficiently close to each other for successful dispersal between them. Matrix between HCAs would be managed on rotations compatible with owl dispersal, as illustrated by the 50-11-40 rule. However, spotted owls and their habitat in Matrix would not be conserved.

A network of LSRs replaced the network of HCAs in the NWFP and a series of policy decisions led to substantial forest areas being reserved outside the LSR network to help conserve the northern spotted owl and other species. FEMAT in 1993 began these additions by calling for small LSRs in coastal forests for the marbled murrelet, a species listed as threatened after completion of the ISC's report. Then the NWFP in 1994 scattered smaller LSRs through Matrix/AMA in the form of 100-acre reserves around existing (as of 1993) spotted owl activity centers and buffers around marbled murrelet nests as they were discovered in Matrix/AMA. The *Revised Recovery Plan* for the northern spotted owl in 2011 further expanded the reserve system with recommendations to retain complex forest (mostly old growth) and also high-value habitat around historical spotted owl activity centers in Matrix/AMA.[48] Finally, critical habitat determinations for the owl in 2012 covered significant areas of Matrix/AMA, which restrict harvest although they do not technically establish reserves. This scattering of reserves and critical habitat designations over Matrix has led to the call for revision of the LSR network to better capture the owl and murrelet reserve system, as have concerns about where to place elements in the network to increase its chance of withstanding wildfire. We discuss here how LSR revision might proceed.

We ground our proposal for LSR revision on key assumptions about goals and pivotal issues. First, the fundamental goal in national forest planning is to maintain and restore the ecological integrity of terrestrial and aquatic ecosystems—to ensure ecological sustainability to the degree possible as described in the 2012 Planning Rule. Within the bounds of achievement of that goal, management of these lands should make important contributions to economic and social sustainability.

Second, we need to apply different management strategies to LSRs in Moist and Dry Forests. The concept of a network of reserves large enough for a self-sustaining population of owls and close enough to each other to enable dispersal is still the starting point for creation of the Moist Forest LSR network. Their placement should be reevaluated, as better information is now available than when they were set almost 30 years ago, and the forests have matured in some places and burned in others. Forest management in the Moist Forest LSRs should focus on restoring the ecological integrity of plantations—the highly artificial

stands that were established after the old-growth forests were cut—
while staying out of mature and old-growth forests. In addition, the
recommended actions for Matrix management in the *Revised Recovery
Plan*, including retention of complex forest, should be integrated into
forest plan revision. Within that context, survival of the northern spotted owl will be the pivotal issue, with control of an aggressive invader—
the barred owl—a central concern.

Dry Forests, on the other hand, face an existential crisis—survival
of the forest itself—and old tree populations within them are a central
concern. Concerns over spotted owls must be embedded in this larger
issue, because at the rate these forests are undergoing catastrophic
wildfires, very little high-quality spotted owl habitat may exist in the
future. Drought and associated insect outbreaks are additional widespread risks in Dry Forests, as shown by the Sierra Nevada forests of
California that suffered catastrophic losses of old trees across millions
of acres during droughts in the first decades of the twenty-first century.

With these understandings as background, we offer some guidelines and thoughts about undertaking revision of LSR locations and
boundaries in the national forests:

The USFWS's scientifically grounded 2012 critical habitat determination for the northern spotted owl provides guidance for LSR revision,
although it must be recognized that recent wildfires have degraded
substantial areas of that habitat. Also, more than spotted owl habitat
considerations will guide LSR selection. For example, LSRs sought
to protect habitat for the marbled murrelet in coastal forests and, of
course, reserve the best remaining old-growth forest ecosystems. Still,
critical habitat provides useful guidance.

Revision of the entire LSR network would best be done simultaneously across the entire range of the northern spotted owl, in cooperation with the USFWS, as the BLM accomplished in its 2016 RMPs for
its part of the range. The LSR network is a range-wide plan, and piece-
meal modifications could easily result in an LSR network that no longer
meets habitat goals for spotted owl survival and recovery.

Given the enormity of range-wide LSR revision and the past lack of
success in attacking this problem at that level, the Forest Service may
bite off a portion of the owl's range in which to revise the LSR network
as part of the simultaneous revision of several national forest plans. Four

national forest plans in northern California combined with one in south-western Oregon, largely covered by Dry Forests, seem likely candidates for such an effort as noted in the 2020 bioregional assessment. If such an LSR revision based on multiple forest plans occurs, the effectiveness of the modified subregional LSR network would need to be analyzed in the context of the LSR network for the rest of the range to make sure that the entire network still provides sufficient habitat for owl recovery.

For Moist Forests, changes in the LSR network that shift more mature and old-growth forest into LSRs could be balanced by shifting some plantations currently in LSRs into Matrix. Such shifts should be an important consideration in any LSR boundary revisions.

For Dry Forests, standards that encourage fuel treatments and forest restoration within the older forests of the LSRs would be as important as LSR boundary changes. Standards need to clarify that such actions are not just allowed but are essential to the survival of older forests there.

More fundamentally, though, is the issue of whether large LSRs make sense for Dry Forests at all. The large LSRs were patterned largely after desired conditions in Moist Forests. Soon after the NWFP was adopted, Jim Agee, the founding father of fire ecology in the Pacific Northwest, visited the Okanogan-Wenatchee National Forest and looked out at the large Dry Forest LSRs that had been created. He almost immediately yelled, "It's going to burn up—don't you all realize that?" Bob Taylor of the California Forestry Association expressed similar thoughts in 1994 when he agonized over the ecological and social risks of utilizing an approach to forest conservation in the Dry Forests of northern California that was designed for the Moist Forests of western Oregon.[49] Wildfires over the last 25 years have confirmed their concerns.

As mentioned before, an alternative approach to large LSRs for Dry Forest landscapes in the NWFP area could focus on maintaining smaller patches or islands of suitable habitat for the owls and their prey within a landscape that is otherwise restored to a fire-resistant and resilient condition. These restored forests of lower density and primarily large, older trees of fire-resistant species would be the dominant element of these restored landscapes (perhaps two-thirds of the area) within which the dense patches (perhaps one-third of the area) are embedded. The dense patches would be placed, to the degree possible, where the associated microclimates make them less susceptible to burn at high severities (so-called fire refuges).[50] Some of them would still get torched from time to time, but since the Matrix retains an overstory of older trees, important elements of replacement patches would already be present. Furthermore, the fact that the dense patches are embedded in a restored landscape should make their probability of burning much lower than otherwise—hence the concept that they would have increased hang time, compared with dense forest in an untreated landscape.

Johnson, Franklin, and Johnson modeled such an approach in *A Plan for the Klamath Tribes' Management of the Klamath Reservation Forest* for a portion of the eastern Cascades in Oregon. Patches of denser, older forest of 300–500 acres were scattered through a landscape dominated by restored mixed-conifer forest. The denser patches would occupy approximately 30 percent of the landscape; the remainder would be shifted to a more historical composition and structure by retaining older trees and reducing competitive pressures on them through thinning and prescribed fire, and also by re-creating a patchy, clumpy spatial structure of much lower density.

Such an approach for Dry Forests would require three key elements. First, the USFWS would need to be deeply involved in their design and in the decision about what mix of restored forest and dense patches would provide sufficient habitat on the national forests to enable survival and recovery of the northern spotted owl. Second, a commitment would be needed to retain old trees across the landscape. Third, it would be important to acknowledge that these patches would not be permanent but would shift over time as they lost their effectiveness.

Ways to Make Needed Changes

How might the forest conservation and management ideas in this chapter be used in national forest planning and management, beyond the important changes suggested to recognizing the goals and rights of tribal nations? We consider three possibilities. Most could be integrated into project-level directions governing specific actions that the agencies take on the ground. Alternatively, the ideas could be incorporated into revision of the NWFP and/or the plans for individual national forests. A third possibility would be to adopt them through congressional action. From the first possibility to the last, the permanence of the decision increases but so does the political struggle to achieve an outcome.

Our recommendations for conserving natural, older Moist Forests can be implemented largely through project selection and silvicultural prescriptions. Reserving remaining natural, older Moist Forest in Matrix/AMA has been the de facto policy for federal forests in the NWFP area for years in many places and could be continued, as could leaving LSRs unsalvaged. Designating large, old trees for retention in Matrix/AMA harvest units and during salvage is permitted under the standards and guidelines for the NWFP and thus can be done in describing the prescriptions that should be applied. A separate decision process is not needed.

Also, the recommendations could be quickly adopted as general guidance for agency actions. They could be issued by the regional forester, chief, secretary of agriculture, or president as recommendations on interim protection of older, natural forests until forest plans, or the NWFP itself, are amended or revised. They could function much like the recommendations in recovery plans for federally listed species—powerful statements that are left to line officers in the field to apply. The president's recent Executive Order on conservation of mature and old-growth forests is an important step in that direction.

Stepping up forest restoration in older Dry Forests would necessitate the close cooperation of the USFWS and others but, in theory, should not require plan revision. However, it is also true that it may take major efforts like the Rogue Basin Strategy to speed things up, along with a modified conservation strategy for the spotted owl that addresses the barred owl conflict. Further, general guidance on fire salvage could be a key to protecting old trees and advancing forest restoration in Dry Forests.

The current standards and guidelines for implementing the NWFP lack strong support for fuel treatments and forest restoration in older Dry Forests in LSRs at high risk. These actions should be of the highest priority to save these older forests and the old-growth trees within them from high-severity fire and also from drought and insect attack. At the very least, the Regional Ecosystem Office should clarify that such actions are critically needed and encouraged under the NWFP. Or the Forest Service could undertake a very narrow amendment to the

NWFP, as was done across multiple forests by the Northern Region in implementing revised grizzly bear guidance from the USFWS.[51]

Relative to the national forests, forest plan revision and associated creation of a barred owl control plan could take a while to pass through all political and regulatory hurdles. In the meantime, it is important not to wait for all that to happen before changes are made, as occurred under the National Forest Management Act, where a 15-year delay in making needed changes until revised forest plans were approved had dire consequences for old-growth forests.

Revising forest plans and updating the NWFP, along the lines suggested here, would be a welcome outcome if it could be achieved. The Forest Service's revised Planning Rule of 2012 brought the agency's planning philosophy, principles, and requirements into line with the special role these forests play in American life. Sustainability is first defined as maintaining and restoring the ecological integrity of terrestrial and aquatic ecosystems and then contributing to the admirable goals of economic and social sustainability to the degree possible. Species whose viability is judged to be at risk under that approach will be protected through species-specific requirements. Short-term estimates of likely timber harvest will be made in that context.[52] A new day for the national forests of the NWFP area will arrive when national forest plans using the 2012 Planning Rule are implemented, but it may be hard slogging to get there. And it will ultimately require decisions at larger spatial scales, such as regions or ranges of keystone species.

Congress could step up and resolve some of these issues, especially on the future of mature and old-growth forests, providing protection with the permanence of national parks and wilderness. However, intense disputes over national forest goals, particularly the role of timber harvest, have generally stymied such efforts in the past.

Of course, lightning could strike at any time when Congress is in session, such as legislation establishing a system of carbon reserves or other mechanisms. If carbon legislation was developed, the fundamental differences between Moist and Dry Forests would need to be recognized. Carbon reserves work for older Moist Forests, but older Dry Forests need actions to restore and sustain them.

Congress could also mandate particular forest management policies and prescriptions, such as requiring that trees or stands over a certain age be left on-site at harvest. Again, the difference between Moist and Dry Forests would need to be recognized. However, legislating silvicultural prescriptions is harder than it might seem, as it is difficult to build in the flexibility needed to deal with all situations, let alone changing conditions and evolving knowledge.[53] In sum, Congress may eventually get into the fray, but that should not stop progress in the meantime.

Epilogue

There has been a dramatic and continuing evolution of federal forest policy concerns since adoption of the NWFP over 25 years ago, and the twenty-first century provides a greatly altered set of challenges and societal priorities for federal forestlands. New approaches are imperative, and stakeholders and resource professionals need to take a fresh look at attitudes left over from the late-twentieth-century owl wars. Our final thoughts on the future follow.

In the Moist Forest landscapes, twenty-first-century management will not be about prioritizing wood production using continuing generations of plantations. While northern spotted owls remain a concern, the focus will be on dealing with an aggressive invasive more than habitat. Replacing plantations with fully functional forest ecosystems should be a dominant issue in Moist Forest landscapes to provide the habitat needed by native biota and immensely increase sequestered carbon stocks. Saving remaining older forests is also critical. These are the only fully functional forest ecosystems left in plantation-fractured landscapes created from clearcut logging of native forests.

The priority in Dry Forest landscapes must be primarily on sustaining functional forest ecosystems in the face of what have obviously become dramatically greater risks from catastrophic fire, drought, and insects. This will be possible only with greatly accelerated programs to bring resistance back to densely overstocked forests using both mechanical and pyric tools. Sustaining habitat for spotted owls will need to be integrated into this goal, for the alternative will be little spotted owl habitat at all.

Purposeful actions are critical to achieving these goals in both the Moist and Dry Forests that we so altered during the twentieth century. This can and, in our view, should include commercial removal of wood products, which can help subsidize the significant costs of managing these lands and provide support for local communities. These communities can also be important sources of woods workers who can carry out the myriad activities associated with management of the federal forests.

Walking away from our federal forests, as some propose, is not an option for any who truly care about their future. They say, "We have badly managed these forests in the past and we should leave their restoration to nature." Yes, the forests have undergone major human-driven changes in the last 100 years with great alterations in their structure, composition, and functional capabilities. We have greatly altered the context within which they exist, including the climate, disturbance regimes, and biota. Essentially both these forests and the environment represent novel conditions, unlike any that existed historically.

We created these conditions; we should use our knowledge to guide and assist these ecosystems in recovering and sustaining their full functional capabilities. We have learned an immense amount about these forest ecosystems and how they work in the last 50 years; we know that they are not simply collections of young green trees and that we depend on them for myriad services beyond wood. For us to fail to use our science and skills to help return them to full functionality (which is what health is really about) and to leave them to deal with the random and novel conditions that we have created would be a fundamental abdication of our responsibilities as forest stewards.

Appendixes

APPENDIX 1

Scientific Teams

Interagency Scientific Committee (ISC)

Jack Ward Thomas, USFS (chair); Eric D. Forsman, USFS; Joseph B. Lint, BLM; E. Charles Meslow, USFWS; Barry B. Noon, Colorado State University; Jared Verner, USFS

Gang of Four (G-4)

K. Norman Johnson, Oregon State University (coordinator); Jerry F. Franklin, University of Washington; John Gordon, Yale University; Jack Ward Thomas, USFS; James R. Sedell, USFS; Gordon H. Reeves, USFS

Scientific Assessment Team (SAT)

Jack Ward Thomas, USFS (leader); Martin G. Raphael, USFS (associate leader); Robert G. Anthony, USFWS; Eric D. Forsman, USFS; A. Grant Gunderson, USFS; Richard S. Holthausen, USFS; Bruce G. Marcot, USFS; Gordon H. Reeves, USFS; James R. Sedell, USFS; David M. Solis, USFS

Forest Ecosystem Management Assessment Team (FEMAT)
(team and group leaders only)

Jack Ward Thomas, USFS (team leader); Martin G. Raphael, USFS (deputy team leader). Terrestrial ecology group: E. Charles Meslow, USFWS (coleader); Richard S. Holthausen, USFS (coleader). Aquatic/ watershed group: James R. Sedell, USFS (coleader); Gordon H. Reeves, USFS (coleader). Resource analysis group: K. Norman Johnson, Oregon State University (leader). Economic assessment group: Brian Greber, Oregon State University (leader). Social assessment group: Roger N. Clark, USFS (leader). Spatial analysis group: Duane R. Dippon, BLM (coleader); John R. Steffenson, USFS (coleader)

Species Assessment Team (SAT II)

Richard S. Holthausen, USFS (team leader); Keith Aubry, USFS; Kelly M. Burnett, USFS; Nancy Fredricks, USFS; Joseph Furnish, BLM; Robin Lesher, USFS; Martin G. Raphael, USFS; Roger Rosentreter, BLM; Robert Anthony, National Biological Survey; E. Charles Meslow, National Biological Survey; Edward E. Starkey, National Biological Survey

APPENDIX 2

The SAT Analysis

Jack Ward Thomas led the Scientific Assessment Team (SAT), which included a core team composed primarily of owl and fish experts,[1] and a support team with a variety of expertise in plant ecology, wildlife biology, and hydrology that provided additional advice and analysis. SAT assembled quickly in December 1992 to identify the products they would develop and to organize members into three groups of expertise: northern spotted owls, fish, and other species. This appendix focuses on the portion of SAT that addressed whether the Forest Service's implementation of the alternative in the Record of Decision associated with the 1992 FEIS (the ISC's conservation strategy for the northern spotted owl) would lead to extirpation of any of the 32 species identified in the FEIS as being closely associated with older forests.[2]

SAT, with the help of the evaluation panels, identified "mitigation options for habitat conditions conducive to providing for high viability" for all species identified as being at viability risk.[3] SAT first called for adoption of the standards and guidelines in the recently adopted or proposed forest plans and the draft recovery plan for the spotted owl (based largely on the ISC's conservation strategy). With that foundation, SAT recommended adoption of four additional mitigation measures:

1) Standards and guidelines for Riparian Habitat Conservation Areas for protecting the 112 fish populations at risk. These standards and guidelines required extensive riparian buffers on both fish-bearing and non-fish-bearing streams, with the buffer widths measured in terms of site-potential tree heights—the maximum average height a tree can attain on a particular site. The SAT noted that a variety of nonfish species identified to be at viability risk would also benefit from increased old-growth protection afforded

397

by the wider buffers. These buffer requirements were embedded in a comprehensive Fish Habitat Conservation Strategy, described below.

2) Standards and guidelines for marbled murrelet nesting habitat that were also expected to benefit other species at viability risk.

3) Standards and guidelines for conservation of rare and locally endemic species, including specific mosses, fungi, and invertebrates (especially land snails and mollusks, and salamanders that occur in narrow ranges). As part of these standards, SAT called for surveys to identify and protect sites occupied by these species prior to ground-disturbing activities.

4) Additional standards and guidelines for other species in the upland forest outside reserves, including snag retention and other provisions for specific woodpeckers and other birds, survey and protection of great gray owl sites, and a variety of requirements for lynx habitat where lynx were present.[4]

Also, SAT identified over 200 species for which a viability analysis was not possible because of insufficient life-history data but noted that the mitigation measures they proposed would provide increased protection for some of these species. In general, though, surveys, research, monitoring, and an adaptive management approach would be necessary to conserve them.

The Fish Habitat Conservation Strategy was a major step forward in aquatic and riparian conservation and deserves further explanation. First, SAT reiterated the finding of the Gang of Four analysis that the recently adopted or proposed forest plans augmented by the ISC's strategy for the spotted owl would not protect anadromous fish populations or resident trout.

Second, Sedell and Reeves, as part of SAT, refined and further developed the conservation strategy they had initiated in the G-4 analysis. As they stated in the SAT report: "The strategy is designed to provide a high probability for maintaining and restoring habitat for fish. Its focus is on maintaining and restoring ecological functions and processes that operate in a watershed to create habitat."[5]

They further stated:

Our strategy is to maintain as close to a 'natural' disturbance regime as is possible within watersheds and landscapes, many of which have already been altered by human activities. It needs to be emphasized, however, that this will require time for this strategy to work. Because it is based on natural disturbance processes, it may require timescales of decades to over a century to accomplish all of its objectives. Significant improvements in fish habitat, however, can be expected on a timescale of 10 to 20 years.[6]

SAT's Fish Habitat Conservation Strategy rested on four components that provided the foundation for the Aquatic Conservation Strategy in the NWFP: (1) identifying a landscape-level network of watershed refugia by refining the Key Watersheds of the G-4 analysis; (2) establishing Riparian Habitat Conservation Areas where the only activities allowed were those benefiting or not adversely affecting fish habitat, with interim buffer widths measured by site-potential tree heights: two site-potential tree heights on each side of fish-bearing streams and one on each side of non-fish-bearing streams; (3) conducting watershed analyses to evaluate geomorphic and ecological processes, revise boundaries of riparian buffers based on site-specific analyses, and plan restoration measures; and (4) implementing comprehensive watershed restoration. Shifting from fixed riparian widths (of national forest plans and the G-4 analysis) to site-potential tree heights as the measure of protective riparian widths was an especially important and lasting change.[7]

APPENDIX 3

The FEMAT Viability Analysis

Leaders of FEMAT's viability assessments explained how they employed these four possible outcomes in estimating species viability under a FEMAT option:

> We asked panelists to assign 100 "likelihood votes" (or points) across the four outcomes in the scale. A panelist could express complete certainty in a single outcome for a species/option combination by allocating all 100 points to a single outcome. The panelist could express uncertainty by spreading votes across the outcomes. An individual panelist could refrain from assessing a species because they simply had too little understanding to venture an informed opinion. The entire panel could also choose not to rate a species if they thought there was inadequate scientific knowledge about the species. These species were marked "not rated" on the assessment forms, but they were of no less concern than rated species. Discussions about the need to study and provide for these species was captured in the panel transcripts and panel leaders' reports. . . .
>
> We adopted the likelihood voting methodology in an effort to quantify scientific and personal uncertainty. . . . We felt that honest expressions of how little or how much was known about species/option interactions could help us and decisionmakers better understand the issues and make more informed tradeoffs. We emphasized to panelists that the likelihoods are not probabilities in the classical notion of frequencies. They represented degrees of belief in future outcomes, expressed in a probability-like scale that could be mathematically aggregated and compared across options and species. This use of the "judgmental probabilities" is

consistent with the theory and practice of decision analysis and decision science.[1]

The viability leaders further noted that their analysis was limited

to assessing sufficiency of habitat on federal land to provide for the persistence of the species. We did not assess population viability per se . . . some species are more heavily influenced than other species by habitat on nonfederal lands or other conditions (such as air pollution). In these cases, our habitat-ratings for federal lands do not reflect the expected consequences of these other influences. The fate of such species is not principally a function of federal forest management.[2]

They compared options by

assessing whether a species (or group) attained an 80 percent or greater likelihood of achieving Outcome A: Habitat of sufficient quality, distribution, and abundance to allow the species population to stabilize, well distributed across federal lands. . . . This basis for comparison represents a relatively secure level of habitat and thus provides a stringent criterion for comparison. The same process was used to assess the likelihood of maintaining a functional, interacting late-successional and old-growth forest ecosystem.

In focusing on the attainment of 80 percent likelihood of achieving outcome A, we are not suggesting that only options attaining this likelihood satisfy the viability regulation. We think it likely that options attaining such a percentage would be viewed as meeting the requirement, but a score of less than 80 should not automatically be regarded as a failing grade. Similarly, in some instances it may be appropriate to look at categories A and B (that is, A plus B) as the benchmark. Indeed, in situations where a species is already restricted to refugia, it may be appropriate to look at A plus B plus C.[3]

APPENDIX 4

Reserve Objectives and Permitted Silviculture under the NWFP

Late-Successional Reserves[1]

OBJECTIVE:

- "Protect and enhance conditions of late successional and old-growth forest ecosystems."

PERMITTED SILVICULTURE:

- "West of the Cascades—Thinning (precommercial and commercial) may occur in stands up to 80 years old regardless of the origin of the stands."
- "East of the Cascades and in the Oregon and California Klamath Provinces—Given the increased risk of fire in these areas . . . additional management activities are allowed in Late-Successional Reserves. . . . Silvicultural activities aimed at reducing risk shall focus on younger stands . . . activities in older stands may be appropriate if: (1) the proposed management activities will clearly result in greater assurance of long-term maintenance of habitat, (2) activities are clearly needed to reduce risks, and (3) activities will not prevent the Late-Successional Reserves from playing an effective role in the objectives for which they were established."
- "Salvage guidelines are intended to prevent negative effects on late-successional habitat, while permitting some commercial wood volume removal" where salvage is "defined as the removal of trees from an area following a stand-replacing event."
- General [salvage] guidelines include the following:
  - "Planning for salvage should focus on long-range objectives, which are based on desired future condition of the forest."
  - "All standing live trees should be retained."

- "Management should focus on retaining snags that are likely to persist until late successional conditions have developed."
- "Management should retain adequate coarse woody debris quantities in the new stand so that in the future it will still contain amounts similar to naturally regenerated stands."
- "In some cases, salvage operations may actually facilitate habitat recovery. For example, excessive amounts of coarse woody debris may interfere with stand regeneration activities following some disturbances."
- "Some salvage that does not meet the preceding guidelines will be allowed when it is essential to reduce the future risk of fire or insect damage to late-successional forest conditions."
- "Thinning or other silvicultural treatments inside reserves are subject to review by the Regional Ecosystem Office to ensure that the treatments are beneficial to the creation of late-successional forest conditions."

Riparian Reserves

OBJECTIVE:

- Help maintain or restore the hydrologic, geomorphic, and ecologic processes that directly affect standing and flowing water bodies such as lakes and ponds, wetlands, and streams. They are part of the Aquatic Conservation Strategy, which was developed to restore and maintain the ecological health of watersheds and aquatic ecosystems.[2]

PERMITTED SILVICULTURE:

- Apply silvicultural practices to control stocking, to reestablish and manage stands, and to acquire desired vegetation characteristics needed to attain Aquatic Conservation Strategy objectives.[3] In practice, the silviculture practiced in the Riparian Reserves, where it was permitted, followed the same approaches described above for Late-Successional Reserves—thinning in young stands and fuel treatments where forests were at high risk of wildfire.

Notes

ACKNOWLEDGMENTS

1 "Voices of the Forests, Voices of the Mills," Oregon State University, http://scarc.library.oregonstate.edu/omeka/exhibits/show/forestryvoices/main/.

2 Scott, *Timber Wars*.

INTRODUCTION

1 Forsman tells his story about tracking the northern spotted owl in his own words at the end of chapter 4.

2 Forsman, interview with Samuel J. Schmieding, December 5, 2016, Corvallis, OR.

CHAPTER 1

1 Boyd, *Indians, Fire, and the Land*.

2 Nesbit, *Collector*.

3 Robbins, *Landscapes of Promise*. See the chapter titled "The Native Ecological Context" for an in-depth discussion of the ideas briefly covered here.

4 Boyd, *Indians, Fire, and the Land*.

5 Mack, "In Pursuit of the Wild *Vaccinium*."

6 Juntunen et al., *World of the Kalapuya*.

7 See White, *"It's Your Misfortune,"* chaps. 3 and 4; and Robbins, "Prescribing the Landscape" in *Landscapes of Promise*, for more on how the US government stole tribal lands in the Pacific Northwest.

8 College of Forestry, *McDonald-Dunn Forest Plan*.

9 Long et al., "Tribal Ecocultural Resources"; Berg, *First Oregonians*.

10 See Long et al., "Tribal Ecocultural Resources," 856–857, for maps showing ancestral lands as cessations within the range of the northern spotted owl, including those of the tribes and bands of western Oregon and northern California.

11 Reilly et al., "Climate, Disturbance, and Vulnerability"; PRISM Climate Group, "30-Year Normals."

12 McCabe et al., "Rain-on-Snow Events."

13 Orr and Orr, *Geology of the Pacific Northwest*.

14 Orr and Orr, *Geology of the Pacific Northwest*. The Oregon Coast Range and Olympic Peninsula, as noted by Orr and Orr, were also formed by tectonic processes operating on sedimentary rock.

15 DellaSala et al., "Global Perspective on the Biodiversity."

16 See J. Franklin et al., *Ecological Forest Management*, chaps. 2 and 3, for more discussion of the ideas presented here.

17 Buotte et al., "Carbon Sequestration"; J. Franklin et al., *Ecological Forest Management*, chap. 2.

18 J. Franklin et al., *Ecological Forest Management*, chap. 3.

19 Abatzoglou et al., "Compound Extremes."

20 Tepley et al., "Post-Fire Tree Establishment"; J. Franklin et al., *Ecological Forest Management*, chap. 3.

21 Merschel, "Historical Forest Succession."

22 J. Franklin et al., *Ecological Forest Management*, chap. 3.

23 J. Franklin et al., chap. 3; Hessburg et al., "Dry Forests and Wildland Fires."

24 See J. Franklin et al., *Ecological Forest Management*, chap. 3, for a detailed description of Moist and Dry Forests.

25 See Hagmann et al., "Evidence for Widespread Changes," for a persuasive summary of the evidence that Dry Forests in the western United States have become much denser over the last 150 years; also Hagmann et al., "Historical and Current Forest Conditions," for similar evidence about the eastern Cascades area in figure 1.6b; and Metlen et al., "Regional and Local Controls," for similar evidence in southwestern Oregon.

26 See Reeves et al., "Aquatic Conservation Strategy—Review of Relevant Science," for more discussion of the ideas presented here.

27 Lichatowich, *Salmon without Rivers*.

28 Lichatowich, *Salmon without Rivers*.

CHAPTER 2

 1 Dana and Fairfax, *Forest and Range Policy*, 82.

 2 White, *"It's Your Misfortune,"* chaps. 2–6.

 3 Dana and Fairfax, *Forest and Range Policy*, 62–63.

 4 Dana and Fairfax, 82.

 5 Clary, *Timber and the Forest Service*, 22.

 6 Clary, 23.

 7 Clary, 23.

 8 Clary, 23; Pinchot, *Use of the National Forest Reserves*. Over time, the tiny "Use Book" became the multivolume *Forest Service Manual*.

 9 Clary, *Timber and the Forest Service*, 85.

10 See J. Franklin, *Ecological Forest Management*, chap. 17, for discussion of the regulated forest.

11 Steen, *U.S. Forest Service*, 224–225, 251.

12 Dana and Fairfax, *Forest and Range Policy*, 167; Steen, *U.S. Forest Service*, 251–252; Lewis, *Forest Service and the Greatest Good*, 107–108.

13 Dana and Fairfax, *Forest and Range Policy*, 168.

14 While the Forest Service would attempt to offer approximately the same volume per year on each national forest under this policy, sometimes the offerings would not sell because of poor markets, and the agency would offer them again in future years so that over five to ten years the full allowable cut would be sold. Upon purchase, the buyer usually had three to five years to complete the sale, so harvest over time was often more varied than offerings or sales.

15 L. Davis and Johnson, *Forest Management*, chap. 8.

16 Robbins, *Landscapes of Conflict*, 169.

17 Brock, *Money Trees*.

18 https://www.blm.gov/or/files/PubLawNo405.pdf

19 Dana and Fairfax, *Forest and Range Policy*, 105–108, 165–167.

20 Dana and Fairfax, 190–194.

21 This famous quote as it applies to the Multiple-Use Sustained-Yield Act comes
    from the opinion by Ninth Circuit judge Goodwin: "This language, partially
    defined in section 531 in such terms as 'that (which) will best meet the needs of
    the American people' and 'making the most judicious use of the land,' can hardly
    be considered concrete limits upon an agency discretion. Rather, it is language
    which 'breathe(s) discretion at every pore.'" Strickland v. Morton, 519 F.2d 467, 469
    (9th Cir. 1975).

22 Clary, *Timber and the Forest Service*, 172.

23 Lewis, *Forest Service and the Greatest Good*, 120–125.

24 Dana and Fairfax, *Forest and Range Policy*, 198–199.

25 Glicksman, "Traveling in Both Directions."

26 Dana and Fairfax, *Forest and Range Policy*, 218; Glicksman, "Traveling in Both
    Directions."

27 Dana and Fairfax, *Forest and Range Policy*, 221.

28 Glicksman, "Traveling in Both Directions."

29 Robbins, *Landscapes of Conflict*, 172–173.

30 Robbins, 177.

31 USFS, *Douglas-Fir Supply Study*.

32 L. Davis and Johnson, *Forest Management*, chap. 16; Hirt, *Conspiracy of Optimism*,
    263–265. The Forest Service made sure that its interpretation of sustained yield
    would be permitted under the National Forest Management Act, which was
    passed in 1976.

33 Dana and Fairfax, *Forest and Range Policy*, 323–327; Hirt, *Conspiracy of
    Optimism*, 256–260.

CHAPTER 3

1 This chapter is drawn primarily from the class notes of a senior course on forest
   policy that Johnson taught at Oregon State University, with the able assistance of
   PhD students James Johnston and Deanne Carlson. Susan Jane M. Brown improved
   the description of the laws discussed in this chapter. See J. Franklin et al., *Ecological
   Forest Management*, chap. 7, for a more in-depth treatment.

2 5 U.S.C. §§ 551 et seq.

3 Pub. L. No. 93-205, 87 Stat. 884, 16 U.S.C. §§ 1531–1544. For an excellent
   comprehensive introduction to NEPA, see US Department of Energy, *Citizen's
   Guide to the NEPA*. Council on Environmental Quality regulations implementing
   NEPA that are binding on all federal agencies can be found at https://ceq.doe.gov.
   The council enacted NEPA regulations that are binding on all federal agencies in
   1978. See 40 C.F.R. Part 1500. Many federal agencies such as the Forest Service
   and BLM also have their own NEPA regulations, with which they must also
   comply. It should be noted, though, that the Trump administration adopted new
   NEPA regulations before leaving office in 2021 that greatly reduced requirements
   for the divulgence of environmental effects of proposed actions, which form the
   heart of the NEPA regulations. The Biden administration is reversing many of
   those changes as this book is being finished.

4   Steen, *U.S. Forest Service*, xxiv.
5   Reid, "Understanding and Evaluating."
6   Robertson v. Methow Valley Citizens Council, 490 U.S. 332 [1989].
7   Glicksman, "Traveling in Both Directions."
8   Glicksman, "Traveling in Both Directions."
9   AuCoin, *Catch and Release*.
10 G. Williams, *U.S. Forest Service*, 293.
11 Pub. L. No. 93-205, 87 Stat. 884 (1973), 16 U.S.C. §§ 1531–1544.
12 USFWS, "Endangered and Threatened Plants and Animals."
13 Doremus, "Story of *TVA v. Hill.*"
14 Tennessee Valley Authority v. Hill, 437 U.S. 153 (1978).
15 Unless nonfederal landowners and managers apply for a federal permit, in which case the federal responsibilities apply.
16 West Virginia Div. of Izaak Walton League of America, Inc. v. Butz, 522 F.2d 945 (4th Cir. 1975).
17 Fairfax and Achterman, "Monongahela Controversy."
18 Fairfax and Achterman, "Monongahela Controversy."
19 122 Cong. Rec. 5619 (1976) (remarks of Sen. Humphrey).
20 Pub. L. No. 94-588; 16 U.S.C. §§ 1600–1614.
21 16 U.S.C. § 1604(e).
22 16 U.S.C. § 1604(g).
23 16 U.S.C. § 1604(g)(3)(B).
24 USDA, "Part 219—Planning," 219.19.
25 16 U.S.C. § 1612(a).
26 16 U.S.C. § 1604(h)(1).
27 USDA, "Part 219—Planning," 219.19.
28 Wellock, "Dicky Bird Scientists," 390.
29 Wellock, "Dicky Bird Scientists," 392. In drafting regulations specific for fish and wildlife, Webb had been influenced by MacArthur and Wilson's *Theory of Island Biogeography*, in particular by their idea that small populations were more vulnerable to extinction. Webb inferred that a requirement for viable populations on national forests would reduce the risk of extinctions. Steve Mealey, personal communication, December 28, 2021.
30 Steve Mealey, personal communication, December 28, 2021.
31 Steve Mealey, personal communication, December 28, 2021.
32 USDA, "Part 219—Planning," 219.19.
33 Wellock, "Dicky Bird Scientists," 390–391.
34 USDA, "Part 219—Planning," 219.8(a).
35 J. Franklin et al., *Ecological Forest Management*, chap. 18.

CHAPTER 4

1   J. Franklin distinguishes between wild science and domesticated science in his preface to Johnson et al., *Bioregional Assessments*.
2   J. Franklin et al., *Ecological Characteristics*.
3   Maser and Sedell, *From the Forest to the Sea*.
4   J. Franklin et al., *Interim Definitions*.
5   Harmon et al., "Ecology of Coarse Woody Debris."

6   F. Swanson et al., *History, Physical Effects*, abstract.
7   J. Franklin and Donato, "Variable Retention Harvesting."
8   Dietrich, *Final Forest*; J. Franklin and Donato, "Variable Retention Harvesting."
9   Gregory and Ashkenas, *Riparian Management Guide.*
10 Kelly and Braasch, *Secrets of the Old Growth Forest.*
11 Fred Swanson, personal communication, December 5, 2019.
12 Forsman, "Preliminary Investigation."
13 Forsman, interview with Samuel J. Schmieding, 31, December 5, 2016, Corvallis, OR.
14 Ruggierro et al., *Wildlife and Vegetation.*
15 Raphael and Marcot, "Validation of a Wildlife-Habitat-Relationships Model."
16 Wagner, *After the Blast*, 6–7.
17 J. Franklin and MacMahon, "Messages from a Mountain."
18 J. Franklin et al., *Ecological Forest Management*, chaps. 3 and 4; J. Franklin and Donato, "Variable Retention Harvesting."
19 M. Swanson et al., "Forgotten Stage of Forest Succession."
20 Landslides triggered by rapid, shallow earth movements that become fluidized and channelized.
21 Naiman et al., "Fundamental Elements."

CHAPTER 5
1   Yaffee, *Wisdom of the Spotted Owl*, 14.
2   Yaffee, 14.
3   Yaffee, 16–18.
4   Yaffee, 18–19.
5   Over time the task force broadened to also include agencies from the state of Washington.
6   Yaffee, *Wisdom of the Spotted Owl*, 21.
7   See Forsman's story at the end of chap. 4.
8   Yaffee, *Wisdom of the Spotted Owl*, 34.
9   Yaffee, 41.
10 LaFollette, *Saving All the Pieces.*
11 Yaffee, *Wisdom of the Spotted Owl*, 47.
12 Yaffee, 47–48.
13 Yaffee, 48.
14 Yaffee, 73.
15 National Wildlife Federation v. U.S. Forest Service, 592 F.Supp. 931,934 (D. Or 1984).
16 Yaffee, *Wisdom of the Spotted Owl*, 75.
17 Jim Lyons, personal communication, August 27, 2019.
18 Heinrichs, "Old Growth Comes of Age."
19 Thomas, interview with Rick Freeman, 1.
20 Forsman, "Habitat Utilization."
21 Forsman, interview with Samuel J. Schmieding, December 5, 2016, Corvallis, OR.
22 Yaffee, *Wisdom of the Spotted Owl*, 81.
23 Yaffee, 62.
24 Yaffee, 64.

25 USFS, *Regional Guide for the Pacific Northwest Region*.
26 Yaffee, *Wisdom of the Spotted Owl*, 77.
27 Yaffee, 77.
28 Yaffee, 81.
29 Scott, *Timber Wars*. See episode 3, "The Owl," for a colorful story about the Stahl-Lande collaboration.
30 Lande, "Demographic Model"; Doak, "Spotted Owls."
31 Lande to Stahl, Portland, June 12, 1985, cited in Yaffee, *Wisdom of the Spotted Owl*, 98.
32 Scott, *Timber Wars*, episode 3; Yaffee, *Wisdom of the Spotted Owl*, 98.
33 Yaffee, *Wisdom of the Spotted Owl*, 87.
34 Yaffee, 87.
35 G. Williams, *U.S. Forest Service*, 259.
36 USFS, "Draft Supplemental to the Final Environmental Impact Statement"; Wellock, "Dicky Bird Scientists."
37 Yaffee, *Wisdom of the Spotted Owl*, 96–101.
38 USFS, "Final Supplemental to the Final Environmental Impact Statement."
39 Yaffee, *Wisdom of the Spotted Owl*, 106.
40 Seattle Audubon Society v. Robertson, No. 89-CV-160-WLD (W.D. Wash. filed Feb. 8, 1989).
41 Lande, "Demographic Model."
42 Yaffee, *Wisdom of the Spotted Owl*, 114.
43 Caldbick, "Dwyer, William, J."
44 Northern Spotted Owl (*Strix occidentalis caurina*) v. Hodel, 716 F.Supp. 479 (W.D. Wash. 1988).
45 Personal communication from Andy Kerr of The Larch Company, who started his conservation career during the Ford administration, January 26, 2021.
46 Associated Press, "Protesters Halt Logging."
47 Associated Press, "Protesters Halt Logging."
48 Associated Press, "Protesters Halt Logging."
49 Associated Press, "Protesters Halt Logging."
50 Scott, *Timber Wars*, episode 1, "The Last Stand."
51 Scott, *Timber Wars*, episode 2, "The Ancient Forest."
52 Scott, *Timber Wars*, episode 2.
53 Durbin, *Tree Huggers*, 108.
54 Section 318 of Pub. L. No. 101-121, October 23, 1989.
55 Thomas et al., *Conservation Strategy*, 47–49; Yaffee, *Wisdom of the Spotted Owl*, 123.
56 Thomas et al., *Elk of North America*.
57 Dietrich, *Final Forest*, 241–242.
58 Thomas et al., *Conservation Strategy*, 1.
59 Thomas et al., 1.
60 Murphy and Noon, "Integrating Scientific Methods," 5.
61 Murphy and Noon, 3.
62 Murphy and Noon, 5–6.
63 Durbin, *Tree Huggers*, 115. See Durbin's book for a lively and illuminating description of the ICS's voyage of discovery.

64 Thomas et al., *Conservation Strategy*, 39.
65 Thomas et al., 3.
66 Thomas et al., 4; "d.b.h." refers to the diameter of a tree 4.5 feet from the ground.
67 Thomas et al., *Conservation Strategy*, 4–5.
68 Yaffee, *Wisdom of the Spotted Owl*, 125.
69 Dietrich, *Final Forest*, 242–243.
70 Dietrich, 243.
71 Steen, *Jack Ward Thomas*, 19.
72 Steen, 16.
73 Dietrich, *Final Forest*, 240.
74 Thomas et al., *Conservation Strategy*, 5.
75 Litigation over timber sales sold under the terms of Section 318 would continue for years.
76 Seattle Audubon Society v. Robertson, 1991 U.S. Dist. LEXIS 10131 (W.D. Wash. 1991).
77 Seattle Audubon Society v. Evans, 771 F.Supp. 1081, 1084 (W.D. Wash. 1991).
78 Seattle Audubon Society v. Evans, 952 F.2d 297 (9th Cir. 1991).
79 Seattle Audubon Society v. Evans, 771 F. Supp. at 1089. The BLM was guilty by association for failing to come up with its own owl plan and would eventually find itself similarly enjoined.
80 Seattle Audubon Society v. Evans, 771 F. Supp. at 1090.
81 Seattle Audubon Society v. Evans, 771 F. Supp. at 1096.
82 Seattle Audubon Society v. Evans, 771 F. Supp. at 1092.

CHAPTER 6

1 The analysis, results, and conclusions from the Gang of Four effort described in this chapter are drawn from Johnson et al., *Alternatives for Management*. Descriptions of working with Congress and with other scientists and specialists on the effort come from the collective memories of the authors.
2 Jim Lyons, personal communication, August 27, 2019.
3 Ruggierro et al., *Wildlife and Vegetation.*
4 A fifth-field watershed covers 40,000–250,000 acres, and a sixth-field watershed covers 10,000–40,000 acres.
5 Nehlsen et al., "Pacific Salmon at the Crossroads."
6 The harvest for the previous decade (1980–1989) is highlighted here, rather than the higher average harvest for the previous five years (1985–1989), because the decadal base is more reflective of the estimated allowable cut in the forest plans of the 1980s. The percentage comparisons for both time spans are shown in figure 6.2.
7 USDA, "Part 219—Planning," 219.19.
8 Johnson et al., *Alternatives for Management.*
9 See FEMAT, *Forest Ecosystem Management*, 6-32 to 6-33, for description of indirect and induced effects.
10 Durbin, *Tree Huggers*, 170.
11 Lewis, *Forest Service and the Greatest Good*, 204–205; Freeman, "Eco-Factory," 636.
12 USFWS, *Final Draft Recovery Plan for the Northern Spotted Owl.*
13 USFWS, *Final Draft Recovery Plan for the Northern Spotted Owl.*

14 Dick Holthausen (coleader of the recovery plan), personal communication, January 11, 2020.
15 Marcot and Thomas, *Of Spotted Owls*, 5.
16 J. Franklin et al., *Ecological Forest Management*, chap. 7; Durbin, *Tree Huggers*, 135–136.
17 The National Marine Fisheries Service is within NOAA.
18 Durbin, *Tree Huggers*, 138–139; Yaffee, *Wisdom of the Spotted Owl*, 246; Marcot and Thomas, *Of Spotted Owls*, 7. David Cottingham, interview with K. Norman Johnson, May 2, 2022.
19 Lane County Audubon Society v. Jamison, 958 F.2d 290 (9th Cir. 1992), 1.
20 USFS, "Final Environmental Impact Statement for Management of the Northern Spotted Owl."
21 Seattle Audubon Society v. Moseley, 798 F. Supp. 1473 (W.D. Wash.), *supplemented*, 798 F. Supp. 1484 (W.D. Wash. 1992).
22 Seattle Audubon Society v. Moseley. Judge Dwyer stated that "the Forest Service has violated NEPA by failing to explain or justify a rating in the FEIS of Alternative B, the plan chosen, as having a 'low to medium-low probability of providing for viable populations of late-successional forest associated wildlife species other than northern spotted owls.' FEIS at 3 & 4-140." The judge further said, "The agency also contends that the statement in the FEIS describing the impact on other species is merely a quotation from comments by the Scientific Panel on Late Successional Forest Ecosystems, not the agency's own view. That is what makes this a NEPA question rather than one under NFMA at this stage. If the 'low to medium-low' viability rating were admittedly the Forest Service's own rating, summary judgment under NFMA would be entered now."
23 Seattle Audubon Society v. Moseley.
24 Seattle Audubon Society v. Moseley.
25 Seattle Audubon Society v. Espy, 998 F.2d 699 (9th Cir. 1993), *and aff'd in part, appeal dismissed in part sub nom.* Seattle Audubon Society v. Espy, 998 F.2d 699 (9th Cir. 1993).

CHAPTER 7

1 Katie McGinty, interview with K. Norman Johnson, October 22, 2018.
2 Katie McGinty, interview with K. Norman Johnson, October 22, 2018.
3 Tom Collier, interview with K. Norman Johnson, October 21, 2018.
4 Tom Collier, interview.
5 Jim Lyons, personal communication, August 27, 2019.
6 Tom Tuchmann, interview with K. Norman Johnson, January 13, 2021.
7 Seattle Audubon Society v. Moseley, 798 F. Supp. 1473 (W.D. Wash.), *supplemented*, 798 F.
8 McGinty, "Forest Conference Discussion."
9 Tom Tuchmann, interview with K. Norman Johnson, January 13, 2021.
10 Video of the three sessions of the Forest Conference can be found at https://www.c-span.org/video/?39331-1/northwest-environmental-issues and also at ?39332-1 and ?39338-1.
11 Katie McGinty, interview with K. Norman Johnson, October 22, 2018.
12 Jim Lyons, personal communication, August 27, 2019.
13 FEMAT, *Forest Ecosystem Management*, i.
14 Yaffee, *Wisdom of the Spotted Owl*, 142–143.

15 Yaffee, *Wisdom of the Spotted Owl*, 143.

16 FEMAT, *Forest Ecosystem Management*, ii.

17 Katie McGinty, interview with K. Norman Johnson, October 22, 2018.

18 Seattle Audubon Society v. Moseley, 798 F. Supp. 1473 (W.D. Wash.), *supplemented*, 798 F. Supp. 1484 (W.D. Wash. 1992).

19 Forest Service letters of direction, July 30, 1992, and August 28, 1992, as noted in appendix 2-A of Thomas et al., *Viability Assessments*.

20 Raphael et al., "Marbled Murrelet."

21 Thomas et al., *Viability Assessments*, 259.

22 Thomas, interview with Rick Freeman, 2.

23 Thomas et al., *Viability Assessments*, 257–258.

24 A mollusk is an invertebrate of a large phylum that includes snails, slugs, mussels, and octopuses. They have a soft, unsegmented body and live in aquatic or damp habitats, and most have an external calcareous shell.

25 Thomas et al., *Viability Assessments*, 264.

26 Thomas et al., 268–270.

27 Emanuel, "Forest Conference Follow-Up."

28 Department of Agriculture, Department of the Interior, Department of Labor, Department of Commerce, Environmental Protection Agency, Office of Environmental Policy, Office of Science and Technology, National Economic Council, Council of Economic Advisers, and Office of Management and Budget.

29 FEMAT, *Forest Ecosystem Management*, i–iv. McGinty, Lyons, and Tuchmann took the lead in negotiating the FEMAT instructions with Thomas.

30 FEMAT, *Forest Ecosystem Management*, ii.

31 FEMAT, ii–iv.

32 FEMAT, iv.

33 Forest Conference Executive Committee, "Forest Conference," 2.

34 FEMAT, iii.

35 Clinton, "Remarks at Conclusion," 388.

36 Jim Pipkin, interview with K. Norman Johnson, October 21–22, 2018.

37 FEMAT, *Forest Ecosystem Management*, ii.

38 Sedell et al., "Development and Evaluation"; Swanson et al., "Natural Variability."

39 Sedell et al., "Development and Evaluation."

40 Kelly Burnett, personal communication, December 28, 2021.

41 FEMAT, *Forest Ecosystem Management*, table 5-5, p. 5-37.

42 Jim Lyons, personal communication, August 27, 2019.

43 Thomas, interview with Rick Freeman, 4.

44 J. Franklin and Donato, "Variable Retention Harvesting."

45 Three other options were considered (2, 6, and 10), but they were not fully evaluated and are not covered here.

46 A portion of the area within LSRs would also be classified as Riparian Reserves, but Riparian Reserves are acknowledged in this accounting only in Matrix/AMA, where they add to the total reserve acreage.

47 FEMAT, 2-28. Estimates of LS/OG forest were made from satellite imagery classified by Pacific Meridian Resources for the national forests, and from aerial photos of the BLM's western Oregon forests.

48 FEMAT, *Forest Ecosystem Management*, 2-25.

49 Marcot and Thomas, *Of Spotted Owls*, 11.

50 Norris, "Science Review," 118.

51 Skillen, *Federal Ecosystem Management*, 192.

52 Bruce Marcot, personal communication, February 8, 2020.

53 FEMAT, *Forest Ecosystem Management*, 4-43.

54 FEMAT, *Forest Ecosystem Management*, 5-70.

55 Steen, "Interview with Jack Ward Thomas," 23.

56 FEMAT, *Forest Ecosystem Management*, 2-36.

57 FEMAT, 2-36.

58 FEMAT, 2-36.

59 FEMAT, 2-36.

60 Greber, "Impacts of Technological Change."

61 Greber, "Economic Assessment."

62 FEMAT, *Forest Ecosystem Management*, 6-20 to 6-22.

63 Daniels, *Rise and Fall*.

64 Johnson, "Sustainable Harvest Levels."

65 While new sales were enjoined under Dwyer's rulings, harvesting of past sales
    could generally continue.

66 Johnson et al., *Sustainable Harvest Levels and Short-Term Timber Sales*, 21–24.

67 In addition to workers displaced, there would be "indirect effects" caused from
    changing business expenditures in the region, such as purchases of equipment,
    and "induced effects" caused by changing personal expenditures. FEMAT, *Forest
    Ecosystem Management*, 6-32 to 6-33.

68 Greber, "Economic Assessment," 39.

69 FEMAT, *Forest Ecosystem Management*, 7-50 to 7-52.

70 FEMAT, 7-51.

71 FEMAT, 7-68.

72 Carroll and Daniels, "Science and Politics."

73 FEMAT, *Forest Ecosystem Management*, 7-69.

74 David Cottingham, interview with K. Norman Johnson, May 2, 2022.

75 Forest Conference Executive Committee, "Forest Conference," 1–2.

76 Forest Conference Executive Committee, 2.

77 Forest Conference Executive Committee, 2.

78 Forest Conference Executive Committee, 5.

79 Forest Conference Executive Committee, 6.

80 Forest Conference Executive Committee, 3.

81 Forest Conference Executive Committee, 6.

82 Forest Conference Executive Committee, 4, table after 6.

83 Forest Conference Executive Committee, 11–12.

84 Forest Conference Executive Committee, 16–17.

85 Forest Conference Executive Committee, 2.

86 Forest Conference Executive Committee, 2.

87 Clinton and Gore, *Forest Plan*.

88 Clinton and Gore, *Forest Plan*.

89 The Environmental Protection Agency administers the Clean Water Act.

90 Clinton and Gore, *Forest Plan*.

91 Power, "Reaction to President."

92 Carroll, *Community and the Northwestern Logger;* Lee et al., "Social Impact of Timber Harvest."
93 FEMAT, *Forest Ecosystem Management,* 7-84.
94 Strong, "Tribal Right to Fish," 34.
95 Anderson, "Prescription for Ecological Disaster," 39.
96 FEMAT, *Forest Ecosystem Management,* 6-34.

CHAPTER 8

1 USDA and USDI, "Draft Supplemental Environmental Impact Statement."
2 The description of the SAT II process in this section benefited from discussions with two members of SAT II (Dick Holthausen and Kelly Burnett) and Jim Lyons.
3 Anderson, "Prescription for Ecological Disaster," 39.
4 Jim Lyons, personal communication, August 27, 2019.
5 USDA and USDI, "Final Supplemental Environmental Impact Statement," appendix J2.
6 This limitation may have been partly to avoid a lawsuit over violations of the Federal Advisory Committee Act (FACA). This act requires that advisory committees with nonfederal members that are established or utilized by the executive branch meet certain requirements such as holding open meetings, keeping notes, and following other protocols. The Northwest Forest Resource Council sued the federal government, arguing that FEMAT was such an advisory committee, that it did not meet FACA requirements for such a committee, and that the Northwest Forest Plan should therefore be invalidated. The court agreed with the plaintiffs but declined to provide injunctive relief, in part because it did not see how adhering to FACA would have changed the outcome. Northwest Forest Resource Council v. Espy, 846 F. Supp. 1009, 1010 (D.D.C. 1994).
7 USDA, "Part 219—Planning," 219.19.
8 Local extinction, also known as extirpation, is the condition of a species (or other taxon) that ceases to exist in a chosen geographic area of study, though it still exists elsewhere.
9 USDA and USDI, "Final Supplemental Environmental Impact Statement," appendix J2; Attorney-Client Work Product, "Habitat for Viable Populations."
10 See chapter 3 for discussion of the diversity clause in NFMA.
11 USDA and USDI, "Final Supplemental Environmental Impact Statement," appendix J2; Attorney-Client Work Product, "Habitat for Viable Populations."
12 USDA and USDI, "Final Supplemental Environmental Impact Statement," appendix J2.
13 Kelly Burnett (represented the aquatics group in SAT II), personal communication, December 13, 2021.
14 USDA and USDI, "Final Supplemental Environmental Impact Statement."
15 The three Survey and Manage categories in the SAT II effort (table 8.2) were turned into four categories in the FEIS and presented in a slightly different order.
16 Dick Holthausen (led SAT II), personal communication, January 11, 2020; Stelle, "Major Issues."
17 USDA and USDI, "Final Supplemental Environmental Impact Statement," S-21. The FEIS (S-21) leaves the distinct impression that all changes that had been made, including the addition of Survey and Manage, were incorporated into the new harvest estimates: "The PSQ figures for Alternative 9 are changed from the Draft SEIS to reflect modifications made to Alternative 9 as a result of public

comments and internal review. The overall result of the revisions to PSQ for Alternative 9 between the Draft SEIS and the Final SEIS is a reduction of 92 MMBF per year; from 1,050 MMBF to 958 MMBF per year, not including the 'other wood.'"

18 USDA and USDI, "Record of Decision," 4.

19 McGinty, "Completion of the Forest Plan," 1.

20 McGinty, 3–4.

21 McGinty, 2.

22 McGinty, 2.

23 McGinty, 2–3.

24 McGinty, 4.

25 McGinty, 4.

26 McGinty, 4.

27 McGinty, 4.

28 McGinty, 4; handwritten note of President Clinton.

29 R. Paul Richard's memo was attached as a forward to McGinty, "Completion of the Forest Plan."

30 Figure 7.5 shows the net LSR area for Option 9 (20 percent) after removal of overlap with Administratively Withdrawn areas.

31 USDA and USDI, "Record of Decision," 2.

32 USDA and USDI, "Standards and Guidelines," B-6.

33 Record of Decision," 6.

34 "Standards and Guidelines," C-21.

35 FEMAT, *Forest Ecosystem Management*, 3-28.

36 USDA and USDI, "Standards and Guidelines," D4–D7.

37 USDA and USDI, "Record of Decision"; USDA and USDI, "Standards and Guidelines"; Adaptive Management Area Work Group, "Standards and Guidelines."

38 Seattle Audubon Society v. Lyons, 871 F. Supp. 1291 (W.D. Wash. 1994), *aff'd*, 80 F.3d 1401 (9th Cir. 1996).

39 Seattle Audubon Society v. Lyons, 871 F. Supp. 1291 (W.D. Wash. 1994).

40 Seattle Audubon Society v. Lyons, 871 F. Supp. 1291 at 1300.

41 Other claims include violation of the Federal Advisory Committee Act, questions about the adequacy of the administrative record, and application of the DC circuit case Sweet Home Chapter of Communities for a Great Oregon v. Babbitt, 17 F.3d 1463 (D.C. Cir. 1994). The court rejected these claims, too.

42 Seattle Audubon Society v. Lyons, 871 F. Supp. 1291 at 1310.

43 Seattle Audubon Society v. Lyons, 871 F. Supp. at 1310.

44 Seattle Audubon Society v. Lyons, 871 F. Supp. at 1310.

45 Seattle Audubon Society v. Lyons, 871 F. Supp. at 1314–1315.

46 Seattle Audubon Society v. Lyons, 871 F. Supp. at 1315.

47 Seattle Audubon Society v. Lyons, 871 F. Supp. at 1315–1316.

48 Seattle Audubon Society v. Lyons, 871 F. Supp. at 1315.

49 Seattle Audubon Society v. Lyons, 871 F. Supp. at 1321.

50 Seattle Audubon Society v. Lyons, 871 F. Supp. at 1322.

51 Seattle Audubon Society v. Lyons, 871 F. Supp. at 1321.

52 Seattle Audubon Society v. Lyons, 871 F. Supp. at 1323.

53 Seattle Audubon Society v. Lyons, 871 F. Supp. at 1312–1313.

54 Seattle Audubon Society v. Lyons, 871 F. Supp. at 1313.

55 Seattle Audubon Society v. Lyons, 871 F. Supp. at 1313–1314.

56 Seattle Audubon Society v. Moseley, 80 F.3d 1401 (9th Cir. 1996).

57 USDA and USDI, "Standards and Guidelines," E-15.

58 Jim Pipkin, interview with K. Norman Johnson, October 21–22, 2019.

59 Paul Henson, memorandum to state of Oregon supervisor, USFWS, February 25, 2020.

60 Marcot and Thomas, *Of Spotted Owls.*

61 Tuchmann et al., *Northwest Forest Plan.*

62 Pipkin, *Northwest Forest Plan Revisited.*

63 FEMAT, *Forest Ecosystem Management*, 5-35.

64 Carroll and Daniels, "Science and Politics," 106.

65 USDA and USDI, "Standards and Guidelines," B13.

CHAPTER 9

1 Assistance on legal analysis provided by Susan Jane M. Brown.

2 See Thomas's wonderful memoir *Forks in the Trail*, 231–245, for a blow-by-blow account of his early interactions with the new Republican Congress.

3 Pub. L. No. 104-19, July 27, 1995, 109 Stat. 241; Goldman and Boyles, "Forsaking the Rule of Law."

4 Elderkin, "What a Difference"; Steen, *Jack Ward Thomas*, 218–220.

5 See Steen's chapter on the year 1995 in *Jack Ward Thomas* for the turmoil and disarray in the Clinton administration as it attempted to limit damage from the Salvage Rider.

6 Clinton, "Memorandum on Timber Salvage," 1192.

7 Elderkin, "What a Difference."

8 Steen, *U.S. Forest Service*, xxvii.

9 Steen, *Jack Ward Thomas*, 294.

10 Neel, "Interagency Office."

11 Tom Tuchmann, interview with K. Norman Johnson, January 13, 2021.

12 Pipkin, *Northwest Forest Plan Revisited.*

13 Sonner, "Gore Calls Salvage Logging."

14 Tuchmann et al., *Northwest Forest Plan*, 155.

15 Tuchmann et al., 156.

16 Tuchmann et al., 156.

17 It may be surprising that private harvest fell at the same time (1990–1994) as federal harvest, since the opposite might have been expected. However, Daniels, in *Rise and Fall*, found that the lucrative Asian log export market for private timber also collapsed during this period.

18 Charnley et al., *Rural Communities*, 156. Thus, Charnley et al. estimated that 7,600 of the 19,000 jobs lost (40 percent) between 1990 and 1994 could be attributed to factors other than reduced timber supplies. By comparison, Helvoigt and Adams, in "Stochastic Frontier Analysis," found that 38 percent of the decline in employment at sawmills between 1988 and 1994 (when lumber production in Oregon and Washington declined by a third) can be attributed to technological change that reduced labor requirements.

19 Charnley, "Northwest Forest Plan as a Model."

20 Tuchmann et al., *Northwest Forest Plan*, 166–172.

21 Thompson, "Forest Communities," 4.
22 Charnley, "Northwest Forest Plan as a Model," 335.
23 Charnley et al. "Socioeconomic Well-Being and Forest Management," 643.
24 Charnley, "Northwest Forest Plan as a Model," 334.
25 Grinspoon et al., *Northwest Forest Plan.*
26 Grinspoon et al., *Northwest Forest Plan.*
27 Carlton, "First It Takes."
28 Tuchmann and Davis, *O&C Lands Report*, 20.
29 Tuchmann and Davis, 20.
30 Charnley et al., "Rural Communities and Economies."
31 Lehner, "What Replaces Timber?"
32 Thompson, "Forest Communities," 1; see Charnley et al., "Rural Communities and Economies," for a detailed analysis.
33 Haynes, "Contribution of Old Growth Timber."
34 Skillen, *Federal Ecosystem Management*; Robbins, *Place for Inquiry.*
35 Molina et al., "Protecting Rare, Old-Growth Forest," 313.
36 Molina et al., "Protecting Rare, Old-Growth Forest."
37 Charnley et al., *Northwest Forest Plan*, 14–16.
38 Northwest Ecosystem Survey Team, https://nestcascadia.wordpress.com/nest-faq/.
39 Charnley and Donoghue, *Public Values and Forest Management*, 5.
40 Jehl, "Clinton Forest Chief Acts."
41 Dombeck and Thomas, "Declare Harvest of Old-Growth."
42 USFWS, "Final Biological Opinion."
43 Gifford Pinchot Task Force et al. v. United States Fish and Wildlife Service, No. 00-CV-05462 (W.D. Wash. filed Aug. 11, 2000).
44 Gifford Pinchot Task Force v. U.S. Fish & Wildlife Serv., 378 F.3d 1059 (9th Cir. 2004), *amended*, 387 F.3d 968 (9th Cir. 2004) (*GPTF v. FWS*).
45 *GPTF v. FWS* at 1069–1070.
46 *GPTF v. FWS* at 1075.
47 American Forest Resource Council v. Clarke (D. DC 2003).
48 Northwest Ecosystem Alliance v. Rey, 380 F. Supp. 2d at 1175 (W.D. Wash. 2005).
49 Northwest Ecosystem Alliance v. Rey, 380 F. Supp. 2d at 1190.
50 Northwest Ecosystem Alliance v. Rey, 380 F. Supp. 2d at 1192–1193; see also Organized Vill. of Kake v. U.S. Dep't of Agric., 795 F.3d 956, at 966–967 (9th Cir. 2015).
51 Northwest Ecosystem Alliance v. Rey, 380 F. Supp. 2d at 1175.
52 The Bush administration in 2007 moved forward with one last attempt to entirely eliminate the Survey and Manage program and issued an FEIS and ROD that purportedly addressed the flaws the court had identified in its previous decision. Environmental groups again challenged the decision, which was struck down in Conservation NW. v. Rey, 674 F. Supp. 2d 1232 (W.D. Wash. 2009), on many of the same grounds.
53 Molina et al., "Protecting Rare, Old-Growth Forest"; Marcot et al., "Other Species and Biodiversity."
54 A programmatic biological opinion is one that addresses an agency's multiple actions in a program, region, or other basis.

55 Pacific Coast Federation of Fishermen's Assoc. v. NMFS, Civ. No. C97-775R (W.D. Wash. 1998).

56 Pac. Coast Fed'n of Fishermen's Ass'ns v. Nat'l Marine Fisheries Serv., 265 F.3d 1028 (9th Cir. 2001) at 7.

57 Pac. Coast Fed'n of Fishermen's Ass'ns v. Nat'l Marine Fisheries Serv., 265 F.3d 1028 (9th Cir. 2001) at 8.

58 Northwest Ecosystem Alliance v. Rey, 380 F. Supp. 2d, at 1175, 1192 (W.D. Wash 2005).

59 Pac. Coast Fed'n of Fishermen's Ass'ns v. Nat'l Marine Fisheries Serv., 482 F. Supp. 2d 1248, at 3 (W.D. Wash 2007).

60 Pacific Rivers Council et al. v. Shepard, 2012 WL 950032 (D. Or. 2012).

61 Stidham et al., "Role of Economic Emergency."

62 Siskiyou Reg'l Educ. Project v. Goodman (Biscuit Fire), No. 04-3058-CO, 2004 U.S. Dist. LEXIS 15575 (D. Or. 2004).

63 Stidham et al., "Role of Economic Emergency."

64 Alliance for the Wild Rockies v. Cottrell, 622 F.3d 1045 (9th Circuit 2010).

65 J. Franklin et al., *Ecological Forest Management*, chap. 1.

66 See Tappeiner et al., "Densities, Ages, and Growth Rates," 638: "Regeneration of old-growth stands [in coastal Oregon] occurred over a prolonged period, and trees grew at low density with little self-thinning; in contrast, after timber harvest, young stands may develop with high density of trees with similar ages and considerable self-thinning. The results suggest that thinning may be needed in dense young stands where the management objective is to speed development of old-growth characteristics."

67 USDA and USDI, "Standards and Guidelines," B-6.

68 Tappeiner et al., "Densities, Ages, and Growth Rates."

69 USDA and USDI, "Standards and Guidelines," C-12.

70 See appendix 4 for details.

71 Hessburg et al., "Dry Forests and Wildland Fires."

72 Baker and Ehle, "Uncertainty in Surface Fire History"; M. Williams and Baker, "Spatially Extensive Reconstructions." See Hagmann et al., "Evidence for Widespread Changes," for an effective refutation of the view that Dry Forests were generally dense historically.

73 Raphael et al., *Assessing the Compatibility*.

74 Raphael et al., 11.

75 Raphael et al., 2.

76 J. Franklin, "Adaptive Management Areas."

77 Ritchie, *Ecological Research*.

78 Wigglesworth, "It Is Just a Drop in the Bucket."

79 Cissel et al., "Landscape Management." As promising as the Blue River Project seemed to the scientists who undertook it, the project did not escape the conflict over harvest of LS/OG forest: two of three initial Blue River treatments were successfully implemented, but only one-third of the last treatment was implemented after a decade of protest and litigation, and the Blue River effort came to a close.

80 Duncan, "Too Early to Tell"; Stankey et al., "Adaptive Management."

81 Duncan, "Too Early to Tell"; Bormann et al., "Adaptive Management."

CHAPTER 10

1 R. Davis et al., *Northwest Forest Plan.*
2 Dugger et al., "Effects of Habitat."
3 McIver et al., *Status and Trend.*
4 Raphael et al., "Marbled Murrelet," 337.
5 Reeves et al., "Aquatic Conservation Strategy—Review of Relevant Science."
6 USFS, "Lowell Country Project." A story map for the project can be found at https://usfs.maps.arcgis.com/apps/Cascade/index.html?appid=ca34fb77a35f4210a 27848d14e554514.
7 Much of the required road work needed for the Lowell Country Project was built into the project itself, as tasks that the timber buyer would need to accomplish.
8 J. Franklin and Johnson, "2012 Planning Rule."
9 USFS, "Flat Country Project."
10 Spies et al., "Twenty-Five Years."
11 Thomas, "Sustainability of the Northwest."
12 Patricia A. Grantham (supervisor, Klamath National Forest), interview with K. Norman Johnson, September 10, 2018.
13 Carlton, "First It Takes."
14 J. Franklin et al., *Ecological Forest Management*, chap. 9.
15 Jessica Halofsky et al., "Changing Wildfire, Changing Forests," 4. For these estimates, they employed Representative Concentration Pathway 4.5 or 8.5 from Vose et al., "Temperature Changes in the United States."
16 Jessica Halofsky et al., "Changing Wildfire, Changing Forests."
17 J. Franklin et al., "Effects of Global Climatic Change."
18 Jessica Halofsky et al., "Changing Wildfire, Changing Forests."
19 J. Franklin et al., *Ecological Forest Management*, chap. 2.
20 J. Franklin et al., chap. 3.
21 Reilly et al., "Climate, Disturbance, and Vulnerability."
22 R. Davis et al., "Normal Fire Environment."
23 J. Franklin et al., "Effects of Global Climatic Change"; Spies et al., "Climate Change Adaptation Strategies."
24 Westerling et al., "Warming and Early Spring."
25 J. Franklin et al., "Effects of Global Climatic Change."
26 McCord et al., "Earl Seral Pathways"; Serra-Diaz et al., "Disequilibrium of Fire-Prone Forests."
27 Jessica Halofsky et al., "Changing Wildfire, Changing Forests."
28 Raffa et al., "Cross-Scale Drivers"; Jessica Halofsky et al., "Changing Wildfire, Changing Forests."
29 Reeves et al., "Aquatic Conservation Strategy—Review of Relevant Science."
30 Reeves et al., "Aquatic Conservation Strategy: Relevance to Management."
31 Buotte et al., "Carbon Sequestration."
32 Murray et al., "Estimating Leakage." For more discussion, also see J. Franklin et al., *Ecological Forest Management*, chap. 15.
33 Joshua Halofsky et al., "Nature of the Beast"; J. Franklin et al., *Ecological Forest Management*, chap. 3.
34 J. Franklin et al., *Ecological Forest Management*, chap. 3.
35 Hessburg et al., "Wildfire and Climate Change"; North et al., "Pyrosilviculture Needed."

36 Reeves et al., "Aquatic Conservation Strategy—Relevance to Management."

37 M. Swanson et al., "Forgotten Stage of Forest Succession."

38 See extensive description of this approach in J. Franklin et al., *Ecological Forest Management*, chap. 3; J. Franklin and Johnson, "2012 Planning Rule."

39 See J. Franklin et al., *Ecological Forest Management*, chap. 6, for more discussion.

40 USFWS, "Revised Recovery Plan"; USFWS, "Endangered and Threatened Wildlife and Plants: Revised Critical Habitat."

41 USFWS, "Revised Recovery Plan," 3-43, 3-49, 3-67.

42 Henson et al., "Improving Implementation," 867.

43 USFWS, "Endangered and Threatened Wildlife and Plants: Revised Critical Habitat."

44 Diller, "To Shoot or Not to Shoot."

45 Wiens et al., "Competitive Interactions."

46 Courtney et al., *Scientific Evaluation.*

47 Hamer et al., "Home Range Attributes."

48 USFWS, "Barred Owl Study Update."

49 Diller, "To Shoot or Not to Shoot."

50 Diller, 2.

51 Diller, 3.

52 A. Franklin et al., "Range-Wide Declines," abstract.

53 Horn et al., "Potential Trophic Cascades."

54 USFWS, "Experimental Removal of Barred Owls."

55 Diller, "To Shoot or Not to Shoot."

56 Wiens et al., "Invader Removal."

57 43 U.S.C. § 2601 (Oregon and California Revested Lands Sustained Yield Management *Act* of *1937*).

58 The BLM did commit to manage consistent with the concepts underlying the NWFP in a Memorandum of Understanding. However, the agency would choose the land allocations and management standards to attain that goal, paying close attention to the limits set by the USFWS and NMFS. The BLM also still participates in the Interagency Monitoring Program. Ray Davis, personal communication, December 15, 2021.

59 BLM, "Resource Management Plans."

60 BLM, "Resource Management Plans."

61 BLM, "Resource Management Plans."

62 BLM, "Resource Management Plans."

63 Using this same approach to modeling allowable cuts on the land allocations for the BLM under the NWFP, the BLM reported (in its NEPA documents for the 2016 RMPs) an allowable cut approximately 30 percent higher than that projected at the time of creation of the NWFP.

64 BLM, "Resource Management Plans."

65 BLM, "Resource Management Plans."

66 Gilman, "BLM Moves Away."

67 Pacific Rivers, et al. v. Bureau of Land Management, 815 Fed. Appx 107 (9th Circ. 2020).

68 Richard Hardt, BLM, personal communication, December 19, 2019.

69 American Forest Resource Council v. Hammond, 422 F. Supp. 3d 184, (CD. DC. 2019).

70 National Association of Home Builders v. Defenders of Wildlife, 551 U.S. 669 (2007).

71 American Forest Resource Council v. Hammond, 422 F. Supp 3d at 191.

72 American Forest Resource Council v. Hammond, 422 F. Supp 3d at 191.

73 American Forest Resource Council v. Hammond, 422 F. Supp 3d at 191.

74 See J. Franklin et al., *Ecological Forest Management*, chaps. 3 and 4; and Palik et al., *Ecological Silviculture*, chaps. 6 and 8, for more details on an ecological forestry approach.

75 J. Franklin et al., *Ecological Forest Management*, chap. 20.

76 See Johnson et al., "Sustaining the People's Lands," for a summary of the report, and Committee of Scientists, *Sustaining the People's Lands*, for the full report.

77 Cooper, "Second Committee of Scientists."

78 USDA, 36 C.F.R. Part 219.

79 USFS, *Land Management Planning Handbook*, chap. 20, p. 58.

80 USDA, 36 C.F.R. Part 219; USFS, *Land Management Planning Handbook*; J. Franklin et al., *Ecological Forest Management*, chap. 20.

81 Spies et al., *Synthesis of Science*.

82 Spies et al., *Synthesis of Science*, 181.

83 Spies et al., "Twenty-Five Years," 5.

84 USFS, *Bioregional Assessment*, 3.

85 USFS, 60–61.

86 USFS, 60.

87 USFS.

88 Pihulak et al., "Oregon's 2020 Wildfires."

89 Numerous reports in the *Oregonian* and *Eugene Register-Guard* during September 2020.

90 Reilly, et al, 2022. "Cascadia Burning."

91 J. Franklin et al., "Setting the Stage."

92 Joshua Halofsky et al., "Nature of the Beast"; J. Franklin et al., *Ecological Forest Management*, chap. 3.

93 Abatzoglou et al., "Compound Extremes."

94 Zald and Dunn, "Severe Fire Weather."

95 Ray Davis, US Forest Service PNW Research Station, Corvallis, personal communication, July 15, 2021.

96 USFWS, "Endangered and Threatened Wildlife and Plants: 12-Month Finding."

CHAPTER 11

1 Clinton and Gore, *Forest Plan*.

2 FEMAT, *Forest Ecosystem Management*, chap. 4.

3 Skillen, in *Federal Ecosystem Management*, identifies all five principles as providing this scaffolding, but this one principle stands above the others in setting the framework for the NWFP.

4 The Free Dictionary, www.thefreedictonary.com/worldview.

5 Johnson, "Will Linking Science to Policy."

6 Cashore and Howlett, "Behavioral Thresholds."

7 Johnson, "Will Linking Science to Policy."

8 Clinton, "Remarks at the Conclusion," 388.

9 Jim Pipkin, personal communication.

10 Dietrich, *Final Forest*, 240.

11 Simard, *Finding the Mother Tree.*

12 Perry and Jones, "Summer Streamflow Deficits."

13 Steen, "Interview with Jack Ward Thomas," 23.

14 Johnson et al., *Bioregional Assessments.*

15 Seattle Audubon Society v. Lyons, 871 F. Supp. 1311.

16 USFS, *Bioregional Assessment*, 9.

17 J. Franklin et al., *Ecological Forest Management*, chap. 20.

18 Schultz et al., "Wildlife Conservation Planning."

19 Greeley et al., "Timber: Mine or Crop?"; Pinchot, *Breaking New Ground.*

20 Brock, *Money Trees.*

21 USDA, 36 C.F.R. Part 219.

22 BLM, "Resource Management Plans."

23 Dana and Fairfax, *Forest and Range Policy*, 65.

24 USDA, "36 C.F.R. Part 219; USFS, *Land Management Planning Handbook.*

25 J. Franklin et al., *Ecological Forest Management*, chap. 20.

26 Skillen, *Federal Ecosystem Management*, 220.

27 Hays, *Conservation and the Gospel of Efficiency*, 3.

28 Shindler and Brunson, "Social Acceptability."

29 Shindler and Brunson, 1.

30 Holling, "Resilience and Stability"; Gunderson, "Stepping Back"; Gunderson and Holling, *Panarchy.*

31 Gunderson, in "Stepping Back," also emphasized that effective monitoring can moderate the adaptive cycle.

CHAPTER 12

1 Thomas et al., "Northwest Forest Plan."

2 Downing et al., "Human Ignitions on Private Lands."

3 Executive Office of the President, "Strengthening the Nation's Forests," Sec. 1.

4 Executive Office of the President, "Strengthening the Nation's Forests," Sec. 2.

5 J. Franklin and Johnson, "2012 Planning Rule."

6 USFS, "Flat Country Project."

7 Phalan et al., "Successes and Limitations."

8 See Hagmann et al., "Evidence for Widespread Changes," for a comprehensive review of historical evidence for the change in composition, structure, and density in Dry Forests of the West.

9 Churchill et al., "Restoring Forest Resilience"; Palik et al., *Ecological Silviculture*, chap. 12.

10 See Hessburg et al., "Wildfire and Climate Change," on the crucial role of prescribed and managed wildfire, in addition to thinning, in saving Dry Forests.

11 Spies et al., "Synthesis of Science," 181.

12 USDA and USDI, "Standards and Guidelines," C-13.

13 USDA and USDI, C-13.

14 Spies et al., "Twenty-Five Years," 5. It should be noted that "the purpose should not be to eliminate high severity fire but rather to reduce the proportion of the fire

acreage that burns at high severity to a level most consistent with history. Wildfires, including high-severity fire, produce large woody debris, regenerate hardwoods, and promote productivity and landscape resilience for many aquatic species, including salmon (Reeves et al., . . . ["Aquatic Conservation Strategy— Review of Relevant Science"])."

15 Henson et al., "Improving Implementation."

16 Metlen et al., *Rogue Basin Cohesive Forest Restoration.*

17 Downing et al., "Human Ignitions on Private Lands."

18 USDA and USDI, "Standards and Guidelines," C-13 to C-14.

19 Johnston et al., "Does Conserving Roadless Wildland."

20 Oregon Nat. Res. Council Fund v. Brong, 492 F.3d 1120 (9th Cir. 2007).

21 Larson et al., "Tamm Review."

22 BLM, "Resource Management Plans."

23 Larson et al., "Tamm Review."

24 Larson et al., "Tamm Review."

25 Paul Henson, personal communication, March 2021.

26 USFS, *Bioregional Assessment.*

27 USFWS, "Endangered and Threatened Wildlife and Plants: 12-Month Finding."

28 Diller, "To Shoot or Not to Shoot," 4.

29 Reeves et al., "Aquatic Conservation Strategy—Review of Relevant Science."

30 Poage et al., "Long-Term Patterns."

31 USFS, "Lowell Country Project."

32 NMFS, *Recovery Plan for Oregon Coast Coho.*

33 Reeves et al., "Aquatic Conservation Strategy—Review of Relevant Science."

34 Kelly Burnett, personal communication, December 13, 2021.

35 See Reeves et al., *Initial Evaluation,* for evidence of the differing capacity of stream segments on federal and nonfederal lands.

36 USDA, 36 C.F.R. Part 219; USFS, *Land Management Planning Handbook.*

37 J. Franklin and Johnson, "2012 Planning Rule."

38 Larson et al., "Tamm Review."

39 For more on indigenous law, including case law, and discussions of sovereignty rights, removal to reservations, government-to-government relationships, the doctrine of trust, the 1934 Indian Reorganization Act, the 1958 Termination Act, and the 1974 Self-Determination Act, see Wilkins and Lomawaima, *Uneven Ground*; and Deloria, *American Indians, American Justice.*

40 Klamath Tribes v. United States of America, No. 96-381-HA, October 2, 1996. (D. Or. 1996).

41 Seminole Nation v. United States, 316 U.S. 286, 296–97, 62 S.Ct. 1049, 86 L.Ed. 1480 (1942).

42 Klamath Tribes v. United States of America. No. 96-381-HA, October 2, 1996.(D. Or. 1996).

43 USFS, *Bioregional Assessment,* 11.

44 USFS, 17.

45 USFS, 46.

46 USFS, 61.

47 See Long et al., "Tribal Ecocultural Resources," 856–857, for detailed maps of the ancestral lands of tribes in the NWFP area.

48 USFWS, "Revised Recovery Plan," 3-43, 3-67.

49 Taylor, "Rejecting a Reserve Approach," 22.

50 Meddens et al., "Fire Refugia."

51 USFS, *Bioregional Assessment*, 38.

52 J. Franklin et al., *Ecological Forest Management*, chap. 20.

53 Cheng et al., "Is There a Place for Legislating."

APPENDIX 2

1 See appendix 1 for a list of SAT team members.

2 Thomas et al., *Viability Assessments*, 8.

3 Thomas et al., 268.

4 Thomas et al., 268–300.

5 Thomas et al., 438.

6 Thomas et al., 440.

7 Thomas et al., 441–442.

APPENDIX 3

1 FEMAT, *Forest Ecosystem Management*, 4-43-44.

2 Meslow et al., "Assessment of Terrestrial Species," 26.

3 FEMAT, *Forest Ecosystem Management*, 2-28.

APPENDIX 4

1 USDA and USDI, "Standards and Guidelines," C-11 to C-14.

2 USDA and USDI, "Standards and Guidelines," B-12 to B-13.

3 USDA and USDI, "Standards and Guidelines," C-32.

Literature Cited

Abatzoglou, John T., David E. Rupp, Larry W. O'Neill, and Mojtaba Sadegh. "Compound Extremes Drive the Western Oregon Wildfires of September 2020." *Geophysical Research Letters* 48, no. 8 (April 2021).

Adaptive Management Area Work Group. "Standards and Guidelines and the Adaptive Management Area System." Paper no. 1. Portland, OR: Regional Ecosystem Office, 2000. https://www.blm.gov/or/plans/surveyandmanage/files/14-policy_paper_on_standards.pdf.

Agrawal, A., R. S. Schick, E. P. Bjorkstedt, R. G. Szerlong, M. N. Goslin, B. C. Spence, T. H. Williams, and K. M. Burnett. *Predicting the Potential for Historical Coho, Chinook, and Steelhead Habitat in Northern California.* NOAA-TM-NMFS-SWFSC-379. US Department of Commerce, June 2005.

Anderson, Michael. "A Prescription for Ecological Disaster." *Journal of Forestry* 92, no. 4 (April 1994): 39.

Archives West. "Warner Creek Fire Collection, 1991–2001." http://archiveswest.orbiscascade.org/ark:/80444/xv93031.

Associated Press. "Protesters Halt Logging in Friendly Standoff; 13 Arrested." *AP News*, March 26, 1989. https://apnews.com/article/00dad8e93972012895d1fbf1310e279c.

Attorney-Client Work Product. "Habitat for Viable Populations and Risk of Extinction." November 19, 1993. Folder: Forest Conference—Forest Viability. OA/ID 4302. Pp. 97–103. FOIA collection 2012-0769-F (Seg. 1). Records of Kathleen "Katie" McGinty, 1993–1994. William J. Clinton Presidential Library, Little Rock, AR. https://clinton.presidentiallibraries.us/files/original/9294ecdb6d05b5ce1bc06f6a624d4bfa.pdf.

AuCoin, Les. *Catch and Release: An Oregon Life in Politics.* Corvallis: Oregon State University Press, 2019.

Baker, William L., and Donna Ehle. "Uncertainty in Surface-Fire History: The Case of Ponderosa Pine Forests in the Western United States." *Canadian Journal of Forest Research* 31, no. 7 (July 2001): 1205–1226.

Berg, Laura, ed. *The First Oregonians.* 2nd ed. Corvallis: Oregon State University Press, 2007.

BLM (Bureau of Land Management). "Resource Management Plans (RMPs) for Western Oregon." 2016. https://www.blm.gov/programs/planning-and-nepa/near-you/oregon-washington/rmps-westernoregon.

Blumm, Michael C., Susan Jane M. Brown, and Chelsea Stewart-Fusek. "The World's Largest Ecosystem Management Plan: The Northwest Forest Plan after a Quarter-Century." *Environmental Law Review* 52, no. 2. 2002.

Bormann, Bernard T., Richard W. Haynes, and Jon R. Martin. "Adaptive Management of Forest Ecosystems: Did Some Rubber Hit the Road?" *BioScience* 57, no. 2 (February 2007): 186–191.

Boyd, Robert, ed. *Indians, Fire, and the Land in the Pacific Northwest.* 2nd ed. Corvallis: Oregon State University Press, 2022.

Brock, Emily M. *Money Trees: The Douglas Fir and American Forestry: 1900–1994.* Corvallis: Oregon State University Press, 2015.

Buotte, Polly C., Beverly E. Law, William J. Ripple, and Logan T. Berner. "Carbon Sequestration and Biodiversity Co-benefits of Preserving Forests in the Western United States." *Ecological Applications* 30, no. 2 (December 2019): e02039. https://doi.org/10.1002/eap.2039.

Burnett, Kelly M., Gordon H. Reeves, Daniel J. Miller, Sharon Clarke, Ken Vance-Borland, and Kelly Christiansen. "Distribution of Salmon-Habitat Potential Relative to Landscape Characteristics and Implications for Conservation." *Ecological Applications* 17, no. 1 (2007): 66–80.

Burrow, John. *A History of Histories: Epics, Chronicles, and Inquiries from Herodotus and Thucydides to the Twentieth Century.* New York: Vintage Books, 2007.

Caldbick, John. "Dwyer, William J. (1929–2002)." HistoryLink.org, January 31, 2013. https://www.historylink.org/File/5338.

Carlton, Jim. "First It Takes, Then It Gives: The Environmental Movement Almost Killed Hayfork, Calif. Now, It's the Town's Main Hope." *Wall Street Journal*, March 24, 2008. https://www.wsj.com/articles/SB120605512503353159.

Carroll, Matthew S. *Community and the Northwestern Logger: Continuities and Changes in the Era of the Spotted Owl.* Boulder, CO: Westview Press, 1995.

Carroll, Matthew S., and Steven E. Daniels. "The Science and Politics of Forest Management: President Clinton's Northwest Forest Plan." In *New Strategies for Wicked Problems*, edited by Edward P. Weber, Denise Lach, and Brent S. Steel, 95–110. Corvallis: Oregon State University Press, 2017.

Carroll, Matthew S., Robert G. Lee, and Rebecca J. McLain. "Occupational Community and Forest Work: Three Cases from the Pacific Northwest." In *Communities and Forests: Where People Meet the Land*, edited by Robert G. Lee and Donald R. Field, 159–175. Corvallis: Oregon State University Press, 2005.

Carroll, Matthew S., Charles W. McKetta, Keith A. Blatner, and Con Schallau. "A Response to 'Forty Years of Spotted Owls? A Longitudinal Analysis of Logging Industry Job Losses.'" *Sociological Perspectives* 42, no. 2 (Summer 1999): 325–333.

Carson, Rachel. *Silent Spring.* Boston: Houghton Mifflin Harcourt, 1962.

Cashore, Benjamin, and Michael Howlett. "Behavioral Thresholds and Institutional Rigidities as Explanations of Punctuated Equilibrium Processes in Pacific Northwest Forest Policy Dynamics." In *Punctuated Equilibrium and the Dynamics of U.S. Environmental Policy*, edited by Robert Repetto, 137–161. New Haven, CT: Yale University Press, 2006.

Charnley, Susan. "The Northwest Forest Plan as a Model for Ecosystem Management: A Social Perspective." *Conservation Biology* 20, no. 2 (2006): 330–340.

Charnley, Susan, Candace Dillingham, Claudia Stuart, Cassandra Moseley, and Ellen Donoghue. *Northwest Forest Plan—The First 10 Years (1994–2003): Socioeconomic Monitoring of the Klamath National Forest and Three Local Communities.* General Technical Report PNW-GTR-764. Portland, OR: US Forest Service, Pacific Northwest Research Station, 2008. https://doi.org/10.2737/PNW-GTR-764.

Charnley, Susan, and Ellen M. Donoghue. *Public Values and Forest Management.* Vol. 5 of *Northwest Forest Plan—The First 10 Years (1994–2003): Socioeconomic Monitoring Results*, technical coordinator Susan Charnley. General Technical Report PNW-GTR-649. Portland, OR: US Forest Service, Pacific Northwest Research Station, 2006. https://doi.org/10.2737/PNW-GTR-649.

Charnley, Susan, Ellen M. Donoghue, Claudia Stuart, Candace Dillingham, Lita P. Buttolph, William Kay, Rebeca J. McLain, Cassandra Moseley, Richard H. Phillips, and Lisa Tobe. *Rural Communities and Economies.* Vol. 3 of *Northwest Forest Plan—The First 10 Years (1994–2003): Socioeconomic Monitoring Results,* technical coordinator Susan Charnley. General Technical Report PNW-GTR-649. Portland, OR: US Forest Service, Pacific Northwest Research Station, 2006. https://doi. org/10.2737/PNW-GTR-649.

Charnley, Susan, Jeffrey D. Kline, Eric M. White, Jesse Abrams, Rebecca J. McLain, Cassandra Moseley, and Heidi Huber-Stearns. "Socioeconomic Well-Being and Forest Management in Northwest Forest Plan–Area Communities." In *Synthesis of Science to Inform Land Management within the Northwest Forest Plan Area,* vol. 3, technical coordinators Thomas A. Spies, Peter A. Stine, Rebecca Gravenmier, Jonathan W. Long, Matthew J. Reilly, and Rhonda Mazza, 625–715. General Technical Report PNW-GTR-966. Portland, OR: US Forest Service, Pacific Northwest Research Station, 2018. https://doi.org/10.2737/PNW-GTR-966.

Chen, Jiquan, Jerry F. Franklin, and Thomas A. Spies. "Contrasting Microclimates among Clearcut, Edge, and Interior of Old-Growth Douglas-Fir Forest." *Agricultural and Forest Meteorology* 63, no. 3–4 (1993): 219–237.

Cheng, Antony S., R. J. Gutiérrez, Scott Cashen, Dennis R. Becker, John Gunn, Amy Merrill, David Ganz, Michael Liquori, David S. Saah, and William Price. "Is There a Place for Legislating Place-Based Collaborative Forestry Proposals? Examining the Herger-Feinstein Quincy Library Group Forest Recovery Act Pilot Project." *Journal of Forestry* 114, no. 4 (July 2016): 494–504. https://doi.org/10.5849/ jof.15-074.

Churchill, Derek J., Andrew J. Larson, Matthew C. Dahlgreen, Jerry F. Franklin, Paul F. Hessburg, and James A. Lutz. "Restoring Forest Resilience: From Reference Spatial Patterns to Silvicultural Prescriptions and Monitoring." *Forest Ecology and Management* 291, no. 1 (2013): 442–457.

Cissel, John H., Frederick J. Swanson, and Peter. J. Weisberg. "Landscape Management Using Historical Fire Regimes: Blue River, Oregon." *Ecological Applications* 9, no. 4 (1999): 1217–1231.

Clary, David A. *Timber and the Forest Service.* Lawrence: University Press of Kansas, 1986.

Clinton, William J. "Memorandum on Timber Salvage Legislation." August 1, 1995. In *Public Papers of the Presidents of the United States: William J. Clinton (1995, Book II),* 1192. Washington, DC: Office of the Federal Register, National Archives and Records Administration, [1995]. https://www.govinfo.gov/app/details/PPP-1995-book2/PPP-1995-book2-doc-pg1192.

———. "Remarks at the Conclusion of the Forest Conference in Portland, April 2, 1993." In *Public Papers of the Presidents of the United States: William J. Clinton (1993, Book I),* 387–389. Washington, DC: Office of the Federal Register, National Archives and Records Administration, [1993]. https://www.govinfo.gov/app/ details/PPP-1993-book1/PPP-1993-book1-doc-pg387.

Clinton, William J., and Albert A. Gore Jr. *The Forest Plan: For a Sustainable Economy and a Sustainable Environment.* July 1, 1993. Folder: Forest Conference-Press Package, Forest Plan. OA/ID Number: 4300. William J. Clinton Presidential Library Archives, Little Rock, AR. https://clinton.presidentiallibraries.us/files/orig inal/5f7047b9e5e3675fb002be08b5d0db76.pdf.

College of Forestry, Oregon State University. *McDonald-Dunn Forest Plan.* 2005. https://cf.forestry.oregonstate.edu/sites/default/files/mcdunn_plan.pdf.

Committee of Scientists. *Sustaining the People's Lands: Recommendations for Stewardship of the National Forests and Grasslands into the Next Century.*

Washington, DC: US Department of Agriculture, March 15, 1999. https://www. fs.fed.us/emc/nfma/includes/cosreport/Committee%20of%20Scientists%20 Report.htm.

Cooper, Arthur W. "The Second Committee of Scientists: Moving Forward While Looking Backward." *Journal of Forestry* 97, no. 5 (May 1999): 16–18.

Courtney, S. P., J. A. Blakesley, R. E. Bigley, M. L. Cody, J. P. Dumbach, R. C. Fleischer, A. B. Franklin, et al. *Scientific Evaluation of the Status of the Northern Spotted Owl*. Portland, OR: Sustainable Ecosystems Institute, 2004. https://www.fws.gov/ oregonfwo/species/data/northernspottedowl/BarredOwl/Documents/ CourtneyEtAl2004.pdf.

Dana, Samuel T., and Sally K. Fairfax. *Forest and Range Policy: Its Development in the United States*. New York: McGraw Hill, 1980.

Daniels, Jean M. *The Rise and Fall of the Pacific Northwest Log Export Market*. General Technical Report PNW-GTR-624. Portland, OR: US Forest Service, Pacific Northwest Research Station, 2005. https://doi.org/10.2737/PNW-GTR-624.

Davis, D., T. Hibbitts, and P. McCaig. *A Forestry Program for Oregon: Public Opinion about Forests and Forest Management in Oregon*. 2001. On file with Davis, Hibbitts, and McCaig, Inc., 1100 NW Glisan, Suite 300-B, Portland, OR, 97209.

Davis, Lawrence S., and K. Norman Johnson. *Forest Management*. 3rd ed. New York: McGraw-Hill, 1987.

Davis, Raymond J., David M. Bell, Matthew J. Gregory, Zhiqiang Yang, Andrew N. Gray, Sean P. Healey, and Andrew E. Stratton. *Northwest Forest Plan—The First 25 Years (1994–2018): Status and Trends of Late-Successional and Old-Growth Forests*. General Technical Report PNW-GTR-1004. Portland, OR: Forest Service, Pacific Northwest Research Station, 2020. https://www.fs.usda.gov/treesearch/ pubs/65070.

Davis, Raymond, Zhiqiang Yang, Andrew Yost, Cole Belongie, and Warren Cohen. "The Normal Fire Environment—Modeling Environmental Suitability for Large Forest Wildfires Using Past, Present, and Future Climate Normals." *Forest Ecology and Management* 390 (April 2017): 173–186. https://doi.org/10.1016/j. foreco.2017.01.027.

DellaSala, Dominick A., S. B. Reed, T. J. Frest, and James Strittholt. "A Global Perspective on the Biodiversity of the Klamath-Siskiyou Ecoregion." *Natural Areas Journal* 19, no. 4 (1999): 300–319.

Deloria, Vine, Jr. *American Indians, American Justice*. Austin: University of Texas Press, 1983.

Dietrich, William. *The Final Forest: The Battle for the Last Great Trees of the Pacific Northwest*. New York: Simon and Schuster, 1992.

Diller, Lowell V. "To Shoot or Not to Shoot: The Ethical Dilemma of Killing One Raptor to Save Another." *Wildlife Professional* 7 (2013): 54–57. http://www.orww. org/Elliott_Forest/References/Academic/Wildlife_Society/Diller_20130100.pdf.

Doak, D. "Spotted Owls and Old Growth Logging in the Pacific Northwest." *Conservation Biology* 3, no. 4 (December 1989): 389–396.

Dombeck, Mike, and Jack Ward Thomas. "Declare Harvest of Old Growth Off-Limits and Move On." *Seattle Post-Intelligencer*, August 23, 2003. https://www.seattlepi. com/news/article/P-I-Focus-Declare-harvest-of-old-growth-forests-1122206.php.

Doremus, Holly. "The Story of *TVA v. Hill*: A Narrow Escape for a Broad New Law." In *Environmental Law Stories: An In-Depth Look at Ten Leading Cases on Environmental Law*, edited by Oliver A. Houck and Richard James Lazarus, 109– 140. New York: Foundation Press, 2005.

Downing, William M., Christopher J. Dunn, Matthew P. Thompson, Michael D. Caggiano, and Karen C. Short. "Human Ignitions on Private Lands Drive USFS Cross-Boundary Wildfire Transmission and Community Impacts in the Western US." *Scientific Reports* 12 (2022). https://doi.org/10.1038/s41598-022-06002-3.

Dugger, Katie M., Eric D. Forsman, Alan B. Franklin, Raymond J. Davis, Gary C. White, Carl J. Schwarz, Kenneth P. Burnham, et al. "The Effects of Habitat, Climate, and Barred Owls on the Long-Term Population Demographics of Northern Spotted Owls." *Condor* 118, no. 1 (February 2016): 57–116. https://doi.org/10.1650/CONDOR-15-24.1.

Dumont, Clayton W. "The Demise of Community and Ecology in the Pacific Northwest: Historical Roots of the Ancient Forest Conflict." *Sociological Perspectives* 39, no. 2 (Summer 1996): 277–300.

Duncan, Sally. "Too Early to Tell, or Too Late to Rescue? Adaptive Management Under Scrutiny." *Science Findings* 33 (April 2001). Portland, OR: US Forest Service, Pacific Northwest Research Station. https://www.fs.usda.gov/pnw/publications/too-early-tell-or-too-late-rescue-adaptive-management-under-scrutiny.

Duncan, Sally, and Jonathan R. Thompson. "Forest Plans and Ad Hoc Scientist Groups in the 1990s: Coping with the Forest Service Viability Clause." *Forest Policy and Economics* 9, no. 1 (2006): 32–41.

Durbin, Kathie. *Tree Huggers: Victory, Defeat, and Renewal in the Northwest Ancient Forest Campaign.* Seattle, WA: The Mountaineers, 1996.

Elderkin, Susan. "What a Difference a Year Makes." *High Country News*, September 2, 1996. https://www.hcn.org/issues/89/2748.

Emanuel, Rahm. "Forest Conference Follow-Up." April 19, 1993. Memorandum for Mack McLarty. Folder: Forest Summit [1]. OA/ID 2530. 2009-0140-F. CPE: Political Affairs—Rahm E. William J. Clinton Presidential Library, Little Rock, AR.

Executive Office of the President. "Strengthening the Nation's Forests, Communities, and Local Economies." April 27, 2022. https://www.federalregister.gov/documents/2022/04/27/2022-09138/strengthening-the-nations-forests-communities-and-local-economies.

Fairfax, Sally K., and Gail L. Achterman. "The Monongahela Controversy and the Political Process." *Journal of Forestry* 75, no. 8 (1977): 485–487.

FEMAT (Forest Ecosystem Management Assessment Team). *Forest Ecosystem Management: An Ecological, Economic, and Social Assessment.* Washington, DC: US Government Printing Office, 1993. https://www.blm.gov/or/plans/nwfpnepa/FEMAT-1993/1993_%20FEMAT_Report.pdf.

Ferris, Ann E., and Eyal G. Frank. "Labor Market Impacts of Land Protection: The Northern Spotted Owl." *Journal of Environmental Economics and Management* 109 (September 2021).

Forest Conference Executive Committee. "Forest Conference." June 18, 1993. Memorandum for the president. Folder: Forest Conference-White House Memos. OA/ID 4302. 2-24. FOIA collection 2012-0769-F (Seg. 1). Records of Kathleen "Katie" McGinty, 1993–1994. William J. Clinton Presidential Library, Little Rock, AR. https://clinton.presidentiallibraries.us/files/original/54eabddc15299779c788e d42f30d0615.pdf.

Forsman, Eric D. "Habitat Utilization by Spotted Owls in the West-Central Cascades of Oregon." PhD diss., Oregon State University, 1980.

———. Interview with Samuel J. Schmieding, December 5, 2016. Northwest Forest Plan Oral History Collection (OH 48). Special Collections and Archives Research Center, Oregon State University Libraries, Corvallis.

——. "A Preliminary Investigation of the Spotted Owl in Oregon." Master's thesis, Oregon State University, 1975.

Franklin, Alan B., Katie M. Dugger, Damon B. Lesmeister, Raymond J. Davis, J. David Wiens, Gary C. White, James D. Nichols, et al. "Range-Wide Declines of Northern Spotted Owl Populations in the Pacific Northwest: A Meta-Analysis." *Biological Conservation* 259 (July 2021 suppl.). https://doi.org/10.1016/j.biocon.2021.109168.

Franklin, Jerry F. "Adaptive Management Areas." *Journal of Forestry* 92, no. 4 (April 1994): 50.

——. Preface to *Bioregional Assessments: Science at the Crossroads of Management and Policy,* edited by K. Norman Johnson, Frederick Swanson, Margaret Herring, and Sarah Greene, xi–xiii. Washington, DC: Island Press, 1999.

——. "Preserving Biodiversity: Species, Ecosystems, or Landscapes?" *Ecological Applications* 3, no. 2 (1993): 202–205.

Franklin, Jerry F., Kermit Cromack Jr., William Denison, Arthur McKee, Chris Maser, James Sedell, Fred Swanson, and Glen Juday. *Ecological Characteristics of Old-Growth Douglas-Fir Forests.* General Technical Report PNW-GTR-118. Portland, OR: US Forest Service, Pacific Northwest Research Station, 1981. https://doi.org/10.2737/PNW-GTR-118.

Franklin, Jerry F., and Daniel C. Donato. "Variable Retention Harvesting in the Douglas-Fir Region." *Ecological Processes* 9, no. 8 (2020). https://doi.org/10.1186/s13717-019-0205-5.

Franklin, Jerry F., F. Hall, W. Laudenslayer, C. Maser, J. Nunan, J. Poppino, C. J. Ralph, and T. Spies (Old-Growth Definition Task Group). *Interim Definitions for Old-Growth Douglas-Fir and Mixed Conifer Forests in the Pacific Northwest and California.* Research Note PNW-RN-447. Portland, OR: US Forest Service, Pacific Northwest Research Station, 1986. https://doi.org/10.2737/PNW-RN-447.

Franklin, Jerry F., and K. Norman Johnson. "The 2012 Planning Rule and Ecological Integrity: Maintaining and Restoring the National Forests of the Douglas-Fir Region." In *193 Million Acres: Toward a Healthier and More Resilient US Forest Service,* edited by Steve Wilent, 487–519. Bethesda, MD: Society of American Foresters, 2018.

——. "A Restoration Framework for Federal Forests in the Pacific Northwest." *Journal of Forestry* 110, no. 8 (2012): 429–439.

Franklin, Jerry F., K. Norman Johnson, and Debora L. Johnson. *Ecological Forest Management.* Long Grove, IL: Waveland Press, 2018.

Franklin, Jerry F., and James A. MacMahon. "Messages from a Mountain." *Science* 288, no. 5469 (2000): 1183–1184.

Franklin, Jerry. F., Thomas Spies, and Frederick J. Swanson. "Setting the Stage: Vegetation Ecology and Dynamics." In *People, Forests, and Change: Lessons from the Pacific Northwest,* edited by Deanna H. Olson and Beatrice Van Horne, 16–32. Washington, DC: Island Press, 2017.

Franklin, Jerry F., Frederick J. Swanson, Mark Harmon, David A. Perry, Thomas A. Spies, Virginia H. Dale, Arthur McKee, et al. "Effects of Global Climatic Change on Forests in North America." *Northwest Environmental Journal* 7, no 2 (1991): 233–254.

Freeman, Richard. "The EcoFactory: The United States Forest Service and the Political Construction of Ecosystem Management." *Environmental History* 7, no. 4 (2002): 632–658.

Furnish, James. 2015. *Toward a Natural Forest.* Corvallis: Oregon State University Press.

Gilman, Sarah. "BLM Moves Away from Landmark Northwest Forest Plan." *High Country News*, July 25, 2016. https://www.hcn.org/issues/48.12/blm-moves-away-from-landmark-northwest-forest-plan.

Glicksman, Robert L. "Traveling in Both Directions: Roadless Area Management under the Clinton and Bush Administrations." *Environmental Law* 34, no. 4 (2004): 101–166.

Goldman, Patti A., and Kristen L. Boyles. "Forsaking the Rule of Law: The 1995 Logging without Laws Rider and Its Legacy." *Environmental Law* 27, no. 4 (1997): 1035.

Greber, Brian J. "Economic Assessment of FEMAT Options." *Journal of Forestry* 92, no. 4 (1994): 36–39.

———. "Impacts of Technological Change on Employment in the Timber Industries of the Pacific Northwest." *Western Journal of Applied Forestry* 8, no. 1 (1993): 34–37.

Greeley, W. B., Earle H. Clapp, Herbert A. Smith, Raphael Zon, W. N. Sparhawk, Ward Shepard, and J. Kittredge Jr. "Timber: Mine or Crop?" *Yearbook of Agriculture*, 1922, 83–180.

Gregory, Stan, and Linda Ashkenas. *Riparian Management Guide.* Department of Fisheries and Wildlife, Oregon State University, December 4, 1990. https://andrewsforest.oregonstate.edu/sites/default/files/lter/pubs/pdf/pub1214.pdf.

Grinspoon, Elisabeth, Delilah Jaworski, and Richard Phillips. *Northwest Forest Plan—The First 20 Years (1994–2013): Social and Economic Status and Trends.* Report FS/R6/PNW/2015/0006. Portland, OR: US Forest Service, Pacific Northwest Region, 2016.

Gunderson, Lance H. "Stepping Back: Assessing for Understanding in Complex Regional Systems." In *Bioregional Assessments: Science at the Crossroads of Management and Policy*, edited by K. Norman Johnson, Fred Swanson, and Margaret Herring, 27–42. Washington, DC: Island Press, 1999.

Gunderson, Lance H., and C. S. Holling, eds. *Panarchy: Understanding Transformations in Human and Natural Systems.* Washington, DC: Island Press, 2002.

Hagmann, R. K., P. Hessburg, S. Prichard, N. A. Povak, P. M. Brown, P. Z. Fulé, R. E. Keane, et al. "Evidence for Widespread Changes in Structure, Composition, and Fire Regimes of Western North American Forests." *Ecological Applications* 31, no. 8 (December 2021). https://doi.org/10.1002/eap.2431.

Hagmann, R. Keala, Debora L. Johnson, and K. Norman Johnson. "Historical and Current Forest Conditions in the Range of the Northern Spotted Owl in South Central Oregon, USA." *Forest Ecology and Management* 389 (April 2017): 374–385.

Halofsky, Jessica E., David L. Peterson, and Brian J. Harvey. "Changing Wildfire, Changing Forests: The Effects of Climate Change on Fire Regimes and Vegetation in the Pacific Northwest, USA." *Fire Ecology* 16, no. 4 (2020). https://doi.org/10.1186/s42408-019-0062-8.

Halofsky, Joshua S., Daniel C. Donato, Jerry F. Franklin, Jessica E. Halofsky, David L. Peterson, and Brian J. Harvey. "The Nature of the Beast: Examining Climate Adaptation Options in Forests with Stand-Replacing Fire Regimes." *Ecosphere* 9, no. 3 (2018). https://doi.org/10.1002/ecs2.2140.

Hamer, Thomas E., Eric D. Forsman, and Elizabeth M. Glenn. "Home Range Attributes and Habitat Selection of Barred Owls and Spotted Owls in an Area of Sympatry." *Condor* 109, no. 4 (November 2007): 750–768.

Harmon, M. E., J. F. Franklin, F. J. Swanson, P. Sollins, S. V. Gregory, J. D. Lattin, N. H. Anderson, et al. "Ecology of Coarse Woody Debris in Temperate Ecosystems." In

Advances in Ecological Research, vol. 15, edited by A. MacFadyen and E. D. Ford, 133–302. Orlando, FL: Academic Press, 1986.

Haynes, Richard W. "Contribution of Old-Growth Timber to Regional Economies in the Pacific Northwest." In *Old Growth in a New World*, edited by Thomas A. Spies and Sally L. Duncan, 83–94. Washington, DC: Island Press, 2009.

Hays, Samuel. *Conservation and the Gospel of Efficiency: The Progressive Conservation Movement, 1890–1920.* Cambridge, MA: Harvard University Press, 1959.

Heinrichs, Jay. "Old Growth Comes of Age." *Journal of Forestry* 81, no. 12 (December 1983): 776–779.

Helvoigt, Ted L., and Darius M. Adams. "A Stochastic Frontier Analysis of Technical Progress, Efficiency Change and Productivity Growth in the Pacific Northwest Sawmill Industry." *Forest Economics and Policy* 11, no. 4 (July 2009): 280–287.

Helvoigt, Ted L., Darius M. Adams, and A. L. Ayre. "Employment Transitions in Oregon's Wood Products Sector during the 1990s." *Journal of Forestry* 101, no. 4 (June 2003): 42–46.

Henson, Paul, Rollie White, and Steven P. Thompson. "Improving Implementation of the Endangered Species Act: Finding Common Ground through Common Sense." *BioScience* 68, no. 11 (November 2018): 861–872. https://doi.org/10.1093/biosci/biy093.

Hessburg, Paul F., James K. Agee, and Jerry F. Franklin. "Dry Forests and Wildland Fires of the Inland Northwest USA: Contrasting the Landscape Ecology of the Pre-settlement and Modern Eras." *Forest Ecology and Management* 211 (June 2005): 117–139.

Hessburg, Paul F., Susan J. Prichard, R. Keala Hagmann, Nicholas A. Povak, and Frank K. Lake. "Wildfire and Climate Change Adaptation of Western North American Forests: A Case for Intentional Management." *Ecological Applications* 31, no. 8 (December 2021). https://doi.org/10.1002/eap.2432.

Hirt, Paul W. *A Conspiracy of Optimism: Management of the National Forests since World War Two.* Lincoln: University of Nebraska Press, 1994.

Holling, C. S. "Resilience and Stability of Ecological Systems." *Annual Review of Ecology and Systematics* 4 (November 1973): 1–23.

Horn, Samantha R., Barry R. Noon, J. David Wiens, and William J. Ripple. "Potential Trophic Cascades Triggered by the Barred Owl Range Expansion." *Wildlife Society Bulletin* 40, no. 4 (2016): 615–624.

Janssen, Marco A. "A Future Full of Surprises." In *Panarchy: Understanding Transformations in Human and Natural Systems*, edited by Lance H. Gunderson and C. S. Holling, 241–260. Washington, DC: Island Press, 2002.

Jehl, Douglas. "Clinton Forest Chief Acts to Stop Logging of the Oldest Trees." *New York Times*, January 9, 2001. https://www.nytimes.com/2001/01/09/us/clinton-forest-chief-acts-to-stop-logging-of-the-oldest-trees.html.

Johnson, K. Norman. "Sustainable Harvest Levels and Short-Term Sales." *Journal of Forestry* 92, no. 4 (1994): 41–43.

———. "Will Linking Science to Policy Lead to Sustainable Forestry? Lessons from the Federal Forests of the United States." In *Sustainable Forestry: From Monitoring and Modelling to Knowledge Management and Policy Science*, edited by Keith Reynolds, Alan Thomson, Margaret Shannon, Michael Kohl, Duncan Ray, and K. Rennolls, 15–29. New York: Oxford University Press, 2007.

Johnson, K. Norman, James Agee, Robert Beschta, Virginia Dale, Linda Hardesty, James Long, Larry Nielsen, et al. "Sustaining the People's Lands: Recommendations for Stewardship of the National Forests and Grasslands into the Next Century." *Journal of Forestry* 97, no. 5 (May 1999): 6–12.

Johnson, K. Norman, S. Crim, K. Barber, M. Howell, and C. Cadwell. 1993. *Sustainable Harvest Levels and Short-Term Timber Sales for Options Considered in the FEMAT Report: Methods, Results, and Interpretations.* On file with Susan Charnley, Portland Forestry Sciences Laboratory, P.O. Box 3890, Portland, OR 97208.

Johnson, K. Norman, Jerry F. Franklin, and Debora L. Johnson. *A Plan for the Klamath Tribes' Management of the Klamath Reservation Forest.* 2008. https:// klamathtribes.org/wp-content/uploads/2020/02/Klamathtribes_ forestmanagement_plan.pdf.

Johnson, K. Norman, Jerry F. Franklin, Jack Ward Thomas, and John Gordon. *Alternatives for Management of Late-Successional Forests of the Pacific Northwest.* A report to the Agricultural Committee and the Merchant Marine and Fisheries Committee of the US House of Representatives by the Scientific Panel on Late-Successional Forest Ecosystems. 1991.

Johnson, K. Norman, Frederick Swanson, Margaret Herring, and Sarah Greene, eds. *Bioregional Assessments: Science at the Crossroads of Management and Policy:* Washington, DC: Island Press, 1999.

Johnson, Kirk. "A Town That Thrived on Logging Is Looking for a Second Growth." *New York Times*, November 14, 2014. https://www.nytimes.com/2014/11/16/us/ town-that-thrived-on-logging-is-looking-for-a-second-growth.html.

Johnston, James D., John B. Kilbride, Garrett W. Meigs, Christopher J. Dunn, and Robert E. Kennedy. "Does Conserving Roadless Wildland Increase Wildfire Activity in Western U.S. National Forests?" *Environmental Research Letters* 16, no. 8 (2021): 084040. https://doi.org/10.1088/1748-9326/ac13ee.

Juntunen, Judy Rycraft, May D. Dasch, and Ann Bennett Rogers. *The World of the Kalapuya.* Philomath, OR: Benton County Historical Society and Museum, 2005.

Kelly, David, and Gary Braasch. *Secrets of the Old Growth Forest.* Kaysville, UT: Gibbs and Smith, 1988.

LaFollette, Cameron. *Saving All the Pieces: Old Growth Forest in Oregon.* Oregon Student Public Interest Research Group, 1978. https://static1.squarespace.com/ static/573a143a746fb9ea3f1376e5/t/59ce7ed880bd5eb1e334433b/1506705156714/ SavingAllthePiecesVLQ.pdf.

Lande, Russell. "Demographic Model of the Northern Spotted Owl." *Oecologia* 75 (1988): 601–607.

Larson, Andrew J., Sean M. A. Jeronimo, Paul F. Hessburg, James A. Lutz, Nicholas A. Povak, C. Alina Cansler, Van R. Kane, and Derek J. Churchill. "Tamm Review: Ecological Principles to Guide Post-Fire Forest Landscape Management in the Inland Pacific and Northern Rocky Mountain Regions." *Forest Ecology and Management* 504 (January 2022): 119680.

Lee, R. G., M. S. Carroll, and K. K. Warren. "The Social Impact of Timber Harvest Reductions in Washington State." In *Revitalizing the Timber Dependent Regions of Washington*, edited by Paul Sommers and Helen Birss, 3–19. Seattle: Northwest Policy Center, University of Washington, 1991.

Lehner, Josh. "What Replaces Timber?" Oregon Office of Economic Analysis, July 25, 2019. https://oregoneconomicanalysis.com/2019/07/25/what-replaces-timber.

Lewis, James G. *The Forest Service and the Greatest Good: A Centennial History.* Durham, NC: Forest History Society, 2005.

Lichatowich, James A. *Salmon without Rivers: A History of the Pacific Salmon Crisis.* Washington, DC: Island Press, 1999.

Long, Jonathan, Frank K. Lake, Kathy Lynn, and Carson Viles. "Tribal Ecocultural Resources and Engagement." In *Synthesis of Science to Inform Land Management within the Northwest Forest Plan Area*, vol. 3, technical coordinators Thomas A.

Spies, Peter A. Stine, Rebecca Gravenmier, Jonathan W. Long, Matthew J. Reilly, and Rhonda Mazza, 851–897. General Technical Report PNW-GTR-966. Portland, OR: US Forest Service, Pacific Northwest Research Station, 2018. https://doi.org/10.2737/PNW-GTR-966.

MacArthur, Robert, and Edward O. Wilson. *Theory of Island Biogeography*. Princeton, NJ: Princeton University Press, 1967.

Mack, Cheryl A. "In Pursuit of the Wild *Vaccinium*: Huckleberry Processing Sites in the Southern Washington Cascades." *Archaeology in Washington* 4 (1992): 3–16.

Marcot, Bruce G., Karen L. Pope, Keith Slauson, Hartwell H. Welsh, Clara A. Wheeler, Matthew J. Reilly, and William J. Zielinski. "Other Species and Biodiversity of Older Forests." In *Synthesis of Science to Inform Land Management within the Northwest Forest Plan Area*, vol. 2, technical coordinators Thomas A. Spies, Peter A. Stine, Rebecca A. Gravenmier, Jonathan W. Long, Matthew J. Reilly, and Rhonda Mazza, 371–459. General Technical Report PNW-GTR-966. Portland, OR: US Forest Service, Pacific Northwest Research Station, 2018. https://doi.org/10.2737/PNW-GTR-966.

Marcot, Bruce G., and Jack Ward Thomas. *Of Spotted Owls, Old Growth, and New Policies: A History since the Interagency Scientific Committee Report*. General Technical Report PNW-GTR-408. Portland, OR: US Forest Service, Pacific Northwest Research Station, 1997. https://doi.org/10.2737/PNW-GTR-408.

Maser, Chris, and James R. Sedell. *From the Forest to the Sea: The Ecology of Wood in Streams, Rivers, Estuaries, and Oceans*. Boca Raton, FL: CRC Press, 1994.

McCabe, Gregory J., Martyn P. Clark, and Lauren E. Hay. "Rain-on-Snow Events in the Western United States." *Bulletin of the American Meteorological Society* 88, no. 3 (2007): 319–328. https://doi.org/10.1175/BAMS-88-3-319.

McCord, Millen, Matthew J. Reilly, Ramona J. Butz, and Erik S. Jules. "Early Seral Pathways of Vegetation Change Following Short Interval, Repeated Wildfire in Low Elevation Mixed-Conifer-Hardwood Forest Landscape of the Klamath Mountains, Northern California." *Canadian Journal of Forest Research* 50, no. 1 (January 2020): 13–23..

McDade, M. H., F. J. Swanson, W. A. McKee, J. F. Franklin, and J. VanSickle. "Source Distances for Coarse Woody Debris Entering Small Streams in Western Oregon and Washington." *Canadian Journal of Forest Research* 20, no. 3 (March 1990): 326–330.

McGinty, Katie. "Completion of the Forest Plan." April 7, 1994. Memorandum for the president. Folder: Forest Conference-Decision Memos/Record of Decision (RoD). OA/ID 4301. 17-23. FOIA collection 2012-0769-F (Seg. 1). Records of Kathleen "Katie" McGinty, 1993–1994. William J. Clinton Presidential Library, Little Rock, AR. https://clinton.presidentiallibraries.us/files/original/4fbe437615b30fc3d2b8d7 9aed20c36.pdf.

———. "Forest Conference Discussion with President at Thursday Lunch," March 17, 1993. Memorandum to the vice president. Folder: Memo to the Vice President from Katie McGinty. OA/ID 4572. 2-5. FOIA collection 2012-0769-F (Seg. 1). Records of Kathleen "Katie" McGinty, 1993–1994. William J. Clinton Presidential Library, Little Rock AR. https://clinton.presidentiallibraries.us/files/original/ cc99fc53318898b3cede9488e3bff67a.pdf.

McIver, William R., Scott F. Pearson, Craig Strong, Monique M. Lance, Jim Baldwin, Deanna Lynch, Martin G. Raphael, Richard D. Young, and Nels Johnson. *Status and Trend of Marbled Murrelet Populations in the Northwest Plan Area, 2000 to 2018*. General Technical Report PNW-GTR-996. Portland, OR: US Forest Service, Pacific Northwest Research Station, 2021.

Meddens, Arjan J. H., Crystal A. Kolden, James A. Lutz, Alistair M. S. Smith, C. Alina Cansler, John T. Abatzoglou, Garrett W. Meigs, William M. Downing, and Meg A. Krawchuk. "Fire Refugia: What Are They and Why Do They Matter for Global Change?" *BioScience* 68, no. 12 (2018): 944–954. https://doi.org/10.1093/biosci/biy103.

Merschel, Andrew. "Historical Forest Succession and Disturbance Dynamics in Coastal Douglas-Fir Forests in the Southwest Cascades." PhD diss., Oregon State University, 2021.

Meslow, E. Charles. "Spotted Owl Protection: Unintentional Evolution toward Ecosystem Management." *Endangered Species Update* 10, no. 3–4 (1993): 34–38.

Meslow, E. Charles, Richard S. Holthausen, and David A. Cleaves. "Assessment of Terrestrial Species and Ecosystems." *Journal of Forestry* 92, no. 4 (April 1994): 24–27.

Metlen, Kerry L., Darren Borgias, Bryce Kellogg, Michael Schindel, Aaron Jones, George McKinley, Derek Olson, et al. *Rogue Basin Cohesive Forest Restoration Strategy: A Collaborative Vision for Resilient Landscapes and Fire Adapted Communities.* Medford, OR: Nature Conservancy, 2017.

Metlen, Kerry L., Carl N. Skinner, Derek R. Olson, Clint Nichols, and Darren Borgias. "Regional and Local Controls on Historical Fire Regimes of Dry Forests and Woodlands in the Rogue River Basin, Oregon." *Forest Ecology and Management* 430 (2018): 43–58.

Molina, Randy, Bruce G. Marcot, and Robin Lesher. "Protecting Rare, Old-Growth Forest Associated Species under the Survey and Manage Guidelines of the Northwest Forest Plan." *Conservation Biology* 20, no. 2 (2006): 306–318.

Murphy, Dennis D., and Barry R. Noon. "Integrating Scientific Methods with Habitat Conservation Planning Reserve Design for Northern Spotted Owls." *Ecological Applications* 2, no. 1 (February 1992): 3–17.

Murray, Brian C., Bruce A. McCarl, and Heng-Chi Lee. "Estimating Leakage from Forest Carbon Sequestration Programs." *Land Economics* 80, no. 1 (2004): 109–124.

Naiman, Robert J., Timothy J. Beechie, Lee E. Benda, Dean R. Berg, Peter A. Bisson, Lee H. Macdonald, Matthew D. O'Connor, Patricia L. Olson, and E. Ashley Steel. "Fundamental Elements of Ecologically Healthy Watersheds in the Pacific Northwest Coastal Ecoregion." In *Watershed Management: Balancing Sustainability and Environmental Change*, edited by Robert J. Naiman, 127–188. New York: Springer-Verlag, 1992.

Neel, Roy. "Interagency Office of Forestry and Economic Development." November 29, 1993. Memorandum to Secretary Babbitt, Secretary Espy, and others. Folder: Forest Conference—Office of Forestry and Economic Development. OA/ID 4300. 30-32. FOIA collection 2012-0769-F (Seg. 1). Records of Kathleen "Katie" McGinty, 1993–1994. William J. Clinton Presidential Library, Little Rock AR. https://clinton.presidentiallibraries.us/files/original/c287eef0835aafad2708c66ed7cb7d55.pdf.

Nehlsen, Willa, Jack E. Williams, and James Lichatowich. "Pacific Salmon at the Crossroads: Stocks at Risk from California, Oregon, Washington, and Idaho." *Fisheries* 16, no. 2 (1991): 4–21.

Nesbit, Jack. *The Collector: David Douglas and the Natural History of the Northwest.* Seattle: Sasquatch Books, 2010.

NMFS (National Marine Fisheries Service). *Recovery Plan for Oregon Coast Coho Salmon Evolutionarily Significant Unit.* Portland, OR: National Marine Fisheries

Service, West Coast Region, 2016. https://repository.library.noaa.gov/view/noaa/15986.

Norris, Logan A. "Science Review of the Northwest Forest Plan." In *Bioregional Assessments: Science at the Crossroads of Management and Policy*, edited by K. Norman Johnson, Frederick Swanson, Margaret Herring, and Sarah Greene, 117–120. Washington, DC: Island Press, 1999.

North, Malcolm P., Rob A. York, Brandon M. Collins, Matthew David Hurteau, Gavin M. Jones, Eric Knapp, Leda Nikola Kobziar, et al. "Pyrosilviculture Needed for Landscape Resilience of Dry Western United States Forests." *Journal of Forestry* 199, no. 5 (September 2021): 520–544. https://doi.org/10.1093/jofore/fvab026.

Orr, William N., and Elizabeth L. Orr. *Geology of the Pacific Northwest*. Long Grove, IL: Waveland Press, 2006.

Palik, Brian J., Anthony W. D'Amato, Jerry F. Franklin, and K. Norman Johnson. *Ecological Silviculture: Foundations and Applications*. Long Grove, IL: Waveland Press, 2021.

Perry, Timothy D., and Julia A. Jones. "Summer Streamflow Deficits from Regenerating Douglas-Fir Forest in the Pacific Northwest, USA." *Ecohydrology* 10, no. 2 (2017): 1–13.

Phalan, Benjamin T., Joseph M. Northrup, Zhiqiang Yang, Robert L. Deal, Josée S. Rousseau, Thomas A. Spies, and Matthew G. Betts. "Successes and Limitations of an Iconic Policy: Impacts of the Northwest Forest Plan on Forest Composition and Bird Populations." *PNAS* 116, no. 8 (2019): 3322–3327. https://doi.org/10.1073/pnas.1813072116.

Pihulak, Jillian, Jenessa Stemke, Jordan Lindsey, and Daniel Leavell. "Oregon's 2020 Wildfires." OSU Extension Fire Program, Oregon State University, 2021. Updated January 18, 2022. https://osugisci.maps.arcgis.com/apps/MapSeries/index.html?appid=6629651002db435d9df188003d790847.

Pinchot, Gifford. *Breaking New Ground*. New York: Harcourt, Brace, 1947.

———. *The Use of the National Forest Reserves*. Washington, DC: Department of Agriculture, 1905. https://foresthistory.org/wp-content/uploads/2017/10/1905_use_book.pdf.

Pipkin, James. *The Northwest Forest Plan Revisited*. Washington, DC: US Department of the Interior, Office of Policy Analysis, 1998.

Poage, Nathan J., and John C. Tappeiner II. "Long-Term Patterns of Diameter and Basal Area Growth of Old-Growth Douglas-Fir Trees in Western Oregon." *Canadian Journal of Forest Research* 32, no. 7 (July 2002): 1232–1243.

Power, Jay. "Reaction to President Clinton's Forest Plan." Video conference on C-SPAN, July 1, 1993. https://www.c-span.org/video/?44083-1/us-forest-policy.

PRISM Climate Group. 2020. "30-Year Normals." Northwest Alliance for Computational Science and Engineering, Oregon State University. Accessed March 10, 2022. http://prism.oregonstate.edu/normals.

Raffa, Kenneth F., Brian H. Aukema, Barbara J. Bentz, Allan L. Carroll, Jeffrey A. Hicke, Monica G. Turner, and William H. Romme. "Cross-Scale Drivers of Natural Disturbances Prone to Anthropogenic Amplification: The Dynamics of Bark Beetle Eruptions." *BioScience* 58, no. 6 (2008): 501–517. https://doi.org/10.1641/B580607.

Raphael, Martin G., Gary A. Falxa, and Alan E. Burger. "Marbled Murrelet." In *Synthesis of Science to Inform Land Management within the Northwest Forest Plan Area*, vol. 1, technical coordinators Thomas A. Spies, Peter A. Stine, Rebecca Gravenmier, Jonathan W. Long, Matthew J. Reilly, and Rhonda Mazza, 301–337. General Technical Report PNW-GTR-966. Portland, OR: US Forest Service,

Pacific Northwest Research Station, 2018. https://doi.org/10.2737/
PNW-GTR-966.

Raphael, Martin G., Paul Hessburg, Rebecca Kennedy, John Lehmkuhl, Bruce G.
Marcot, Robert Scheller, Peter Singleton, and Thomas Spies. *Assessing the
Compatibility of Fuel Treatments, Wildfire Risk, and Conservation of Northern
Spotted Owl Habitats and Populations in the Eastern Cascades: A Multi-scale
Analysis.* JFSP Research Project Reports 31 (2013). https://digitalcommons.unl.
edu/jfspresearch/31/.

Raphael, Martin G., and Bruce G. Marcot. "Validation of a Wildlife-Habitat-
Relationships Model: Vertebrates in a Douglas-Fir Sere." In *Wildlife 2000:
Modeling Habitat Relationships of Terrestrial Vertebrates*, edited by Jared Verner,
Michael L. Morrison, and C. John Ralph, 129–138. Madison: University of
Wisconsin Press, 1986.

Reeves, Gordon H., Deanna H. Olson, Steven M. Wondzell, Peter A. Bisson, Sean
Gordon, Stephanie A. Miller, Jonathan W. Long, and Michael J. Furniss. "The
Aquatic Conservation Strategy of the Northwest Forest Plan: Relevance to
Management of Aquatic and Riparian Ecosystems." In *Synthesis of Science to
Inform Land Management within the Northwest Forest Plan Area: Executive
Summary*, technical coordinators Thomas A. Spies, Peter A. Stine, Rebecca
Gravenmier, Jonathan W. Long, Matthew J. Reilly, and Rhonda Mazza, 93–106.
General Technical Report PNW-GTR-970. Portland, OR: US Forest Service,
Pacific Northwest Research Station, 2018. https://doi.org/10.2737/
PNW-GTR-970.

———. "The Aquatic Conservation Strategy of the Northwest Forest Plan—A Review of
Relevant Science after 23 Years." In *Synthesis of Science to Inform Land
Management within the Northwest Forest Plan Area*, vol. 2, technical coordinators
Thomas A. Spies, Peter A. Stine, Rebecca Gravenmier, Jonathan W. Long,
Matthew J. Reilly, and Rhonda Mazza, 461–607. General Technical Report PNW-
GTR-966. Portland, OR: US Forest Service, Pacific Northwest Research Station,
2018. https://doi.org/10.2737/PNW-GTR-966.

Reeves, Gordon H., Brian R. Pickard, and K. Norman Johnson. *An Initial Evaluation of
Potential Options for Managing Riparian Reserves of the Aquatic Conservation
Strategy of the Northwest Forest Plan.* General Technical Report PNW-GTR-937.
Portland, OR: US Forest Service, Pacific Northwest Research Station, 2016.
https://doi.org/10.2737/PNW-GTR-937.

Reid, Leslie M. "Understanding and Evaluating Cumulative Watershed Impacts." In
Cumulative Watershed Effects of Fuel Management in the Western United States,
edited by William J. Elliot, Ina Sue Miller, and Lisa Audin, 277–298. General
Technical Report RMRS-GTR-231. Fort Collins, CO: US Forest Service, Rocky
Mountain Research Station, 2010. https://doi.org/10.2737/RMRS-GTR-231.

Reilly, Matthew J., Thomas A. Spies, Jeremy Littell, Ramona Butz, and John B. Kim.
"Climate, Disturbance, and Vulnerability to Vegetation Change in the Northwest
Forest Plan Area." In *Synthesis of Science to Inform Land Management within the
Northwest Forest Plan Area*, vol. 1, technical coordinators Thomas A. Spies, Peter
A. Stine, Rebecca Gravenmier, Jonathan W. Long, Matthew J. Reilly, and Rhonda
Mazza, 29–94. General Technical Report PNW-GTR-966. Portland, OR: US Forest
Service, Pacific Northwest Research Station, 2018. https://doi.org/10.2737/
PNW-GTR-966.

Reilly, Matthew J., Aaron Xuspan, Joshua S. Halofsky, Crystal Raymond, Andy
McEvoy, Alex W. Dye, Daniel C. Donato, John B. Kim, et. al. "Cascadia Burning:
The Historic, but Not Historically Unprecedented, 2020 Wildfires in the Pacific
Northwest." *Ecosphere* 13 (6) (2022). https://doi.org/10.1002/ecs2.4070.

Ritchie, Martin W. *Ecological Research at the Goosenest Adaptive Management Area in Northeastern California.* General Technical Report PSW-GTR-192. US Forest Service, Pacific Northwest Research Station, 2005. https://doi.org/10.2737/PSW-GTR-192.

Robbins, William G. *Landscapes of Conflict: The Oregon Story, 1940–2000.* Seattle: University of Washington Press, 2004.

———. *Landscapes of Promise: The Oregon Story, 1800–1940.* Seattle: University of Washington Press, 1997.

———. *A Place for Inquiry, a Place for Wonder: The Andrews Forest.* Corvallis: Oregon State University Press, 2020.

Ruggierro, Leonard F., Keith B. Aubry, Andrew B. Carey, and Mark H. Huff, technical eds. *Wildlife and Vegetation of Unmanaged Douglas-Fir Forests.* General Technical Report PNW-GTR-285. Portland, OR: US Forest Service, Pacific Northwest Research Station, 1991.

Sass, Emma M., Brett J. Butler, and Marla Markowski-Lindsay. *Distribution of Forest Ownerships across the Conterminous United States, 2017.* Res. Map NRS-11. Madison, WI: US Forest Service, Northern Research Station, 2020. https://doi.org/10.2737/NRS-RMAP-11.

Schick, Tony, Rob Davis, and Lylla Younes. "Big Money Bought Oregon's Forests. Small Timber Communities Are Paying the Price." Oregon Public Broadcasting, July 11, 2020. https://www.opb.org/news/article/oregon-investigation-timber-logging-forests-policy-taxes-spotted-owl/.

Schultz, Courtney A., Thomas D. Sisk, Barry R. Noon, and Martine A. Nie. "Wildlife Conservation Planning under the United States Forest Service's 2012 Planning Rule." *Journal of Wildlife Management* 77, no. 3 (2013): 428–444.

Scott, Aaron. *Timber Wars.* Podcast. Produced by Aaron Scott, Peter Frick-Wright, and Robbie Carver. Oregon Public Broadcasting, 2020. https://www.opb.org/show/timberwars.

Sedell, James R., Gordon H. Reeves, and Kelly M. Burnett. "Development and Evaluation of Aquatic Conservation Strategies." *Journal of Forestry* 92, no. 4 (1994): 28–31.

Serra-Diaz, Josep M., Charles Maxwell, Melissa S. Lucash, Robert M. Scheller, Danelle M. Laflower, Adam D. Miller, Alan J. Tepley, et al. "Disequilibrium of Fire-Prone Forests Sets the Stage for a Rapid Decline in Conifer Dominance during the 21st Century." *Scientific Reports* 8, no. 1 (2018): 1–12. https://doi.org/10.1038/s41598-018-24642-2.

Shindler, Bruce A., and Mark Brunson. "Social Acceptability in Forest and Range Management." In *Society and Natural Resources: A Summary of Knowledge*, edited by Michael J. Manfredo, Jerry J. Vaske, Brett L. Bruyere, Donald R. Field, and Perry J. Brown. Jefferson, MO: Modern Litho, 2004.

Shindler, Bruce A., Mark Brunson, and George H. Stankey. *Social Acceptability of Forest Conditions and Management Practices: A Problem Analysis.* General Technical Report PNW-GTR-537. Portland, OR: US Forest Service, Pacific Northwest Research Station, 2002. https://doi.org/10.2737/PNW-GTR-537.

Simard, Suzanne. *Finding the Mother Tree: Discovering the Wisdom of the Forest.* New York: Alfred. A. Knopf, 2021.

Skillen, James R. *Federal Ecosystem Management: Its Rise, Fall, and Afterlife.* Lawrence: University Press of Kansas, 2015.

Sonner, Scott. "Gore Calls Salvage Logging 'Biggest Mistake'—Admission Comes in TV Interview Scheduled to Air Tonight." *Spokesman Review*, September 27, 1996.

https://www.spokesman.com/stories/1996/sep/27/gore-calls-salvage-logging-biggest-mistake.

Spies, Thomas A., Thomas W. Giesen, Frederick J. Swanson, Jerry F. Franklin, Denise Lach, and K. Norman Johnson. "Climate Change Adaptation Strategies for Federal Forests of the Pacific Northwest, USA: Ecological, Policy, and Socio-Economic Perspectives." *Landscape Ecology* 25, no. 8 (2010): 1185–1199. https://doi.org/10.1007/s10980-010-9483-0.

Spies, Thomas A., Paul F. Hessburg, Carl N. Skinner, Klaus J. Puettmann, Matthew J. Reilly, Raymond J. Davis, Jane A. Kertis, Jonathan W. Long, and David C. Shaw. "Old Growth, Disturbance, Forest Succession, and Management in the Area of the Northwest Forest Plan." In *Synthesis of Science to Inform Land Management within the Northwest Forest Plan Area: Executive Summary*, technical coordinators Thomas A. Spies, Peter A. Stine, Rebecca Gravenmier, Jonathan W. Long, Matthew J. Reilly, and Rhonda Mazza, 31–46. General Technical Report PNW-GTR-970. Portland, OR: US Forest Service, Pacific Northwest Research Station, 2018. https://doi.org/10.2737/PNW-GTR-970.

Spies, Thomas A., Jonathan Long, Susan Charnley, Paul Hessburg, Bruce Marcot, Gordon Reeves, Damon Lesmeister, et al. "Twenty-Five Years of the Northwest Forest Plan: What Have We Learned?" *Frontiers of Ecology and the Environment* 17, no. 9 (2019): 511–520. https://doi.org/10.1002/fee.2101.

Spies, Thomas A., Peter A. Stine, Rebecca Gravenmier, Jonathan W. Long, Matthew J. Reilly, and Rhonda Mazza, technical coordinators. *Synthesis of Science to Inform Land Management within the Northwest Forest Plan Area: Executive Summary*. General Technical Report PNW-GTR-970. Portland, OR: US Forest Service, Pacific Northwest Research Station, 2018. https://doi.org/10.2737/PNW-GTR-970.

Stankey, George H., Bernard T. Bormann, Clare Ryan, Bruce Schindler, Victoria Sturtevant, Roger N. Clark, and Charles Philpot. "Adaptive Management and the Northwest Forest Plan: Rhetoric and Reality." *Journal of Forestry* 101, no. 1 (January 2003): 40–46.

Steen, Harold K. "An Interview with Jack Ward Thomas." Durham, NC: Forest History Society, 2002. https://foresthistory.org/wp-content/uploads/2017/011/Thomas-JW-OHI-Final.pdf.

———, ed. *Jack Ward Thomas: The Journals of a Forest Service Chief*. Durham, NC: Forest History Society, 2004.

———. *The U.S. Forest Service: A History*. Durham, NC: Forest History Society, 2004.

Stelle, Will. "Major Issues for the ISC Meeting." March 12, 1994. Memorandum to Katie McGinty. Folder: Forest Conference—Forest Viability. OA/ID: 4302. 87-90. FOIA collection 2012-0769-F (Seg. 1). Records of Kathleen "Katie" McGinty, 1993–1994. William J. Clinton Presidential Library, Little Rock, AR. https://clinton.presidentiallibraries.us/files/original/9294ecdb6d05b5ce1bc06f6a624d4bfa.pdf.

Stephens, Scott L., Leda N. Kobziar, Brandon M. Collins, Raymond Davis, Peter Z. Fulé, William Gaines, Joseph Ganey, et al. "Is Fire 'For the Birds?' How Two Rare Species Influence Fire Management across the United States." *Frontiers of Ecology and the Environment* 17, no. 7 (September 2019): 391–399.

Stidham, Melanie, Gwen Busby, and K. Norman Johnson. "The Role of Economic Emergency Situation Determinations in Expediting Fire Salvage." *Environmental Law Reporter* 38 (2008): 10741–10751.

Strong, Ted. "Tribal Right to Fish." *Journal of Forestry* 92, no. 4 (1994): 34. https://doi.org/10.1093/jof/92.4.34.

Swanson, F. J., J. A. Jones, D. O. Wallin, and J. H. Cissel. "Natural Variability—Implications for Ecosystem Management." In *Ecosystem Management: Principles and Applications*, vol. 2, *Eastside Forest Ecosystem Health Assessment*, technical editors M. E. Jensen and P. S. Bourgeron, 80–94. General Technical Report PNW-GTR-318. Portland, OR: US Forest Service, Pacific Northwest Research Station, 1994. https://doi.org/10.2737/PNW-GTR-318.

Swanson, Frederick, George W. Lienkaemper, and James R. Sedell. *History, Physical Effects, and Management Implications of Large Organic Debris in Western Oregon Streams*. General Technical Report PNW-GTR-056. Portland, OR: US Forest Service, Pacific Northwest Research Station, 1976. https://www.fs.usda.gov/treesearch/pubs/22618.

Swanson, Mark E., Jerry F. Franklin, Robert L. Beschta, Charles M. Crisafulli, Dominick A. DellaSala, Richard L. Hutto, David B. Lindenmayer, and Frederick J. Swanson. "The Forgotten Stage of Forest Succession: Early-Successional Ecosystems on Forest Sites." *Frontiers of Ecology and the Environment* 9, no. 2 (2011): 117–125..

Tappeiner, John C., David Huffman, David Marshall, Thomas A. Spies, and John D. Bailey. "Density, Ages, and Growth Rates in Old-Growth and Young-Growth Forests in Coastal Oregon." *Canadian Journal of Forest Research* 27, no. 5 (1997): 638–648.

Taylor, Robert J. "Rejecting a Reserve Approach." *Journal of Forestry* 92, no. 4 (1994): 22. https://doi.org/10.1093/jof/92.4.22.

Tepley, Alan J., Frederick J. Swanson, and Thomas A. Spies. "Post-Fire Tree Establishment and Early Cohort Development in Conifer Forests of the Western Cascades of Oregon, USA." *Ecosphere* 5, no. 7 (July 2017): 1–23. https://doi.org/10.1890/ES14-00112.1.

Thomas, Jack Ward. "Forest Management Assessment Team: Objectives, Process, and Options." *Journal of Forestry* 92, no. 4 (1994): 12–19.

———. *Forks in the Trail: A Conservationist's Trek to the Pinnacles of Natural Resource Leadership*. Missoula, MT: Boone and Crockett Club, 2015.

———. Interview with Rick Freeman, September 29, 1998. OH 370-007. Archives and Special Collections, Mansfield Library, University of Montana. https://scholarworks.umt.edu/forestserviceecosystemmanagement/1/

———. "Sustainability of the Northwest Forest Plan—Dynamic vs. Static Management." Draft report. US Forest Service, Pacific Southwest Region, 2003.

Thomas, Jack Ward, Eric D. Forsman, Joseph B. Lint, E. Charles Meslow, Barry R. Noon, and Jared Verner. *A Conservation Strategy for the Northern Spotted Owl*. Washington, DC: US Government Printing Office, 1990. https://www.fws.gov/arcata/es/birds/nso/documents/ConservationStrategyForTheNorthernSpottedOw_May1990.pdf.

Thomas, Jack Ward, Jerry F. Franklin, John Gordon, and K. Norman Johnson. "The Northwest Forest Plan: Origins, Components, Implementation Experience, and Suggestions for Change." *Conservation Biology* 20, no. 2 (2006): 277–287.

Thomas, Jack Ward, Martin G. Raphael, Robert G. Anthony, Eric D. Forsman, A. Grant Gunderson, Richard S. Holthausen, Bruce G. Marcot, Gordon H. Reeves, James R. Sedell, and David M. Solis. *Viability Assessments and Management Considerations for Species Associated with Late-Successional and Old-Growth Forests of the Pacific Northwest*. [Portland, OR]: US Department of Agriculture, National Forest System, Forest Service Research, 1993.

Thomas, Jack Ward, Dale E. Toweill, and Daniel P. Metz. *Elk of North America: Ecology and Management*. Harrisburg, PA: Stackpole Books, 1982.

Thompson, Jonathan. "Forest Communities and the Northwest Forest Plan: What Socioeconomic Monitoring Can Tell Us." *Science Findings* 95 (August 2007).

Portland, OR: US Forest Service, Pacific Northwest Research Station. https://www.fs.fed.us/pnw/sciencef/scifi95.pdf.

Tuchmann, E. Thomas, Kent P. Connaughton, Lisa E. Freedman, and Clarence B. Moriwaki. *The Northwest Forest Plan: A Report to the President and Congress*. Portland, OR: USDA Office of Forestry and Assistance, 1996.

Tuchmann, E. Thomas, and Chad T. Davis. *O&C Lands Report*. 2013. Prepared for Oregon Governor John Kitzhaber. http://media.oregonlive.com/environment_impact/other/OCLandsReport.pdf.

USDA (US Department of Agriculture). 36 C.F.R. Part 219: National Forest System Land Management Planning. 77 (68) Fed. Reg. 21262–21276 (April 9, 2012).

———. "Part 219—Planning." In 36 C.F.R. Part 219: National Forest System Land and Resource Management Planning. 47 (190) Fed. Reg. 43037–43052 (September 30, 1982). https://www.fs.usda.gov/Internet/FSE_DOCUMENTS/stelprdb5349150.pdf.

USDA and USDI (US Department of Agriculture and US Department of the Interior). "Draft Supplemental Environmental Impact Statement on Management of the Habitat for Late-Successional and Old-Growth Forest Related Species within the Range of the Northern Spotted Owl." Washington, DC: US Department of Agriculture, Forest Service, and US Department of the Interior, Bureau of Land Management, 1993.

———. "Final Supplemental Environmental Impact Statement on Management of the Habitat for Late-Successional and Old-Growth Forest Related Species within the Range of the Northern Spotted Owl." Washington, DC: US Department of Agriculture, Forest Service, and US Department of the Interior, Bureau of Land Management, 1994. https://www.blm.gov/or/plans/nwfpnepa/FSEIS-1994/FSEIS-1994-I.pdf.

———. "Record of Decision for Amendments to Forest Service and Bureau of Land Management Planning Documents within the Range of the Northern Spotted Owl and Appendices." Washington, DC, 1994. https://www.fs.usda.gov/Internet/FSE_DOCUMENTS/stelprd3843201.pdf.

———. "Standards and Guidelines for Management of Habitat for Late-Successional and Old-Growth Forest Related Species within the Range of the Northern Spotted Owl." 1994. https://www.fs.usda.gov/Internet/FSE_DOCUMENTS/stelprd3841091.pdf.

US Department of Energy. *A Citizen's Guide to the NEPA: Having Your Voice Heard (CEQ, 2007; revised 2021)*. Office of NEPA Policy and Compliance, December 1, 2007. https://www.energy.gov/nepa/downloads/citizens-guide-nepa-having-your-voice-heard-ceq-2007-revised-2021.

USFS (US Forest Service). "Bioregional Assessment of Northwest Forests." 2020. https://www.fs.usda.gov/detail/r6/landmanagement/planning/?cid=fseprd677501.

———. *Douglas-Fir Supply Study: Alternative Programs for Increasing Timber Supplies from National Forest Lands*. Portland, OR: US Forest Service, Regions Five and Six, Pacific Northwest Forest and Range Experiment Station, 1969.

———. "Draft Supplemental to the Final Environmental Impact Statement for an Amendment to the Pacific Northwest Regional Guide." Portland OR: US Forest Service, Pacific Northwest Regional Office, 1986.

———. "Final Environmental Impact Statement for Management of the Northern Spotted Owl." Portland, OR: US Forest Service, Pacific Northwest Regional Office, 1992.

———. "Final Supplemental to the Final Environmental Impact Statement for an Amendment to the Pacific Northwest Regional Guide." Portland OR: US Forest Service, Pacific Northwest Regional Office, 1988.

———. "Flat Country Project: Record of Decision, 36 C.F.R. 218 Objection Process." January 2021. https://www.fs.usda.gov/nfs/11558/www/nepa/109274_FSPLT3_5578654.pdf.

———. *Land Management Planning Handbook.* FSH 1909.12. Washington, DC: US Forest Service, 2015. http://www.fs.usda.gov/detail/planningrule/home/?cid=stelprd3828310.

———. "Lowell Country Project." 2020. https://www.fs.usda.gov/project/?project=52868.

———. *Regional Guide for the Pacific Northwest Region.* Portland, OR: US Forest Service, Pacific Northwest Regional Office, 1984.

USFWS (US Fish and Wildlife Service). "Barred Owl Study Update." Portland, OR: US Fish and Wildlife Service, 2019. https://www.fws.gov/oregonfwo/articles.cfm?id=149489616.

———. "Endangered and Threatened Plants and Animals." *Code of Federal Regulations,* 50 C.F.R. 17.3 (October 1, 2001).

———. "Endangered and Threatened Wildlife and Plants: 12-Month Finding for the Northern Spotted Owl." 85 (241) Fed. Reg. 81144–81149 (December 15, 2020).

———. "Endangered and Threatened Wildlife and Plants: Revised Critical Habitat for the Northern Spotted Owl." 77 (46) Fed. Reg. 14062–14165 (March 8, 2012).

———. "Experimental Removal of Barred Owls to Benefit Threatened Northern Spotted Owls—Final Environmental Impact Statement." Portland, OR: US Fish and Wildlife Service, 2013. https://www.fws.gov/oregonfwo/Documents/BarredOwl-FinalEIS.pdf.

———. "Final Biological Opinion for the Preferred Alternative of the Supplemental Environmental Impact Statement on Management of Habitat for Late-Successional and Old-Growth Forest Related Species within the Range of the Northern Spotted Owl." Portland, OR: US Fish and Wildlife Service, 1994.

———. *Final Draft Recovery Plan for the Northern Spotted Owl (Strix occidentalis caurina).* Portland, OR: US Fish and Wildlife Service, 1992.

———. *Revised Recovery Plan for the Northern Spotted Owl (Strix occidentalis caurina).* Portland, OR: US Fish and Wildlife Service, 2011.

Vose, R. S., D. R. Easterling, K. E. Kunkel, A. N. LeGrande, and M. F. Wehner. "Temperature Changes in the United States." In *Climate Science Special Report: Fourth National Climate Assessment,* vol. 1, edited by D. J. Wuebbles, D. W. Fahey, K. A. Hibbard, D. J. Dokken, B. C. Stewart, and T. K. Maycock, 185–206. Washington, DC: US Global Change Research Program, 2017. doi:10.7930/J0N29V45.

Wagner, Eric. *After the Blast: The Ecological Recovery of Mount St. Helens.* Seattle: University of Washington Press, 2020.

Wellock, Thomas R. "The Dicky Bird Scientists Take Charge: Science, Policy, and the Northern Spotted Owl." *Environmental History* 15, no. 3 (July 2010): 381–414.

Westerling, A. L., H. G. Hidalgo, D. R. Cayan, and T. W. Swetnam. "Warming and Earlier Spring Increase Western US Forest Wildfire Activity." *Science* 313, no. 5789 (2006): 940–943. doi:10.1126/science.1128834.

White, Richard. *"It's Your Misfortune and None of My Own": A New History of the American West.* Rev. ed. Norman: University of Oklahoma Press, 1993.

Wiens, J. David, Robert G. Anthony, and Eric D. Forsman. "Competitive Interactions and Resource Partitioning between Northern Spotted Owls and Barred Owls in Western Oregon." *Wildlife Monographs* 185, no. 1 (2014): 1–50.

Wiens, J. David, Katie M. Dugger, J. Mark Higley, and Stan G. Sovern. "Invader Removal Triggers Competitive Release in a Threatened Avian Predator." *PNAS* 118, no. 31 (2021). https://doi.org/10.1073/pnas.2102859118.

Wigglesworth, Alex. "It Is Just a Drop in the Bucket." *Los Angeles Times*, November 9, 2021. https://www.latimes.com/california/story/2021-11-08/us-forest-service-struggles-to-complete-prescribed-burns.

Wilkins, David E., and K. Tsianina Lomawaima. *Uneven Ground: American Indian Sovereignty and Federal Law*. Norman: University of Oklahoma Press, 2002.

Wilkinson, Charles F., and H. Michael Anderson. *Land and Resource Planning in the National Forests*. Washington, DC: Island Press, 1987.

Williams, Gerald W. *The U.S. Forest Service in the Pacific Northwest: A History*. Corvallis: Oregon State University Press, 2009.

Williams, Mark A., and William L. Baker. "Spatially Extensive Reconstructions Show Variable-Severity Fire and Heterogeneous Structure in Historical Western United States Dry Forests." *Global Ecology and Biogeography* 21, no. 10 (2012): 1042–1052.

Yaffee, Steven L. *The Wisdom of the Spotted Owl: Policy Lessons for a New Century*. Washington, DC: Island Press, 1994.

Zald, Harold S. J., and Christopher J. Dunn. "Severe Fire Weather and Intensive Forest Management Increase Fire Severity in a Multi-ownership Landscape." *Ecological Applications* 28, no. 4 (2018): 1068–1080.

Index

Interagency Steering Committee, 246
Izaak Walton League, 51

J

Jamison, Cy, 156
Jamison Plan, 156–157
Jansen, Marco, 356
Johnson, Debora, 387, Plate 6.1
Johnson, Kathy, 99
Johnson, Kirk, 300–301
Johnson, Norm
  2001 congressional testimony, 272
  ecological forestry and, 308–310
  FEMAT and, 179–180, 189–192, 205,
    219, 229
  FORPLAN and, 57
  G-4 and, 122–123, 127, 138, 140, 146,
    148
  ISC and, 117
  photograph of, Plate 10.5
  *A Plan for the Klamath Tribes' Man-
    agement of the Klamath Reservation
    Forest*, 387
  second Committee of Scientists and,
    324
  wildfires and, 309, 329
Johnston, James, 256
Jones, Julia, 344
Journal of Forestry, 308
Juliana, Catia, 105
Jungwirth, Jim, 301
Jungwirth, Lynn, 301

K

Kalapuya Tribe, 12–13
Karuk, 298
Kerr, Andy, 103
Kerrick, Mike, 67
Kershner, Jeff, 151–152
Key Watersheds, 135–137, 185, 188,
  190–191, 198–199, 227–228, 237
  map of, 238
Klamath National Forest (KNF),
  269–270, 300
*Klamath Tribes v. United States of
  America*, 380–381
Knauss, John, 158

L

La Grande, Oregon, 109
LaFollette, Cameron, 92
Lampi, Al, 99

Lande, Russel, 98–100
landslides, 77–79, 94, 137
Lane County Audubon Society v. Jamison,
  158
Larch Mountain salamander, Plate 7.1
Late-Successional Reserves (LSRs), 183,
  236, 239, 402–403
  FEMAT and, 183–185, 191–195
  maps of, Plate 6.2, Plate 8.1
  planned redundancy and, 304
  revisiting of, 383–387
  Salvage Rider and, 255, 258
  SAT II and, 230–231
late-successional/old-growth (LS/OG)
  forests, 128–132, 137–146, 154, 239
  FEMAT and, 178–181, 183–185, 191–
    196, 199–200, 206–207, 212–213
  maps of, Plate 1.2, Plate 6.2
  SAT II and, 225–226, 231
  Survey and Manage and, 231, 269,
    271–272, 274–277
  See also old-growth forests
Layzer, Judith, 353
Leahy, Patrick, 107
Leon, Richard, 321–322
Leopold, Aldo, 36
Lesher, Robin, 179
Lewis & Clark College, 273
Lichatowich, James A., 133–134
Lint, Joe, 109
Long, Jonathan, 16, 382–383
Lowell Country Project, 299, Plate 10.2
Luoma, Jon, 72
Lyons, Jim
  FEMAT and, 180, 190
  Forest Conference and, 162–164,
    167–168
  G-4 and, 122–123, 127, 145–146, 152
  SAF Task Force and, 94–95
  Salvage Rider and, 257–258
  second Committee of Scientists and,
    324

M

MacCleery, Doug, 98–99
MacKaye, Benton, 30
MacMahon, James A., 75
marbled murrelets
  ESA and, 121, 172, 275
  FEMAT and, 184–185, 198
  G-4 and, 121, 129, 139–144
  photograph of, 172